Tourism in Development

Reflective Essays

Tourism in Development

Reflective Essays

Edited by

Peter U.C. Dieke

Brian E.M. King

and

Richard A.J. Sharpley

CABI

CABI is a trading name of CAB International

CABI
Nosworthy Way
Wallingford
Oxfordshire OX10 8DE
UK

Tel: +44 (0)1491 832111
Fax: +44 (0)1491 833508
E-mail: info@cabi.org
Website: www.cabi.org

CABI
WeWork
One Lincoln St
24th Floor
Boston, MA 02111
USA

Tel: +1 (617) 682 9015
E-mail: cabi-nao@cabi.org

A catalogue record for this book is available from the British Library, London, UK.

Library of Congress Cataloging-in-Publication Data

Names: Dieke, Peter U. C., editor. | King, Brian, 1957- editor. | Sharpley, Richard, 1956- editor.
Title: Tourism in development : reflective essays / edited by Peter U. C. Dieke, Brian E. M. King, and Richard Sharpley.
Description: First Edition. | Boston : CAB International, 2020. | Includes bibliographical references and index. | Summary: "This thought-provoking book challenges the orthodoxy of tourism and development, highlights past and current events as well as the repercussions and further developments well into the future"-- Provided by publisher.
Identifiers: LCCN 2020023968 (print) | LCCN 2020023969 (ebook) | ISBN 9781789242812 (Hardback) | ISBN 9781789242829 (eBook) | ISBN 9781789242836 (ePub)
Subjects: LCSH: Tourism. | Sustainable development.
Classification: LCC G155.A1 T59213 2020 (print) | LCC G155.A1 (ebook) | DDC 338.4/791--dc23
LC record available at https://lccn.loc.gov/2020023968
LC ebook record available at https://lccn.loc.gov/2020023969

ISBN: 9781789242812 (hardback)
9781789242829 (ePDF)
9781789242836 (ePub)

Commissioning Editor: Claire Parfitt
Editorial Assistant: Lauren Davies
Production Editor: Tim Kapp

Typeset by Exeter Premedia Services Pvt Ltd, Chennai, India
Printed and bound in the UK by CPI Group (UK) Ltd, Croydon, CR0 4YY

Contents

Contributors

David Airey was among the first group to graduate with a Master's degree in tourism at the University of Strathcyde, taught by Kit Jenkins. Currently he is Emeritus Professor of Tourism at the University of Surrey. Before retiring as pro vice-chancellor of the university in 2014, he had spent most of his career involved in education for tourism – as an academic, as a civil servant and as a consultant. Since retiring he has continued his academic work as a teacher, researcher and adviser. He is known for his work in developing tourism as a subject of study and is the author of a number of publications on tourism and related issues. He is a member of national and international organizations concerned with tourism including elected Fellow of the International Academy for the Study of Tourism and the Academy of Social Sciences. In 2006 he received the Ulysses Prize from the United Nations World Tourism Organization for his work on developing tourism knowledge. SHTM, University of Surrey, Guildford, Surrey GU2 7XH, UK. E-mail: d.airey1@btinternet.com

Tom Baum is Professor of Tourism Employment in the Department of Work, Employment and Organization, University of Strathclyde, Glasgow. His research is focused on policy and practice dimensions of frontline service work, particularly within tourism and hospitality. He completed his PhD under the supervision of Professor C.L. Jenkins. Department of Work, Employment and Organisation, University of Strathclyde Business School. E-mail: t.g.baum@strath.ac.uk

Richard Butler is Emeritus Professor at Strathclyde University, Glasgow, and has taught at universities in the UK, Canada, Australia, Hong Kong, The Netherlands and Italy. A geographer by training, his research interests include destination development, sustainability, insular and remote regions, and links between tourism and political change and religion. A past president of the International Academy for the Study of Tourism and the Canadian Association for Leisure Studies, he has published over 20 books and 100 papers on tourism. He has been an adviser to and conducted research for UNWTO and governments in Canada, UK and Australia. In 2016 he was awarded the Ulysses Medal by UNWTO for excellence in the creation and dissemination of knowledge. E-mail: richard.butler@strath.ac.uk

Donna Chambers is Professor of Tourism and Chair of the Faculty of Business, Law and Tourism Research Institute at the University of Sunderland, UK. She has been employed in Higher Education in the UK since 2003 and she previously worked with the ministry responsible for tourism in Jamaica for five years. She has published books and articles on tourism and decolonization, tourism and gender, visual research methods in tourism, and tourism policy, and she adopts a very critical and interdisciplinary approach to her research. She is an Associate Editor of *Annals of Tourism Research*, a Managing Editor of *Leisure Studies* and is also on the Editorial Board of *Tourism Critiques: Practice and Theory*. Faculty of Business, Law and Tourism, University of Sunderland, St Peter's Campus, Sunderland, SR6 0DD, UK. E-mail: donna.chambers@sunderland.ac.uk

Dora Dongzhi Chen is currently a PhD candidate of the College of Tourism Service and Management at Nankai University. Her research mainly focuses on tourism policy and tourism destination governance. No. 38 Tongyan Road, Haihe Education Park, Jinnan District, Tianjin,300350, P.R.China. E-mail: chendongzhi@mail.nankai.edu.cn

Erik Cohen is the George S. Wise Professor of Sociology (emeritus) at the Hebrew University of Jerusalem, where he taught between 1959 and 2000. He has conducted research in Israel, Peru, the Pacific Islands and, since 1977, Thailand. He is the author of about 250 publications, including several books, including *Contemporary Tourism: Diversity and Change* (Elsevier, 2004) and *Explorations in Thai Tourism* (Emerald, 2008), and he recently edited (with Sam Spector) a volume on *Space Tourism* (Emerald 2019). Erik Cohen is a founding member of the International Academy for the Study of Tourism. He was awarded the UN World Tourism Organization's Ulysses Prize for 2012. He presently lives in Thailand. E-mail: erik.cohen@mail.huji.ac.il

Peter Dieke is Visiting Professor of Tourism at Makerere University Business School, Kampala, Uganda. He has held prior academic positions at a number of institutions including the University of Nigeria, University of Strathclyde (UK), Emirates Academy of Hospitality Management, Dubai (UAE), and George Mason University, Virginia (USA). He completed his undergraduate education at the University of Nigeria in 1975, a Master's degree at the University of Massachusetts in 1981 prior to his pursuit of a doctoral degree in the field of tourism development at the University of Strathclyde in 1988. Professor Carson L. Jenkins was his doctoral research degree supervisor at Strathclyde. His principal research interest is in the developmental aspects of tourism, focusing on policy, planning and implementation strategies for the developing world, with a regional interest in sub-Saharan Africa. He has published widely in these areas. He has also advised governments and undertaken assignments for several international organizations, most notably the United Nations Development Programme (UNDP); UN World Tourism Organization (UNWTO); United Nations Educational, Scientific and Cultural Organization (UNESCO); UN Economic Commission for Africa (ECA); the British Council; and the African Union. 11E Ellisland Road, Cumbernauld, Glasgow, G67 2HG, Scotland, UK. E-mail: pucdieke@gmail.com; pdieke@gmu.edu

Larry Dwyer is Visiting Professor at the University of Technology, Sydney and Griffith Institute for Tourism (GIFT). He completed his PhD in Philosophy at the University of Western Ontario, Canada. Larry was subsequently appointed Professor of Tourism Management at the University of Western Sydney before taking up the position of Qantas Professor of Travel and Tourism Economics at the University of New South Wales. Larry is an elected Fellow and past President of the International Academy for the Study of Tourism, and past President of the International Association for Tourism Economics. His research interests are in the areas of tourism economics and policy, tourism planning and tourism management. Business School, University of Technology, Sydney, Broadway, Sydney Griffith Institute for Tourism, Gold Coast Campus, Queensland. E-mail: larry.dwyer@uts.edu.au

Afamefuna Paul Eyisi is currently a final year PhD student in the Tourism and Events Program, College of Arts, Business, Law and Social Sciences (ABLSS) at Murdoch University in Western Australia. He completed his MA in Tourism and Cultural Resource Management, BA in Archaeology and Tourism and Diploma in Tourism and Museum Studies at the Department of Archaeology and Tourism, University of Nigeria, Nsukka in Enugu State, where he lectured from 2011 to 2017 before leaving for his PhD programme in Australia. His research and teaching interests focus on responsible and sustainable tourism, tourism planning and development in

Africa, residents' attitudes toward the impacts of tourism, environmental and social impact assessment, public archaeology, and ethnographic research methods. He has published articles on archaeology and tourism development in Nigeria. Tourism and Events Program, College of Arts, Business, Law and Social Sciences, Murdoch University, 90 South Street, Murdoch, WA 6150, Perth, Western Australia. E-mail: afamefuna.eyisi@unn.edu.ng

Alison M. Gill is a Professor Emerita at Simon Fraser University, Vancouver, Canada, with a joint appointment in the Department of Geography and the School of Resource and Environmental Management. All of Alison's degrees are in Geography – BA University of Hull, UK; MA University of Alberta, Canada and PhD University of Manitoba, Canada. Her research and publications have centred around growth and change in mountain resort communities with emphasis on issues of governance, community planning, and sustainability. She is a Fellow of the International Academy for the Study of Tourism and the Royal Canadian Geographical Society. Department of Geography, Simon Fraser University, Burnaby, BC, V5A 1S6, Canada. E-mail: agill@sfu.ca

Ulrike Gretzel is a Senior Fellow at the Center of Public Relations, University of Southern California, and serves as the Director of Research at Netnografica, a market research company that extracts insights from online conversations. She received her Ph in Communications from the University of Illinois at Urbana-Champaign. Her research focuses on the impact of technology on human experiences and the structure of technology-mediated communication. She studies social media marketing and destination marketing, influencer marketing and the emerging reputation economy. She has also researched the design of intelligent systems in tourism, smart tourism development, technology adoption and non-adoption in tourism organizations, tourism in technological dead zones, and the quest for digital detox experiences. She is frequently acknowledged as one of the most cited authors in tourism and her work has been quoted in major news outlets such as the *New York Times* and *The Australian*. University of Southern California, 3502 Watt Way, Los Angeles, California, 90089, USA. E-mail: gretzel@usc.edu

Joan C. Henderson studied for an MSc in Tourism at the Scottish Hotel School from 1975 to 1976 when she was taught by Professor Jenkins. Subsequently obtaining a PhD from Edinburgh University, she had periods of employment in UK public and private tourism sectors before becoming a lecturer. She later moved to Singapore and was an Associate Professor at Nanyang Business School, Nanyang Technological University, for over 20 years before returning to the UK in 2019. Her research work is reflected in a range of published articles and other materials and includes aspects of tourism in South-east Asia and the Middle East, tourism and heritage, and the politics of tourism. Now retired from full-time employment and resident in Scotland, she remains a student of tourism with a keen interest in matters of tourism development. Lochearnhead, Perthshire, Scotland FK198PX. E-mail: jch3011@outlook.com

Carson 'Kit' Jenkins is Emeritus Professor of International Tourism at Strathclyde University, Glasgow, Scotland. He is also currently Visiting Professor at the College of Tourism and Services Management, Nankai University, P.R. China and at the International College of Digital Innovation, Chiang Mai University, Thailand. Educated at Hull and Strathclyde Universities, Kit's post-doctoral work was in the field of industrial development and shipbuilding economics. His first overseas position as Research Fellow, Institute of Social and Economic Research, University of the West Indies, Barbados, saw his first involvement in tourism research. His career since then has been devoted to teaching, research and consultancy on various aspects of tourism as a development initiative in developing countries. He has undertaken tourism-related work for most of the international development agencies in over 80 countries. E-mail: kitcarsonjenkins@outlook.com

Brian King is Professor and Associate Dean in the School of Hotel and Tourism Management at The Hong Kong Polytechnic University. He completed his MSc in Tourism at the Scottish Hotel School under the supervision of Professor Kit Jenkins and returned to the school as a lecturer after spending time in the tourism industry. Since moving to Australia (1986) and Hong Kong (2012) his teaching and research has focused on tourism in the Asia-Pacific region with an emphasis on cultural dimensions and leadership. He has published books and articles on tourism marketing, integrated resorts, VFR travel and Asia-Pacific tourism. He is Co-editor-in-Chief of *Tourism, Culture and Communication*. School of Hotel & Tourism Management, The Hong Kong Polytechnic University, 17 Science Museum Road, East TST, Kowloon, Hong Kong SAR, China. E-mail: Brian.king@polyu.edu.hk

Louisa Yee-Sum Lee is Assistant Professor in Lee Shau Kee School of Business and Administration at the Open University of Hong Kong. Louisa earned her PhD in Tourism Management at James Cook University, Australia. Professor Kit Jenkins was one of the external examiners of her PhD dissertation. The teaching and research of Louisa has focused on cultural tourism, urban tourism and service quality management. Recently, Louisa obtained a research grant on the touristification of intangible cultural heritage in city tourism destinations. She is one of the editorial board members of *Journal of Tourism Futures*. Lee Shau Kee School of Business and Administration, The Open University of Hong Kong, 30 Good Shepherd Street, Ho Man Tin, Kowloon, Hong Kong SAR, China. E-mail: Yslee@ouhk.edu.hk

Shaun Mann is a Senior Tourism Development Specialist at the World Bank in Washington, DC. He started his career in tourism in the private sector as a tour operator and lodge developer in his home country of Uganda in 1990. In 1997 he joined the Uganda Tourist Board overseeing the growth of Uganda's diverse tourism offer (e.g. primate viewing, birdwatching, sport fishing, white-water rafting, community tourism). He worked with Kit Jenkins on a new tourism strategy for the country in 2000. Shaun left Uganda in 2003 to complete a Master's in Tourism Administration at George Washington University in Washington, DC, and in 2005 joined the World Bank, just down the street. He has worked on tourism development in 35 countries since then and was instrumental in positioning tourism as a key development sector within the World Bank where the lending portfolio as of 2020 exceeds US$4 billion. The views of the author do not represent the views of the World Bank. The World Bank Group, 1818 H Street, Washington, DC, 20433, USA. E-mail: shaunhmann@yahoo.com

Shelagh Mooney is Associate Professor and Hospitality Postgraduate Program Leader in the School of Hospitality and Tourism, Auckland University of Technology, New Zealand. Shelagh's critical research explores the effects of gender and other dimensions of diversity in employment, focusing on organizational processes within specific societal and industry contexts. Current projects focus on gendered organizations, sustainable human resources management and how universities may best 'future-proof' students in a shifting and dynamic employment landscape. School of Hospitality & Tourism, City Campus AH Building (level 5), 49 Wellesley Street East, Auckland Central, Auckland 1010, New Zealand. E-mail: shelagh.mooney@aut.ac.nz

Alastair M. Morrison, Ph is Research Professor in Marketing, Events and Tourism at the Business School, University of Greenwich. Formerly, he was Distinguished Professor Emeritus at Purdue University, USA, specializing in the area of tourism and hospitality marketing. Alastair holds earned degrees from the University of Strathclyde (BA), Michigan State University (MBA) and Purdue University (Ph). He has lived and worked in five different countries and is an experienced international management consultant. He has published approximately 300 academic articles and conference proceedings and is the author of five books on tourism marketing and development – *Marketing and Managing Tourism Destinations*, 2nd edition (Routledge, 2019); *The Tourism System*, 8th edition, (Kendall/Hunt Publishing Company, 2018); *Global Marketing of China Tourism* (China Architectural & Building Press, 2012); *Hospitality and Travel Marketing*, 4th edition (Delmar Publishers, Inc., 2010); and *Tourism: Bridges across Continents* (McGraw-Hill Australia, 1998) – and Co-Editor of the *Routledge Handbook of Tourism Cities*. Prof. Morrison is the Co-Editor-in-Chief of the *International Journal of Tourism Cities*. Business School, University of Greenwich, Old Royal Naval College, Park View, London SE10 9LS, UK. E-mail: a.morrison@greenwich.ac.uk

Nneoma Grace Ololo is a lecturer in the Humanities Unit of the School of General Studies, with parallel teaching appointment in the Department of Archaeology and Tourism, at the University of Nigeria, Nsukka. Her principal doctoral research interest focused on the developmental aspects of event tourism, and tourism and its sustainability in emerging economies. Nneoma also has research interest in research methodology, planning, policy and responsible/sustainable tourism development in the emerging economies with particular interest in Nigeria. She has published articles and books on tourism, cultural tourism, and philosophy and human existence. Humanities Unit, School of General Studies, University of Nigeria, Nsukka, 410001, Enugu State, Nigeria. E-mail: nneoma.ololo@unn.edu.ng

Philip L. Pearce was the Foundation Professor of Tourism at James Cook University, Australia, and a Distinguished Professor of that University. He has long-standing interests in tourist behaviour and experience. His most recent book, in 2019, was the large edited volume, *Tourist Behaviour: The Essential Companion*. He knew Kit Jenkins for 30 years and shared many agreeable conversations, common views and good times. Louisa Lee is one of Philip's successful 50 PhD students and, like many others, is now flourishing in her post-PhD world. College of Business, Law and Governance, James Cook University, Townsville, Queensland, Australia, 4811. Professor Pearce passed away in 2020 shortly before the publication of this volume and is greatly missed by his various former collaborators amongst the current chapter authors as well as more widely across the tourism scholarly community.

Hanqin Qiu is the Distinguished Professor at Nankai University, and the Dean of College of Tourism and Service Management. She is also the Vice-President of United Nations Educational, Scientific and Cultural Organization (UNESCO) Hong Kong Association, the Editor-in-Chief of *Journal of Quality Assurance in Hospitality & Tourism* and the consultant on education administration to the Sun Yat-sen University. Her research interests focus on the analysis of tourism demand, consumer behaviour, inbound and outbound tourism in China, tourism development and policy issues in Asia Pacific, and tourism education innovations. No. 38 Tongyan Road, Haihe Education Park, Jinnan District, Tianjin,300350, P.R.China. E-mail: qiuhanqin@nankai.edu.cn

Richard Sharpley is Professor of Tourism and Development at the University of Central Lancashire, Preston, UK. He has previously held positions at a number of other institutions, including the University of Northumbria (Reader in Tourism) and the University of Lincoln, where he was professor of Tourism. He is Co-editor of the journal *Tourism Planning & Development* and a member of the editorial boards of a number of other tourism journals. His principal research interests lie within the fields of tourism and development and the sociology of tourism and he has published widely in both. His books include *Tourism and Development: Concepts and Issues, 2nd Edition* (2014, with David Telfer) *Tourism and Development in the Developing World* (2nd edition 2016, with David Telfer), *Tourism, Tourists and Society* (5th edition 2018); *Tourism, Development and Environment: Beyond Sustainability* (2009); Mass *Tourism in a Small World* (2017, with David Harrison) and, most recently, *A Research Agenda for Tourism and Development* (2019, also with David Harrison). Lancashire School of Business & Enterprise, Greenbank Building, University of Central Lancashire, Preston, PR1 2HE, UK. E-mail: rajsharpley@uclan.ac.uk

Terry Stevens is the Founder and MD of the multi-award-winning, international tourism consultancy Stevens & Associates. He has now worked in over 50 countries around the world on a broad range of projects for international development organizations, national and regional tourist boards and many private sector clients. After completing a geography degree in Swansea and an MSc at Reading, he worked for private estates, the Pembrokeshire Coast National Park (during which time he completed his PhD) and a number of public bodies before becoming Professor and Dean in Swansea IHE and Director of the UK Stadia and Arena Management Unit. He was invited by Kit to be a visiting speaker at the Scottish Hotel School and external examiner for the master's and doctoral programmes. In 2019, and again in 2020, Terry was awarded the LUXLife International Destination Development Expert of the Year Award. He has written extensively on all aspects of tourism and destination management. In March 2020 he launched a new website to help build a community of debate about the future recovery of tourism in the post-COVID19 era – www.tourism-futures.com. The Studio, 10 Bwrw Road, Loughor, Swansea, Wales, SA4 6TX. E-mail: terry@stevensassoc.co.uk

Sienny Thio is an Assistant Professor and Head of the Hotel Management Program in the Faculty of Business and Economics at Petra Christian University, Surabaya. She completed her Master of Business in Hospitality Management at Victoria University in Melbourne and obtained her doctoral degree in hotel and tourism management from the Hong Kong Polytechnic University under the supervision of Professor Brian King. Her teaching and research areas are related to the hospitality and tourism field, focusing on human capital, customer and tourist behaviour, and special interest tourism. She has presented papers at conferences both at home and abroad, and published papers in several journals. Faculty of Business and Economics, Petra Christian University, 121–131 Siwalankerto, Surabaya, East Java, Indonesia. E-mail: sienny@petra.ac.id

Geoffrey Wall is Distinguished Professor Emeritus at the University of Waterloo. He was educated mainly in the United Kingdom and has lived in Canada since 1974. He has done research and been involved in multi-year projects in Asia, particularly in Indonesia and China, including Taiwan. He is interested in the impacts of tourism of different types on places with different characteristics and the implications of this for planning and management. Recent studies have addressed issues of sustainability, climate change, and indigenous people and heritage, both natural and built. Department of Geography and Environmental Management, University of Waterloo, Waterloo, Ontario N2L 3G1, Canada. E-mail: gwall@uwaterloo.ca

Stephen Wanhill is the Research Director of Global Tourism Solutions (UK), which is a company that is dedicated to producing tourism facts and figures for many local authorities in Britain, so as to assess tourism's local economic impact. From this it follows that his principal research interests are in the field of tourism destination development. He is currently Visiting Professor at the School of Management, Swansea University, Visiting Professor at the University of Enna "KORE", Sicily, and Emeritus Professor of Tourism Research, Bournemouth University, UK. He has been a board member of the Wales Tourist Board with responsibilities for the development and research divisions. In this capacity, he spent much time reviewing and recommending grant applications for projects that encompass accommodation, restaurants and pubs, attractions and public facilities, such as country parks, visitor centres and infrastructure improvements, as well as contributing to the rolling five-year development plan of the board. West Cross, Swansea, SA3 5BR, Wales, UK. E-mail: stephen@wanhill.force9.co.uk

Allan Williams is Professor of Tourism and Mobility Studies at the University of Surrey where, until recently, he was Research Director at the School of Hospitality and Tourism Management. He studied at the University of Wales and the London School of Economics, and has held posts at the LSE, and the universities of Durham and Exeter, before moving to tourism studies at Surrey in 2011. His research focus is the relationship between mobility and economic development, especially innovation, knowledge, productivity and uncertainty. Previously co-editor of *Tourism Geographies* and *European Urban and Regional Studies*, he first met Kit Jenkins when he became a member of the International Academy for the Study of Tourism. School of Hospitality and Tourism Management Studies, University of Surrey, AP Building, Guildford GU2 7XH, UK. E-mail: allan.williams@surrey.ac.uk

Roy Wood has held professorships at the University of Strathclyde; Breda University of Applied Sciences; the University of Macau (where he was also the Business Faculty Associate Dean for Curriculum and Teaching) and Leeds Beckett University. In addition, he has been Principal and Managing Director of IMI University Centre, Lucerne, Switzerland; Dean of the Oberoi Hotel Group Centre for Learning and Development, India; and Chief Operating Officer of the Gulf Hospitality and Tourism Education Company in Bahrain. Wood is the author, co-author, editor or co-editor of over 200 publications including 16 books and over 50 refereed research papers. In 2014 he was awarded an Honorary Fellowship of the UK Council for Hospitality Management Education in recognition of his contribution to the hospitality field. He is currently a visiting professor of tourism and the visitor economy at the University of Cumbria. E-mail: roy.wood@cumbria.ac.uk

Foreword

It is my great pleasure and privilege to write a foreword to this book. The idea for the book was first proposed by Peter Dieke. Originally it was intended as a platform for the many University of Strathclyde (The Scottish Hotel School) graduates who had developed both careers and reputations in tourism academia and practice. Its focus was to be on the role and potential of tourism as a development option in the developing world. After many discussions which greatly benefitted from Brian King and Richard Sharpley joining the editorial team, the scope of the book was enlarged. We agreed to retain the focus of the book on the developing countries but to invite as potential contributors many of the outstanding academics in the field, together with experienced consultant/practitioners with whom we had worked. We were overwhelmed by the favourable response to our invitation. The potential and actual contributions offered not only a depth of experiential learning and academic standing but also of global experience. The contributors were told that we were not preparing a textbook but were asked to write a reflective and critical essay on the subject of their choice, to offer insights arising from their experiences into the central question of the book; How could tourism contribute to development in the developing countries?

My input to the book was very limited! It was the outstanding editorial team which guided the book from concept to conclusion. We all gratefully acknowledge the guidance of Richard Sharpley in formatting the contributions to provide a systematic structure to the book, and he was our contact with CABI, to which we are grateful for its support and help.

With such an ambitious goal, we are aware that there are many gaps in our coverage of the topic. However, there is enough in the book to, hopefully, further encourage scholarship into the challenging and dynamic tourism and development nexus.

To you all, my sincere thanks.
Kit Jenkins

1 Reflections on 50 Years of Tourism and Development

Carson Jenkins*
Strathclyde University, Glasgow, Scotland

Introduction

The purpose of this chapter is to reflect on over 50 years of observation, engagement and practice of designing, implementing and monitoring tourism projects in developing countries. The focus of this activity was to use tourism to facilitate development objectives in these countries. As an academic, my intent was to explore how significant tourism could be in initiating and sustaining development impacts; as a practitioner, it was to evaluate what techniques and methods best worked to achieve the project outcomes. Over the years, it became apparent that many projects were stand-alone and not demonstrably part of a tourism development strategy. Putting it another way, these projects, although often financially successful, did not necessarily contribute to a vision for the tourism sector. So, as a result of these observations I formed a conclusion that 'policy precedes planning'. At its simplest, if there was no policy for the tourism sector, then what were you planning to achieve? Questions such as these began to widen the debate on the potential role of tourism as a development option. The primary focus on the rate of return on financial investment and economic impacts did not necessarily diminish within this debate; however, broader issues of human welfare, social, cultural and environmental consequences of investment in tourism came to be considered and are now reflected in terms such as responsible tourism and sustainable tourism.

Definitional Ambiguities

Tourism

There is no agreed definition of what constitutes tourism. Attempts have been made to do so but none have won widespread support. Perhaps the problem relates to the understanding of what tourism activity involves – at its most nebulous, it can be described as being 'what tourists do'. It therefore becomes very difficult to provide parameters for tourist activity, and if it is not possible to define the phenomenon, it becomes even more difficult to measure it. For this reason, early tourism economic impact studies using multiplier analysis traced expenditure made by tourists through rounds which eventually become too small or difficult to trace. Even the United Nations World Tourism Organization's (UNWTO) development of Tourism Satellite Accounts, which has standardized impact measurement between countries, is not comprehensive.

*kitcarsonjenkins@outlook.com

DOI: 10.1079/9781789242812.0001

Development

This word is again controversial as it is usually used to refer to an index of Gross National Income (GNI) per capita, indicating how this has increased/decreased annually in a specific country, with an increase signifying development. However, many observers, such as de Kadt (1979) and Brown (1998), recognized that much of the development increase did not seem to improve the quality of life of residents. To consider a wider index of improvements, in 1990 the United Nations Development Programme introduced its annual Human Development Report which uses several social, medical, longevity and related data to measure how residents' lives have improved over the years in member countries (UNDP, 1990). So, the concept of development was broadened from being a specific change in economic output to being understood as a means of improving the lives of people in a country, depending, of course, on how that surplus was used (but see also Chapter 2 in this volume).

Developing countries

This is another controversial term because of the heterogeneity between countries. The World Bank had classified all countries in the lowest two-thirds of its Gross National Income (GNI) per capita tables as developing, while the International Monetary Fund and the United Nations both used different criteria. In 2016, this led the World Bank to abandon its classification between developed and developing countries, stating that 'all countries need development!' (World Bank, 2016). When you consider the disparity in development levels between, for example, Chad, Burundi and China, all of which were previously classified as developing, you can understand the nature of the problem of definition. Another factor was the superseding of the Millennium Development Goals, formulated in 1990 and targeted to help the poorest countries, by the Sustainable Development Goals which are more focused on reducing poverty and inequality of distribution (World Bank, 2019a) The two main targets of the Bank are, by 2030, to end extreme poverty and to promote shared prosperity in every country in a sustainable manner. This is a huge challenge, not least because the Bank's statistics indicate 34 low-income countries with a GNI per capita of US$995 – sometimes referred to as 'least developed' countries – and its lending criteria indicate 59 countries eligible for International Development Agency lending on the most favourable 'soft lending' terms (World Bank, 2019b). What this diversity of statistics suggests is that there are no universal definitions of what constitutes poverty or inequality; each country would have to be looked at as an individual set of circumstances and then attempts made to define measurement parameters. United Nations statistics define least developed countries according to three criteria: low income; human resources weaknesses; and economic vulnerability; of the 46 countries so classified, 34 are in Africa, six are in Asia, five in Australia and the Pacific and one in the Caribbean (UN, 2018).

Despite the challenges set out by the various statistical anomalies, the developing countries (and levels within them) are easier to recognize than to define. Even within groups (for example, Small Island Economies, sub-Saharan Africa) there can be huge variations between countries. This makes it inevitable that this chapter will have to generalize 'developing countries' but, wherever appropriate, provide examples to support the arguments advanced. However, the focus of the chapter – the question: Is tourism a development option? – will not change.

Apparent Links between Tourism and Development

The problems restricting economic growth and development in poor countries are well known and they are both economic and social as well as often embracing a political dimension (Harrison, 2001; Mowforth and Munt, 2003; Huybers, 2007; Sharpley, 2009). As noted by Jenkins (2015, p. 145), 'although developing countries share many problems in common, the intensity of those problems and potential solutions may require different approaches. In this sense it is unrealistic to expect a single model of development; the way forward will be to be aware of the collective experiences of other developing countries but to seek solutions and actions which clearly reflect the circumstances of the individual country.' Efforts to stimulate growth are focused on economic initiatives to provide a surplus to allow some development to take place. Without this economic surplus, the country will be entirely dependent on aid or other forms of external assistance to provide such a stimulus. Except for oil-exporting and former developing countries, such as the United Arab Emirates (UAE), probably the major problem for developing countries is their lack of foreign exchange. With this limitation, they are unable to buy in the foreign expertise, equipment and goods and services to develop required infrastructure. The UAE are a good example of where oil wealth has funded a country's modernizing programme which has created cities such as Dubai and Abu Dhabi and where tourism is firmly situated as a major sector in the economy. Compare this to Nigeria, also oil-rich but suffering continuously from endemic corruption, unstable politics and poor leadership and where tourism potential is seriously neglected. Another current example would be Venezuela, where oil wealth has been dissipated, hyper-inflation exists, and the tourism sector has severely declined. These examples show us that for development to take place, even the availability of resources may not be enough in the absence of political leadership.

When growth declines or stagnates over an extended period, its impacts are often intensified by other problems not uncommon in developing countries, such as population pressures, unemployment, urbanization trends and the lack of a social security network to support the poorest in the society. Where these conditions prevail, then often the only stimulation for growth can come from external assistance, either through development aid or by borrowing from development agencies such as the World Bank and its associates such as the International Finance Corporation (IFC) and the International Development Agency, or through support from government bilateral development organizations such as the German DTZ, the UK's Department for International Development and the American USAID. The European Union, particularly through its Lomé Agreement under which the European Community (now European Union) provides aid and extends trade and tariff protection to 62 African, Caribbean and Pacific (ACP) states, supports tourism and other development projects in these countries. Except for aid provision, which can be inefficient if tied to purchases only from the donor country, borrowing must be repaid in foreign exchange.

Until the mid-1970s, the biggest source of development funding – the World Bank – did not lend directly for tourism projects with indirect support coming from infrastructural projects and through its IFC lending. The Bank now supports lending for economically and financially viable tourism projects, so what has induced this change in approach? Perhaps the answer to that question is found in the nature and characteristics of international tourism trends. It is worth noting that with the exceptions of China and India, both with massive and important domestic tourism sectors, most other developing countries are concentrating on attracting inbound tourists, primarily to earn foreign exchange and other economic benefits such as contributions to government revenues, employment generation, the creation of tourism supply chains and regional development opportunities. In addition to these important economic benefits, there is the undisputable fact that international tourism is a growth industry. Using the United Nations World Tourism Organization's (UNWTO) tourism statistics and those provided by its predecessor, the World Tourism Organization, these show that between 1950 and 2018, international tourist arrivals increased from 25 million to around 1.4 billion, an average annual increase of 4%. This compares to the growth in world trade over the same period of just under 4%.

Moreover, international tourist arrivals in 2018 were 6% higher than in 2017, the best growth rate in the last decade (UNWTO, 2019a). This does not mean that every region or tourist-receiving country has enjoyed this experience but, on a global basis, even in times of economic decline tourism has remained steady as an international export. It also has demonstrated remarkable resilience in recovering from declines in arrivals from various causes, such as the SARS (Severe Acute Respiratory Syndrome) outbreak in China from November 2002 to July 2003 and the global financial crisis of 2007–2008. A further factor is that most international tourists travel from the developed countries, bringing with them hard currencies to spend in the destination countries. To add to this redistribution effect, there is almost a complete absence of barriers to travel for residents of the developed countries making international tourism trips, and within a tariff-free environment. Compare this situation to the one in many developing countries where exports are mainly primary goods, simple manufactured goods and agricultural products. By investing in tourism, they are moving from a tariff and quota restricted trade environment to one where no tariffs apply and there is a potential for earning foreign exchange. It is for these reasons that tourism has been recognized as a development initiative where circumstances permit and where few other options exist. Also, the need to develop infrastructure, amenities and services for international tourists is sometimes described as creating an image of innovation and modernity for a country as opposed to depending entirely on the agricultural and other primary sectors. Most developing countries, of course, are at least dual-economy, with the tourism sector and the agricultural sectors coexisting.

Many early observers of tourism, such as Bryden (1973), Turner and Ash (1975), O'Grady (1981), Britton (1981) and Lea (1988), had noted that the modernization argument can be fragile, as in most developing countries much of the tourism development is financed and maintained from external sources, often referred to as the *dependency* argument. The classic and seminal paper written by Gunder Frank (1966) on the 'Development of Underdevelopment' reflects very well the focus of tourism dependency in most developing countries. Much of tourism funding comes from external sources which can be relatively expensive to secure and must be repaid in foreign currency. In addition, the government may have to offer a range of inducements to foreign investors such as tax rebates, concessions on import duties, tax holidays and other fiscal measures to attract investors. These investment incentives are a form of subsidies to investors and distort the cost of investment. It is often the case that the destination is dependent on the foreign travel trade to attract tourists to the destination and accommodate them, and on foreign carriers for transport, one notable example being Gambia in West Africa. In these circumstances, it is said that the destination is very dependent on external suppliers who can substitute other destinations as they wish. In the case of Gambia, the tourism sector is a major part of the national economy but most of the accommodation is owned by foreign companies and the country has no national airline to transport tourists. Consequently, it is highly dependent on external suppliers and has limited means to mitigate these circumstances. In his early analysis of the economic impact of tourism on developing economies, Cleverdon (1979) calculated that only 20-25% of tourism expenditure remained in-country, with huge financial leakages. In some island countries, such as the Bahamas, the leakage factor may be in excess of 50%. Hence, there are powerful arguments (Burns and Novelli, 2008; Sharpley and Telfer, 2014) to heed the warnings of dependency, particularly where tourism activity is seasonal. However, the question remains: If not tourism, what other diversification options might there be? It also raises a question to be addressed below of what actions can be taken to mitigate the risks of dependency.

The Identification of Tourism Projects

Identification can come from various sources. It could be a government initiative where the government recognizes tourism potential, seeks funding and often manages the facility, acting like an entrepreneur. It can, of course, also fund the project and operate it under a management contract with an international company with expertise in the field. In many cases, such as the Caribbean island countries, tourism activity was already established and potential for growth was seen. Other possibilities are to approach one of the international development funding agencies for technical assistance to investigate and evaluate the potential for tourism development, usually through a comprehensive tourism (sector) development plan. This was the main approach used by the United Nations Development Programme (UNDP) and the United Nations World Tourism Organization (UNWTO); similar approaches were also used by the European Union and, to a certain extent, by the World Bank, the latter funding two huge, long-term tourism projects in Cancun, Mexico and Nusa Dua, Bali, Indonesia. Smaller projects were supported by country technical aid, such as the establishment of the Utallii College in Kenya with Swiss technical assistance. The private sector was also very active in financing schemes which were evaluated as being potentially profitable, as in the case of the Young's Island luxury resort development in St. Vincent in the eastern Caribbean.

In India, the Indian Tourism Development Corporation at one time owned and operated an airline, travel agencies, transport vehicles and hotels. It acted very much like an entrepreneur developing tourism facilities for, for example, skiing, in areas which at that time had potential but not enough to attract private sector funding. When these became profitable, they were sold to the private sector and the funds reinvested in other projects. Many other countries, such as Indonesia, followed this example. In Indonesia, Bali was, and arguably still is, the major tourism destination. When starting to develop near-by Lombok island as primarily an over-spill destination to relieve pressure on Bali, the Lombok Tourism Development Corporation advised farmers not to sell their land to investors but to enter into joint-ownership schemes to benefit from future increases in land values resulting from tourism demand. There were both political and financial reasons for this advice when land ownership is a recurring problem in many areas and particularly when dominated by foreign ownership.

As noted above, there was no single development model. However, given that governments usually possessed insufficient foreign exchange to pay for foreign expertise, there were advantages in negotiating with the international funding agencies where local contributions to project costs would be 'in kind', such as providing office space, transport facilities and secretarial assistance. Another advantage in these projects was that it was usual to take a comprehensive view of development options to include social, cultural and environmental impacts. Although in order to receive funding the projects had to be financially and economically viable, they were more holistic in approach than those from the private sector where profit is the dominant factor. A further consideration in a national plan was that it would include recommendations on tourism legislation, organizational and administrative structures, marketing strategies, training requirements and budgetary considerations. The plan should be encapsulated in a tourism policy. Through counterpart training, the planning team would work with nationals and pass on their knowledge and experience of tourism and facilitate the eventual implementation of the plan which, arguably, was the weakest part of this process.

As part of the tourism development experience, a necessary step was to help initiate a training programme for tourism planners. This was successfully achieved in Indonesia by persuading, initially, two universities to introduce Master's programmes in tourism. This created a foundation of trained people who could be employed in tourism related projects, thus beginning to mitigate the problem of countervailing power. When potential private sector investors came to negotiate with their team of experienced tourism personnel, it was difficult for local officials without experience of tourism to know whether what was being proposed was fair, unfair or completely unbalanced; what were the rewards, what were the problems, how could they evaluate proposals and reach decisions? These problems remain, but there is now available a record of tourism investments in developing countries, as well as international associations such as the UNWTO and the World Travel and Tourism Council to facilitate discussions and exchange experiences. In addition, there is a growing depth to the knowledge pool which can help these countries to better manage the tourism sector.

In any discussion of tourism as a development option, there is inevitably a focus on the roles of government and the private sector in this process. Again, as in many other areas of tourism, there is no particular road map or schema which can determine what the balance of inputs should be; it may depend, for example, on the political ethos and structures of a particular country. Hence, in countries like China, Laos, Vietnam and Cuba, central planning of all development is the norm while the private sector is given a minor role. This typical example of a top-down approach to policies and plans reflects the political structures. In contrast, in other (former) developing countries, such as Singapore, Barbados and Thailand, the role of government has been diminished (or supplemented?) by the greater involvement of the private sector in tourism policies and plans, often culminating in Public-Private Sector (PPS) projects as examples of 'bottom-up' planning. Indeed, there is no doubt that focus of the funding of tourism projects in the developing countries has shifted significantly to 'bottom-up' planning'. But why has this happened?

First, in most developing countries, there is ample evidence to suggest that governments are not particularly successful at operating revenue-generating projects, such as hotels and airlines. The reason typically given to support this argument is that government is not involved in the tourism market place – it is too distant from the customer, and cannot react quickly to changes and crises. Certainly, a comparison of the profitability performance for hotels and airlines owned by either governments or the private sector shows that the latter are usually more profitable. Moreover, when government-owned entities lose money, they are usually subsidized by the tax payer; in contrast, if this happens in the private sector, companies risk bankruptcy, particularly if losses continue over a long period of time. For this reason, international funding agencies now insist on governments creating an 'enabling environment' to facilitate private sector involvement in development initiatives and, of course, by including them in any policy and planning discussions. This does not exclude government from the development process, but gives it a different role.

An indirect but fascinating speculation is that this transformation in the role of governments relates to the collapse of the Union of Soviet Socialist Republics (USSR) in September 1991 following the demolition of the Berlin Wall. Until that time, many western countries, primarily the USA and some in Europe, gave considerable development and aid monies to developing countries in order to secure their votes to support their positions in various United Nations debates. With the emasculation of the USSR and the creation of new, independent countries, aid and development providers began to take a more interventionist role in projects and this did much to encourage private sector involvement and Public–Private partnerships.

Second, the type, scale and location of tourism developments are crucial decisions in creating a sustainable sector. The chosen location must be attractive and suitable for the investor but other stakeholders must also be considered, including the local community where the development is to take place (Murphy, 1985; Tosun, 2000). Much has been written about the topic of community participation in tourism development and it is likely to continue because most observers of tourism believe that one of the inputs to a sustainable development is support from the community where it will take place and operate. However, a limitation of community involvement in the tourism development process is that many communities have no experience of tourism or, as individuals, of being a tourist; consultation might become a form of tokenism – consulting the unknowing? A continuing problem is how to define a community and whom should represent it. The early commentary by Arnstein (1969) continues to have much relevance today.

Third, any development will depend either on the availability of or the provision for developing infrastructure, principally because a lack of adequate infrastructure is often the main barrier to development projects. Traditionally, the private sector regarded infrastructure development as being a responsibility of government; however, governments are increasingly trying to involve the private sector in funding and operating infrastructural projects, such as airport construction and operation, toll roads, utility supplies and educational provisions, the purpose being both to help spread costs and to strengthen Public-Private Sector partnerships. Perhaps the most recent international example of this is China's Belt and Road Initiative which has led to the creation of the Asian Infrastructural Development Bank to help various countries undertake infrastructural projects considered essential to develop trade links, including with important tourism-endowed countries.

Fourth, who should benefit from tourism? This is an issue that has concerned academics since the early days of commentators such as Raul Prebich and Andre Gunder Frank (Telfer, 2015). Dependency school writers argue that tourism is not only inherently dependent and, therefore, vulnerable to external pressures, but also that it has minimal economic impact because of perceived high leakages from tourism revenues. Again, this is to an extent a valid argument. However, the extent of leakages from the tourism sector inevitably depends on a number of factors, such as the country's stage of development, its type of tourism, inter-sectoral linkages and what government policies there may be to limit leakages. It should also be noted that government, through its fiscal policies, is often the main beneficiary of tourism revenues and, hence, how it distributes these revenues is central to any discussion of 'equity and fairness'. Also, in any economic analysis of investment in tourism, it should be recognized that it is essentially an inter-sectoral activity and can and does act as a growth point for incremental development.

Fifth, are there or should there be any limits to the growth of tourism in a country? This is a very difficult question to answer – though one of increasing relevance given contemporary concerns with so-called 'overtourism' (Milano *et al.*, 2019) – as it relates to the very old but relevant concern of a country's or location's carrying capacity. There is no recognized method or formula to answer this question; an answer is often based on intuitive or subjective judgement. As a generalization, however, it may be suggested that it is the tourism absorptive capacity of a location – embracing transport, accommodation, amenities, infrastructure – which are the main determinants of supply. As long recognized, these determinants may, of course, be evaluated differently by visitors and residents (Doxey, 1976). For this reason, it is essential to monitor the impacts and outcomes of tourism developments. A monitoring system involving both government (local) and relevant stakeholders should be in place to relate the measurable outcomes of the tourism plan against stated targets and to make changes when appropriate and necessary. Butler's widely cited Tourism Life Cycle model (Butler, 2009) is useful to plot stage shifts in developments and what impacts these might have. A further problem to recognize here is that tourist-receiving developing countries have very limited control over how tourists choose their destinations; efforts are typically directed towards attracting tourists rather than to optimizing demand (Harrison and Sharpley, 2017).

As the Secretary-General of UNWTO (Zurab Polikashani) recently stated: 'The growth of tourism in recent years confirms that the sector is one of the most powerful drivers of economic growth and development. It is our responsibility to manage it in a sustainable manner and to translate this expansion into real benefits for all countries, and particularly to all local communities, creating opportunities for jobs and entrepreneurship and leaving no one behind. Therefore, UNWTO is focusing in 2019 on education, skills and job creation.' (UNWTO, 2019b). These are, of course, ambitious objectives and help us to discern and focus on the important dichotomy between local and global influences on tourism development.

Current Issues

As argued above, a policy is a prerequisite to the formulation of a tourism plan. Policy itself is a very complex area (Dredge and Jenkins, 2007; Edgell *et al.*, 2008; Airey, 2015) which requires consideration of a wide range of factors at the global, national (macro) and sub-national level (micro) levels. In recognizing tourism as a multi-sector activity, we also should recognize its interconnectedness with the three levels noted. Broadly speaking, policies can be described as an agreed evaluation of alternative options to achieve stated goals. However, it should not be forgotten that policies are aimed at future goals so, in one sense, how accurate the predictions of future outcomes might be is dependent on what information is scrutinized and accepted; they are 'wish lists' of targeted goals. As such, they must be flexible to respond to future, unforeseen changes. Although policy formulation might be tasked to expert individuals and/or companies, policies should be anchored to the client, whether public or private sector, in order to establish guidance for the project and to determine local involvement. In most developing countries, the client will be the government controlling the project through a management committee; moreover, that committee can have a wide representation of other stakeholders from the private sector and communities if appropriate. This is a standard way to manage these types of projects and is usually favoured by international funding agencies. The managing committee can provide a useful channel of support for the project but sometimes is the cause of tension and a source of dissention.

A useful way to sift and input information to the policy process is to use what might be described as the 'Three Box Approach. Figure 1.1 illustrates how this helps to understand both interconnectedness in tourism and the various stages of the policy formulation process.

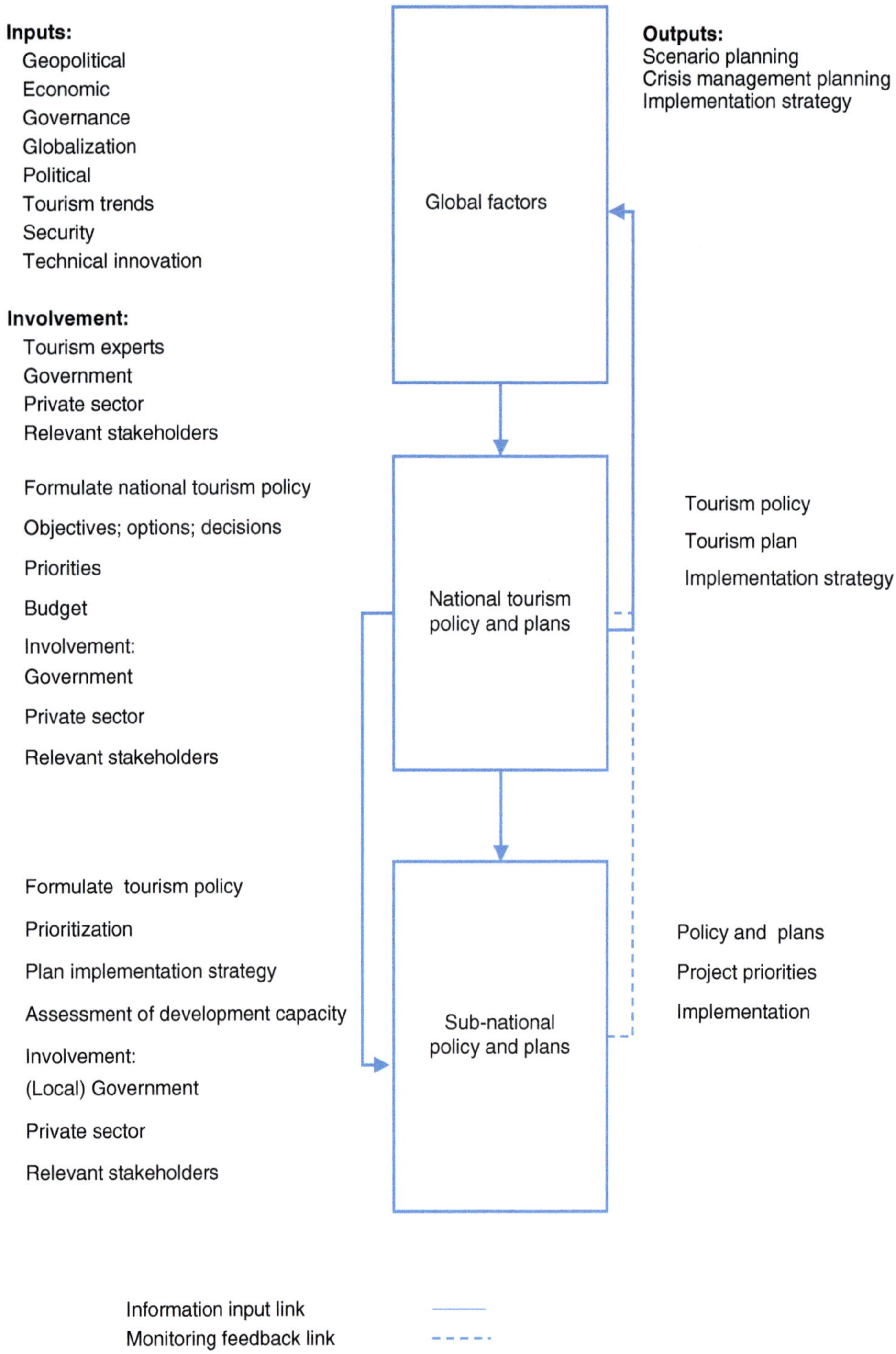

Fig. 1.1. The Three Box approach.

Box 1: Global considerations

These include considerations of global trends and issues which may impact on tourism generally, or be of relevance to specific countries or regions within a country. These factors must be mapped against the tourist-receiving countries' inbound market data – how might their supply markets be affected? There are many such issues, some which are possibly short-term. For example, the grounding of Boeing 737 MAX jets in March 2019 will temporarily affect airline capacity on some routes, dislocate flight schedules and disrupt passenger movements; what were accidents in Indonesia and Ethiopia now have had global impact. Conversely, possible market disruptions caused by the UK's 'Brexit' from Europe may have long-term implications, while the deteriorating trade relations between the USA and China will have implications for travel between and from each country. Similarly, the continuing India–Pakistan conflict (which escalated during 2019) could threaten travel confidence in the Asian region while in Africa, plans to dam the river Nile is threatening water supplies to Egypt and causing tension between that country and up-river countries. In addition, the recent terrorist atrocity in New Zealand (March 2019) is an example of how unforeseen events may impair the confidence of travellers.

There are, of course, many other concerns relevant to tourism at the global level, such as global warming, urbanization trends and the pressures of growing populations. A report from the World Economic Forum, the *Global Risks Report 2019* (WEF, 2019), includes in Chapter 3 a section on 'Future Shocks'. This sets out ten scenarios which may occur in the future and which would have dramatic consequences for global geopolitics. They include considerations such as quantum computing; food becoming a geopolitical weapon; water shortages; cybersecurity; privacy and surveillance issues; and climate change. Not only would it be difficult for any country to escape some of the consequences of these issues if they were to develop, but it is also suggested that these issues could increase and intensify uncertainty, instability and fragility – three of the pillars of the dependency argument. So, through globalization we have seen an intensification of dependency but at a global rather than at a country-to-metropole axis. But the question remains: What should policy makers put into Box 1?

Obviously, these geopolitical factors are important, particularly for some countries.

Some events can unfold quickly while others are more long-term. In tourism, the economic circumstances of countries (and of potential travellers) are important, but analysing global economic data is difficult and often contradictory. Nevertheless, the Secretary-General of the UNWTO is upbeat about the future of tourism and identifies five 'drivers of growth' as being: 'a favourable economic environment; strong outbound demand from major source markets; consolidation of recovery in key destinations affected by previous crises; enhanced connectivity; and enhanced visa facilitation' (UNWTO, 2019b). However, this forecast contrasts with the more pessimistic view expressed in World Bank publications, such as *World Economic Prospects: Darkening Skies* (World Bank, 2019c): 'Global growth is expected to slow to 2.9% in 2019. International trade and investment are moderating, trade tensions remain elevated, and financing conditions are tightening. Amid recent episodes of financial stress, growth in emerging markets and developing economies has lost momentum and is projected to stall.'

The plethora of geopolitical and financial data at global level means that policy makers must be selective in what they see relevant to be considered in Box 1. This problematic decision gives some substance to the assertion that one should think global but act local! Yet, these are essentially external influences which most countries, and certainly not developing countries, cannot influence. However, as tourism is a discretionary purchase and highly substitutable between destinations, policy decisions should nevertheless aim to be proactive rather than reactive to changes in external factors. What should be included is a form of crisis management plan to design a proactive strategy for unpredictable events, such as health scares or natural disasters – for example a volcanic eruption – in areas where these are prevalent. In other words, in the same way that every airline must have a crisis management plan for air crashes or incidents affecting flights which are outside their control, tourist-receiving countries must plan for unpredictable, but possible, scenarios. At the time of writing, for example, both the UK and the European Union are unveiling a raft of measures, including some highly relevant to tourism, to manage a post-Brexit scenario.

Box 2: Formulation of the national (macro) strategy

This box focuses on the formulation of a national tourism strategy. It is informed by the information in Box 1 and by the many other sector policies within the country that can impact on tourism and its development. These might include, for example: land use policies; foreign investment policies; investment incentives policies; transport policies (particularly air transport if this is the main access channel for tourists); employment legislation and work permit rules; and visa and government regulations relevant to tourism. As an inter-sectoral activity, tourism should not be viewed as a 'standalone' sector; it must be integrated to support national development objectives by reflecting these in targets for tourism. The outcome of these considerations and deliberations should be a national tourism development strategy which becomes the guidelines for tourism planning.

Box 3: Formulation of sub-national tourism policies and plans

The action of this box is guided by the national tourism strategy and, hence, any localized (sub-national) projects must reflect and support it. At this stage, the implementation of the strategy is undertaken through specific tourism activity plans; cohesion with the national objectives is necessary but the local circumstances are reflected in the planning process and objectives. One example of this would be training for tourism employees, addressing questions such as: for whom, what areas specifically, numbers, gender balance and so on. A second example would be to prioritize both projects and their locations. No country can develop everything at the same time; development is essentially incremental.

The main purpose of the Three Boxes approach is to reflect the three levels of information necessary to guide and co-ordinate policy formulation and planning for tourism and to recognize that different geographic areas provide different characteristics and challenges. Both policies and plans need to be flexible and this is one approach which offers some degree of integration and conformity for target achievement. Some developments in ICT, such as big data dredging, have better enabled the interpretation of massive data files and have helped the analysis of trends at the global level.

The Future of Tourism in Developing Countries

All the statistical trends suggest that international tourist arrivals will increase at the global level and that a growing number of these travellers will seek destinations in the developing countries. So, most of the developing regions, such as South-east Asia and Africa, are likely to experience variable growth trends in the foreseeable future, but any upward trend should provide more benefits to the developing countries, particularly with regards to the earning of foreign exchange. For many countries where tourism is the major export sector, this can only be good news! However, for reasons already mentioned in this chapter, developing countries are exposed to dependency risks within the tourism market and to global geopolitical and economic events. How might these risks be mitigated if not eliminated?

First, through development assistance, these countries need access to tourism expertise as a form of countervailing power against the experienced tourism investors they are likely to negotiate with within project development. Second, from the growing amount of literature on tourism in developing countries, there is increased support for the idea that tourism should be viewed as a holistic development and not be only concentrated on the potential economic impacts. Third, the organizational, administrative and fiscal measures related to the tourism sector should be evaluated and changed if necessary and appropriate. Fourth, efforts should be made to develop a country's own supply of tourism professional and technical capacity. Fifth, a monitoring system should be developed and implemented to assess or measure the outputs of the tourism sector against policy targets. Sixth, the best available ITC systems and products should be employed to provide a competitive advantage in market surveillance and product development. Seventh, there is a need to ensure that the tourism product is based on quality rather than to maximize numbers. Eighth, to encourage inter-sector linkages with tourism to minimize economic leakages. Ninth, to raise destination market profile to provide increased visibility in a highly competitive environment. Tenth, where possible and appropriate, involve stakeholders in the tourism policy and planning process.

Although these suggestions are reasonable and attainable, it must be recognized that any prioritization of the list will be dependent on the circumstances and development stage of individual countries. In reference to attracting international tourists, the problems of dependency will always be present. This is not a reason to abandon tourism as a development option but rather to be more proactive to the potential problems involved. Perhaps at the core of development is an issue rarely recognized and discussed, namely, political will. Without strong and sustained leadership, development will not take place.

In this chapter there has not been enough space to explore some of the important Box 1 items such as global warming; 'green' tourism'; the nature of 'responsible' tourism; the meaning and practice of tourism 'sustainability'; and issues of 'inter-generation equity', all topics of considerable importance in the development literature. Consideration of the 'triple bottom line' concept has also had to be excluded. However, as a personal overview of 50 years of tourism in development, hopefully the reader will accept these unavoidable limitations.

References

Airey, D. (2015) Developments in understanding tourism policy. *Tourism Review* 70(4), 246–258. DOI: 10.1108/TR-08-2014-0052.

Arnstein, S.R. (1969) A ladder of citizen participation. *Journal of the American Institute of Planners* 35(4), 216–224. DOI: 10.1080/01944366908977225.

Britton, S.G. (1981) *Tourism, Dependency and Development: A Mode of Analysis*, Occasional Paper No. 23. Development Studies Centre, Australian National University, Canberra.

Brown, F. (1998) *Tourism Reassessed: Blight or Blessing?* Butterworth-Heinemann, Oxford, UK.

Bryden, J. (1973) *Tourism and Development: A Case Study of the Commonwealth Caribbean*. Cambridge University Press, Cambridge, UK.

Burns, P. and Novelli, M. (eds) (2008) *Tourism Development: Growth, Myths and Inequalities*. CAB International, Wallingford, UK.

Butler, R.W. (2009) Tourism destination development: cycles and forces, myths and realities. *Tourism Recreation Research* 34(3), 247–254. DOI: 10.1080/02508281.2009.11081600.

Cleverdon, R. (1979) *The Economic and Social Impacts of International Tourism on Developing Countries*. Economist Intelligence Unit, London.

de Kadt, E. (1979) *Tourism: Passport to Development?* Oxford University Press, New York.

Doxey, G. (1976) When enough is enough – the natives are restless in old Niagara? *Heritage Canada* 2(2), 26–27.

Dredge, G. and Jenkins, J. (2007) *Tourism Planning and Policy*. John Wiley & Sons, Milton, Queensland, Australia.

Edgell, D., Allen, M., Smith, G. and Swanson, J. (eds) (2008) *Tourism Policy and Planning: Yesterday, Today and Tomorrow*. Butterworth-Heinemann, Oxford, UK.

Gunder Frank, A. (1966) The development of underdevelopment. *Monthly Review* 18(4), 17–31. DOI: 10.14452/MR-018-04-1966-08_3.

Harrison, D. (ed.) (2001) *Tourism in the Less Developed World: Issues and Case Studies*. CAB International, Wallingford, UK.

Harrison, D. and Sharpley, R. (eds) (2017) *Mass Tourism in a Small World*. CAB International, Wallingford, UK.

Huybers, T. (ed.) (2007) *Tourism in Developing Countries*. Edward Elgar Publishing Limited, Cheltenham, UK.

Jenkins, C.L. (2015) Tourism policy and planning for developing countries: some critical issues. *Tourism Recreation Research* 40(2), 144–156. DOI: 10.1080/02508281.2015.1045363.

Lea, J. (1988) *Tourism and Development in the Third World*. Routledge, London.

Milano, C., Cheer, J. and Novelli, M. (eds) (2019) *Overtourism: Excesses, Discontents and Measures in Travel and Tourism*. CAB International, Wallingford, UK.
Mowforth, M. and Munt, I. (2003) *Tourism and Sustainability: New Tourism in the Third World*. Routledge, London.
Murphy, P. (1985) *Tourism: A Community Approach*. Methuen, New York.
O'Grady, R. (1981) *Third World Stopover*. World Council of Churches, Geneva.
Sharpley, R. (2009) *Tourism Development and the Environment: Beyond Sustainability*. Earthscan, London.
Sharpley, R. and Telfer, D. (eds) (2014) *Tourism and Development: Concepts and Issues*, 2nd edn. Channel View Publications, Bristol, UK.
Telfer, D. (2015) The evolution of development theory and tourism. In: Sharpley, R. and Telfer, D. (eds) *Tourism and Development; Concepts and Issues*, 2nd edn. Channel View Publications, Bristol, UK, pp. 77–117.
Tosun, C. (2000) Limits to community participation in the tourism development process in developing countries. *Tourism Management* 21(6), 613–633. DOI: 10.1016/S0261-5177(00)00009-1.
Turner, L. and Ash, J. (1975) *The Golden Hordes: International Tourism and the Pleasure Periphery*. Constable, London.
UN (2018) List of least developed countries (December 2018). United Nation Committee for Development Policy. Available at: https://www.un.org/development/desa/dpad/wp-content/uploads/sites/45/publication/ldc_list.pdf (accessed 6 September 2019).
UNDP (1990) *Human Development Report 1990*. United Nations Development Programme, New York.
UNWTO (2019a) International tourist arrivals reach 1.4 billion two years ahead of forecasts. UN World Tourism Organization press release, PR 19003. Available at: http://www2.unwto.org/press-release/2019-01-21/international-tourist-arrivals-reach-14-billion-two-years-ahead-forecasts (accessed 25 April 2019).
UNWTO (2019b) *International Tourism Results 2018 and Outlook 2019*. UN World Tourism Organization. Available at: http://cf.cdn.unwto.org/sites/all/files/pdf/unwto_barometer_jan19_presentation_en.pdf (accessed 6 September 2019).
WEF (2019) *The Global Risks Report 2019*. World Economic Forum, Geneva.
World Bank (2016) *World Development Indicators 2016*. World Bank, Washington DC.
World Bank (2019a) *World Development Indicators: Poverty and Inequality*. Available at: http://datatopics.worldbank.org/world-development-indicators/themes/poverty-and-inequality.html (accessed 25 April 2019).
World Bank (2019b) *World Bank Country and Lending Groups*. Available at: https://datahelpdesk.worldbank.org/knowledgebase/articles/906519-world-bank-country-and-lending-groups (accessed 25 April 2019).
World Bank (2019c) *Global Economic Prospects: Darkening Skies*. Washington, D.C.: International Bank for Reconstruction and Development / World Bank. Available at: http://documents.worldbank.org/curated/en/307751546982400534/pdf/133493-PUB-9781464813863.pdf (accessed 6 September 2019).

2 Tourism and (Sustainable) Development: Revisiting the Theoretical Divide

Richard Sharpley*
University of Central Lancashire, Preston, UK

Introduction: Tourism and Unsustainable Growth

Some two decades ago, at a time when it enjoyed almost universal support in academic, industry and policy circles, I published a paper that sought to critique the concept of sustainable tourism development (Sharpley, 2000). The purpose of doing so was quite straightforward. In the late 1980s, the World Commission on Environment and Development, perhaps better known as the Brundtland Commission, had popularized the idea of sustainable development in its report *Our Common Future* (WCED, 1987). Despite immediate criticisms that primarily highlighted its ambiguous and inherently oxymoronic nature (Redclift, 1987), sustainable development subsequently evolved into the dominant development paradigm of the 1990s and beyond, adopted by development agencies, industries and policy makers alike around the world.

The tourism sector was among those quick to join the sustainable development bandwagon. Initially conceptualized and endorsed by academics (whose continuing advocacy has arguably done much to maintain its widespread appeal), the concept of sustainable tourism development in particular had, by the mid-1990s, achieved 'virtual global endorsement as the new [tourism] industry paradigm' (Godfrey, 1996). This was entirely predictable. For academics who had long criticized what was broadly referred to as 'mass tourism' (Harrison and Sharpley, 2017), sustainable tourism development offered what in principle appeared to be a structured, viable alternative approach to tourism development; it was generally considered that, to be sustainable, tourism development should follow strategies other than the mass tourism model (Pigram, 1990) – that is, alternative (to mass) forms of tourism. Similarly, for a tourism industry facing a chorus of criticism over its allegedly unsustainable practices, collectively described by one commentator as 'a spectre haunting our planet' (Croall, 1995, p. 1), sustainable tourism development represented, in a sense, a lifeline, a set of politically acceptable and popularly attractive guidelines within which the tourism sector could continue to operate and, perhaps, allegedly 'good' tourists (Wood and House, 1991) could continue to enjoy guilt-free travel.

However, the ambiguities and contradictions inherent in the parental paradigm of sustainable development were immediately manifested in debates surrounding its application to the specific context of tourism, not least with regards to issues of definition. For example,

---

*rajsharpley@uclan.ac.uk

 Tourism in Development: Reflective Essays (eds P.U.C. Dieke *et al.*)
DOI: 10.1079/9781789242812.0002

while early proponents followed the logical argument that 'those who insert the word "tourism" between "sustainable" and "development"... [should]... ensure that, under all circumstances, the resultant principles of sustainable tourism are also principles of sustainable development' (Hunter, 1995, p. 163), attention became primarily focused on enhancing the sustainability of tourism itself or, more precisely, maintaining the resource base upon which it depends. In short, (sustainable) tourism became an end in itself rather than a means to an end; it became divorced from its *raison d'être*, namely, development (sustainable or otherwise).

These debates, in turn, pointed to the not insignificant fact that the principles and processes of sustainable tourism development espoused in innumerable academic publications and policy documents not only appeared to accept unquestionably the positive and politically correct rhetoric of sustainable development more generally but, perhaps more importantly, were also based upon highly fragile conceptual foundations in particular. More precisely, no attempt had been made to consider from a theoretical perspective the extent to which tourism, as a specific economic sector widely promoted as a vehicle of development, could be mapped onto and follow the (highly contested) principles of sustainable development – hence my paper, *Tourism and sustainable development: exploring the theoretical divide*, the purpose of which was, quite simply, to establish a conceptual framework for critiquing the concept of sustainable tourism development. This clearly demonstrated that, in theory, the nature or 'truths' (McKercher, 1993) of tourism in terms of both production and consumption processes were not commensurate with sustainable development or, alternatively stated, the principles of sustainable development in general did not represent a viable set of objectives for the development of tourism in particular.

Since it was published in 2000, transformations in tourism on a global scale have served to vindicate the arguments put forward in that paper. In other words, despite a plethora of policy statements and industry-based initiatives seeking to enhance its contribution to sustainable development, tourism in practice has followed a well-established, familiar yet ultimately unsustainable trajectory. For example:

- Annual international tourist arrival figures have continued to increase exponentially. In 2000, 687 million international arrivals were recorded; the one billion mark was surpassed in 2012 and, by 2017, arrivals reached a remarkable 1.326 billion (UNWTO, 2018a). In other words, in less than two decades, annual tourist arrivals almost doubled while, over the same period, annual international tourist receipts grew from US$481 billion to US$1340 billion. However, although in some respects this may be considered a success story, what is not known are the absolute environmental costs of this growth – the ever-increasing use of and emissions from fossil fuel-based modes of transport; the resources required to service the needs of tourists; and the social and environmental consequences.
- Much of this growth has been accounted for by the emergence of new markets, notably China. Yet less than one sixth of the global population currently engages in international travel, pointing on the one hand to significant global inequity in the opportunity to travel but, on the other hand, the potential for continuing future growth driven by emerging tourism markets. Indeed, at current growth rates, it would appear likely that the UN World Tourism Organization's (UNWTO) forecast of 1.8 billion international tourist arrivals by 2030 will be realized.
- Not only are new destinations continuing to be developed and becoming more popular, but established destinations are, for the most part, attracting ever-increasing numbers of tourists. Of the 217 nation states listed in the UNWTO's international tourism statistics for 2017, 108 (that is, half of all national destinations) received more than one million tourists in 2017; of these, 73 attracted more than three million tourists and 35 more than ten million tourists (extrapolated from UNWTO, 2018a).
- Relatedly, according to the World Travel and Tourism Council (WTTC), in 2017 not only did travel and tourism contribute 10.4% of global GDP, but the sector's growth rate outpaced overall global economic growth for the seventh successive year (WTTC, 2018). Simply stated, the global economy

Table 2.1. Contribution of travel and tourism to GDP 2017: top 50 countries.

Country	% of GDP	Country	% of GDP	Country	% of GDP
Maldives	76.6	St Kitts & Nevis	26.8	Tonga	18.2
Seychelles	65.3	Albania	26.2	New Zealand	17.9
Macao	61.3	Croatia	25.0	Portugal	17.3
Antigua & Barbuda	51.8	Sao Tome & Principe	24.3	Dominican Republic	17.2
Bahamas	47.8	Mauritius	23.8	Hong Kong (China)	16.7
Vanuatu	46.1	Montenegro	23.7	Madagascar	16.6
Cape Verde	44.9	St Vincent & Grenadines	23.4	Mexico	16.0
St Lucia	41.8	Grenada	23.3	Armenia	15.7
Belize	41.3	Cyprus	22.3	Estonia	15.4
Barbados	40.6	Thailand	21.2	Honduras	15.0
Fiji	40.3	Philippines	21.1	Spain	14.9
Dominica	36.6	Kiribati	20.9	Austria	14.6
Iceland	34.6	Gambia	20.1	Azerbaijan	14.6
Jamaica	32.9	Greece	19.7	Panama	14.5
Cambodia	32.4	Jordan	18.7	Tunisia	14.2
Georgia	31.0	Morocco	18.6	Namibia	13.8
Malta	27.1	Lebanon	18.4		

Adapted from: World Data Atlas, 2018

is becoming (albeit slowly) increasingly dependent upon tourism. However, this global average figure masks what is in some specific cases an alarming dependency on tourism. As can be seen from Table 2.1, travel and tourism's total (direct and indirect) contribution to GDP exceeds 40% in 11 national economies, while it accounts for 15% or more of GDP in a total of 44. It goes without saying that such a degree of dependency (indeed, any level of dependency) contradicts the principles of sustainable development.

These trends in tourism reflect, of course, the increasing demand for tourism experiences in both established and new tourism markets. That demand has largely been stimulated and facilitated not only by increasing wealth, more recently particularly in emerging economies, as well as advances in transport and communication technologies, but also by a number of transformations within the global political economy, not least trade-liberalization policies, with the deregulation of air transport and the consequential emergence of low-cost carriers being a notable influence on international tourism flows (Dobruszkes, 2009). These in turn have led to innovation and efficiencies within the travel and tourism sector in particular, thereby further democratizing access to travel and tourism – that is, making tourism more affordable to ever-increasing numbers of people – as well as underpinning continuing economic growth more generally.

Importantly, however, the economic growth-based model, which has remained a fundamental tenet of global development policies since the mid-twentieth century as well as economic planning in industrialized nations since the early nineteenth century, has increasingly come under scrutiny (Rutherford, 2008; Daly, 2014; Jackson, 2016). Indeed, for many, a major contradiction of the Brundtland Commission's vision of sustainable development was the requirement of up to ten-fold growth in the global economy while, equally curiously, the current Sustainable Development Goals (see UN, 2018) include 'Decent Work and Economic

Growth', based on the arguably naïve belief that technology and 'global resource efficiency' will successfully 'decouple economic growth from environmental degradation'. More specifically, recent concerns over climate change have focused attention on the need to reassess the established belief in economic growth, as powerfully summarized by 15-year-old Greta Thunberg at the climate change talks (CoP24) in Poland in December 2018: 'You only talk about moving forward with the same bad ideas that got us into this mess, even when the only sensible thing to do is pull the emergency brake' (CNN, 2018).

Similarly, the potential or, indeed, rationale for further growth in tourism in particular as an agent of development is being increasingly questioned. While organizations such as the UNWTO and the WTTC, for example, enthusiastically continue to promote further growth in international tourism as a basis for global development (and the United Nations declared 2017 as the 'International Year of Sustainable Tourism for Development'), such growth will inevitably contribute to further environmental degradation. It has long been demonstrated, for example, that a significant reduction in international travel, particularly by air, will be required for the tourism sector to make only a moderate contribution to necessary reductions in carbon emissions over the next 30 years (Dubois *et al.*, 2011), yet current estimates suggest that air passenger numbers will almost double over that period (IATA, 2017). At the same time, many popular tourism destinations are now suffering from what is rather euphemistically referred to as 'overtourism' – essentially the inevitable outcome of unregulated market-led production and consumption; that is, their capacity to support tourism has been exceeded, with significant implications for the well-being of local communities.

This is not to say that, over the last half-century or so, tourism has not made a valuable contribution to building or regenerating economies in innumerable destinations around the world, with consequential economic and social benefits for local communities. On the contrary, tourism has, to date, unquestionably proved to be a powerful and in many cases effective vehicle of what is commonly understood to be development; it has successfully provided the foundation for the economic development of many countries – although it must be acknowledged that notable examples exist, such as Cambodia, where this has not occurred (Sharpley and McGrath, 2016) – as well as making a significant contribution to global employment and income. Moreover, tourism can also be considered to be developmental in the context of the socio-cultural benefits accruing to those fortunate enough to participate in it, to be tourists. However, this has been achieved not only with long-recognized associated costs (Sharpley and Telfer, 2015) but also by following an economic growth model that, as noted above, remains central to tourism-based development policies – recent research has revealed, for example, that a majority of current national tourism development policies in Europe highlight a growth in tourist arrivals as a primary objective (Torkington, S. and Guiver, forthcoming) – but this is also ultimately, by definition, unsustainable. Hence, given the increasing awareness of the need to urgently address the challenge of climate change and other resource issues, it is now timely to revisit not the specific (and arguably outdated) concept of sustainable tourism development but the more general relationship between tourism and development, not least within the framework of contemporary understandings of development objectives, processes and challenges and the necessary parameters of resource sustainability.

Development: from economic growth to prosperity

In 1980, the then World Tourism Organization (WTO, now UNWTO) asserted in its Manila Declaration on World Tourism that:

> ...world tourism can contribute to the establishment of a new international economic order that will help to eliminate the widening economic gap between developed and developing countries and ensure the steady acceleration of economic and social development and progress, in particular in developing countries. (WTO, 1980, p.1)

Although the potential for international tourism to contribute to development had already been recognized and, indeed, enthusiastically embraced for a number of years, leading some far-sighted early commentators to question the assumed developmental benefits of tourism (Turner and Ash, 1975; Young, 1977; de Kadt, 1979), in effect, the WTO statement served to officially sanction tourism's role in development, albeit within an overt political-economic framework. A decade later, tourism had, according to Jenkins (1991, p. 61) become 'an important and integral element of [the] development strategies' of many countries, both developing and developed, and it remains so nowadays. At the same time, the stated reasons for selecting tourism as a development option also remain the same; although the UNWTO, for example, most recently claims its role to be the promotion of 'tourism as a driver of economic growth, inclusive development and environmental sustainability' (UNWTO, 2018b), it is specifically the potential economic benefits of tourism, including income and foreign exchange earnings, job creation and the generation of government revenues, that primarily continue to underpin its popularity as a development strategy.

It is both unsurprising and entirely logical that this should be the case; demonstrating, as already noted, remarkable and continuing growth, tourism is undoubtedly a highly attractive and potentially fruitful economic sector to exploit (Sharpley, 2015), although of course the extent to which wider benefits accrue to destinations is determined by a variety of local political-economic factors. More surprising and less logical, however, is the continuing association between tourism and development. In other words, although tourism may undoubtedly underpin economic growth, claims that it may also contribute to development more generally are less tenable when not only contemporary understandings of development but also what some consider to be the negative consequences of economic growth on development are taken into account. Putting it another way, it is all too easy to assume that tourism might contribute to development but, when the concept and processes of development are scrutinized, that assumption becomes fallacious.

Nevertheless, it is important to point out that, at the outset, economic growth and development were intimately related. Emerging in the 1950s in an era of post-war reconstruction and economic development, development thinking initially reflected the belief that the historical development path followed in the West could be transposed to the rest of the world, particularly to the newly independent poorer countries. Specifically, development policy was based upon classical economic theory of liberalizing trade and stimulating the economy focused around economic growth poles (Potter, 2014, p. 18), a concept that continues to resonate with many contemporary tourism resort developments. In essence, development as a process was guided by a normative, top-down western approach considered to be universally applicable; as an outcome, it was considered to flow from and be commensurate with economic growth (Mabogunje, 1980). It is also interesting to note that, during this period, the academic field of development studies first evolved as a branch of development economics, further entrenching the 'positivist orthodoxy' (Leys, 1995, p. 7) of economic planning as a means of achieving development.

As widely acknowledged, however, the enthusiastic support for the so-called modernization paradigm soon waned as it came to be realized that, in many countries, economic growth policies were not only failing to resolve social and political challenges but, in many cases, actually exacerbating them. Indeed, as argued shortly, the continuing contemporary dependence on and promotion of economic growth is resulting, particularly in developed countries, in what might be referred to as 'de-development' or 'social recession' (Rutherford, 2008). In other words, it soon became evident that development was not only an economic challenge or something that could be achieved through appropriate economic planning, but it was also subject to both wider social and political issues and the structure of the global political economy as a whole; hence the emergence of the dependency or underdevelopment paradigm in the late 1960s (Telfer, 2015). Moreover, it also became evident that there could be no universal process of or policy for development, contributing to the subsequent argument of some – reinforced by tangible evidence that, after more than half a century of development policies, many developmental challenges remain entrenched – that development as a global project has failed and should be abandoned (Rahnema, 1997; Sidaway, 2007).

With regards to the meaning of development, then, although economic growth remained a prerequisite, not least for those living in poverty, it came to embrace the broader social objectives of justice and equality. Specifically, it was argued that the fundamental challenges of development were (and indeed still are) poverty, unemployment and inequality, and, therefore, if just one of these has become worse, even if per capita income has grown significantly, development cannot be considered to have occurred (Seers, 1969). In short, development came to be defined not by economic growth but by socio-economic progress. Despite this broader conceptualization of development, however, development policy in practice continued to remain firmly rooted in neo-liberal economic planning, manifested most acutely in the so-called Structural Adjustment Lending Programmes (SALPs) of the 1980s Reagan-Thatcher era. These promoted lending to developing countries based on the condition that they implemented policies to reduce state intervention in the economy, such as encouraging market liberalization, the privatization of state enterprises and so on, and although initially popular among recipient countries, they ultimately proved to be not only ineffective but in some cases destructive, enhancing rather than solving development challenges (Harrigan and Mosley, 1991). SALPs were, therefore, subsequently replaced by Poverty Reduction Strategy Papers which sought to encourage local participation in development strategy. These too have proved to be unsuccessful (Lazarus, 2008), primarily because, as with the preceding SALPs, they assumed the transferability of western economic policies to developing countries while failing to take account of national political-economic factors. This is, perhaps, one of the reasons underpinning the suggestion that the study of development should no longer fall under the umbrella of economics but should sit 'squarely within the intellectual project of political economy and the diverse theoretical traditions associated with it' (Payne and Phillips, 2010, p. 181).

To the three challenges of development referred to above – poverty, unemployment and inequality – Seers (1977) subsequently added a fourth, namely, self-reliance, thereby confronting the prevailing top-down western economic policy approach to development by introducing the requirement of self-determination into the development 'mix'. In so doing, he sowed the seeds of a fundamental shift in development thinking or, more precisely, in what it means to be 'developed' that has occurred over the last two to three decades. Specifically, the focus of development has shifted from the economy and economic growth (essentially a top-down macro perspective), first, towards societies as both the instigators and beneficiaries of development (a 'bottom-up' approach). For example, fundamental to the alternative development paradigm, unashamedly adapted into the principles of so-called alternative tourism, is the belief that development belongs to and should emanate from societies themselves:

> Development is a whole: it is an integral, value-loaded, cultural process; it encompasses the natural environment, social relations, education, production, consumption and well-being. The plurality of roads to development answers to the specificity of cultural or natural situations; no universal formula exists. Development is endogenous; it springs from the heart of each society. (Dag Hammarskjöld Institute, 1975)

Second, the focus of development has now settled on the individual, a micro approach referred to as human development (Knutsson, 2009) that is concerned with promoting individual as well as societal well-being.

The question is, then, what is well-being and does it remain commensurate with 'development' or being 'developed'? For Sen (1999), such well-being is dependent on the possession of capabilities to fulfil what he refers to as functionings, giving individuals the freedom – that is, freedom of choice and freedom from servitude to ignorance, nature, other societies, beliefs and institutions (Sharpley, 2015) – to live the life they wish, to fulfil their potential as individuals and members of societies. A similar approach is adopted in the UNDP's rather broad definition of human development as:

> the expansion of people's freedoms to live long, healthy and creative lives; to advance other goals they have reason to value; and to engage actively in shaping development equitably and sustainably on a shared planet. (UNDP, 2010, p. 22).

Such a conceptualization implicitly embraces tangible measures or indicators of what is conventionally understood to be development

– longevity, education, income, healthcare, access to resources and so on (many of which are explicitly identified in the current SDGs). However, it does not get to the heart of what well-being is, to what well-being means to the individual. Hence, according to Jackson (2016), a more appropriate term (and objective) is prosperity; to live a good life is to live a prosperous life, in the more traditional meaning of the word. That is, whereas prosperity is now typically thought of in material terms – to be prosperous is to be doing well financially – the etymological roots of the word endow it with the meaning of doing well more generally, to be living *pro spere*, or according to one's hopes and expectations and, hence, to be happy and fulfilled.

In this sense, prosperity inevitably involves a material dimension, just as Goulet (1989) suggests that there is an economic component in development. That is, basic needs, such as water, food and shelter, must be satisfied, as well as sufficient income. Moreover, it also involves other tangible dimensions normally associated with development, such as access to or the provision of education, health care, employment and other essential services, all fundamental to living long and healthy lives. However, as Jackson (2016) suggests, there are also social, psychological and perhaps even spiritual dimensions to prosperity; to be prosperous is, for example, to be and feel part of and contribute to society, to respect and be respected by others, to have a sense of hope, purpose and identity, and to feel and fulfil a moral responsibility for others in the sense that personal well-being may emanate from contributing to the well-being of others, both now and in the future. It is no coincidence, then, that the counterpoint to prosperity, namely poverty, is also considered to be multi-dimensional; to be poor is not only to suffer financial and material deprivation (the metrics by which poverty is usually measured), but also to lack social identity or esteem, to lack voice, to be unable to contribute to society and to suffer the feelings or beliefs associated with poverty (Misturelli and Heffernan, 2008).

Therefore, if, as is generally acknowledged, human development is concerned with enhancing people's well-being or prosperity as defined above, it becomes evident that the term 'development', referring to both a process and a goal that is implicitly universal, is perhaps no longer appropriate to describe at least the goal, if not the process. Certainly, as a process, development implies change, betterment, progress and (in economic terms) growth; as an outcome, however, development, or to be developed, implies that a measureable, tangible condition contrasting with underdevelopment has been attained. Yet, not only are societies and social structures, values and norms infinitely variable and dynamic, meaning there can be no universal benchmarks of when development has been attained, but also well-being or prosperity as an individual expectation, experience or condition is unmeasurable. And it also becomes evident that tourism, as a specific economic sector, cannot contribute directly to the prosperity of individuals and societies. Certainly, it may provide the economic foundations for prosperity, such as employment and income, but the extent to which individuals and societies may become prosperous is dependent on numerous endogenous factors independent of the economic benefits accruing (or not) from tourism. Hence, it is perhaps necessary to untie the knot between tourism and development, to recognize that tourism may contribute to economic development or growth but not directly to wider well-being or prosperity.

But therein lies another question: Does economic growth itself remain a viable objective and, hence, do policies that encourage the continuing growth of tourism in particular also remain viable?

Prosperity and Tourism Growth: Which Way Now?

As already observed, since the Industrial Revolution and the emergence of modern capitalist economies, economic growth as measured by an increase in per capita GDP has remained the objective of economic planning and a yardstick of economic health; a growing economy is considered to be successful, an economy in recession is 'failing'. Moreover, growth in average incomes is conventionally understood to improve people's prosperity. In very simplistic terms, as explained by Jackson (2016), if GDP measures the total exchange value of goods and services, then an increase in GDP reflects higher spending on goods and services which, in turn, suggests that those goods and services are not only valued by consumers, but also that they are valued because they improve their lives. Hence, in short, economic growth increases prosperity. And it is for this reason, in part, that economic growth has long been, and remains, fundamental to development policies. Indeed, a principal target of SDG 8 is to 'Sustain per capita economic growth in accordance with national circumstances and, in particular, at least 7% gross domestic product growth per annum in the least developed countries' (UN, 2018).

The significance of growth to capitalistic, market-led economic systems is complex, involving as it does issues of investment, employment, productivity, innovation and so on, which are considered by some to reflect Schumpeter (1976) critique of capitalism as a process of 'creative destruction'. Moreover, the significance of economic growth to wider development or prosperity is equally complex, and a discussion of the dynamics between capitalism, development and growth is well beyond the scope of this chapter (for example, Sklair, 1994). For the present purposes, however, two points demand attention. First, as has long been argued, indefinite economic growth in general (and implicitly of tourism in particular) is unsustainable in a world of finite resource and assimilative capacities, threatening the prosperity of future generations; and second, continuing economic growth is increasingly considered to have negative social impacts.

With regards to the first issue, civilizations throughout history have recognized the need for natural resource conservation. However, in more modern times, the fragile balance between natural and non-renewable resource exploitation, the ability of the ecosystem to assimilate the outputs of human activity (the ecosystem's 'source' and 'sink' functions) and, in particular, increasing production and consumption related to population growth have come more sharply into focus. Indeed, while Malthus's predictions some two centuries ago proved to be inaccurate, it is now argued by some that a 'substantial and rapid reduction in consumptive behaviour' (Turner, 2008, p. 410) will be required to sustain the global system, with the need to reduce carbon emissions to slow down, if not reverse, global warming considered to be the most pressing issue. Those with a less pessimistic outlook, conversely, follow the argument that sustainable development can be achieved through green growth; that is, maintaining an economic growth policy while 'decoupling' (UNEP, 2011) production and consumption from environmental resources or, in more simple terms, reducing their environmental impacts through efficiencies and technological innovation. In other words, it is claimed that economic growth can be sustained through absolute decoupling (reducing the total impact of all economic activity), although, in practice, even in some advanced industrial nations only limited relative decoupling has, to date, been achieved – the environmental impact per unit of production has decreased, but continuing growth in production and consumption means that the total environmental costs are also growing (van der Voet *et al.*, 2005).

Nevertheless, there remains a lack of consensus over the viability of achieving economic growth and decoupling given that the economy–environment nexus varies according to a variety of factors (Jorgenson and Clark, 2012). However, in the specific context of tourism, the implications are perhaps more straightforward, not least because of the limited opportunities for further environmental efficiencies in fossil fuel-based travel, particularly air transport – although it should be noted that, according to Cadarso *et al.* (2016), adding non-transport-related emissions increases tourism's total carbon footprint by 34%. Specifically, from a development perspective, policies to maintain growth in international tourism to those developing countries requiring economic growth may be appropriate but, even to achieve relative decoupling, a reduction is required in all fossil fuel-based tourism transport modes to all other destinations, particularly cruise ships, which not only, in some circumstances, produce more harmful emissions per passenger than aircraft, but are also credited with contributing significantly to 'over-tourism' in many destinations while generating limited economic benefit.

At the same time, cruise ship tourism, which over the last two decades has increased in terms of passenger numbers at twice the rate of tourism more generally (Brida and Zapata, 2010) also points to the second issue related to the problem associated with continuing economic growth, namely the potential for 'de-development' or social recession. Simply stated, under the economic growth paradigm, increasing material wealth (having and consuming more commodities, whether goods or services such as tourism) is a proxy for development or increasing prosperity. However, overall economic growth disguises a number of challenges, not least increasing inequality; the world may be getting richer, but not everyone is benefitting. Recent research reveals that over the last 30 years 'income inequality has increased rapidly in North America and Asia, grown moderately in Europe, and stabilized at an extremely high level in the Middle East, sub-Saharan Africa, and Brazil' (World Inequality Report, 2018). More specifically, the top ten per cent's share of national income in the USA and Canada increased from 34% to 47% over this period. Such income inequality may cause resentment and potential unrest – arguably, for example, recent protests against tourism in many destinations around the world may be based as much on what tourism represents – in essence, a luxury form of consumption – as on the actual presence of large numbers of tourists. And it has long been argued, of course, that tourism development enhances inequality in destinations (Britton, 1982).

In addition, not only is increasing materialism being associated with social trends being manifested in many developed nations, such as increasing individualism, the breakdown of community cohesion and an increasing incidence of mental health issues, collectively referred to as 'social recession' (Rutherford, 2008), but also it has long been argued that greater wealth and levels of consumption, far from increasing happiness, may actually reduce it and increase levels of anxiety (DeLeire and Kalil, 2010). Interestingly, recent work has revealed that the consumption of experiences (such as tourism) may result in longer-lasting happiness than material consumption (Gilovich and Kumar, 2015), although tourism-specific studies suggest that taking more than one holiday does not result in greater happiness (Nawijn, 2011). Also, it has been suggested that greater meaning – reflecting in part the definition of prosperity above – might be achieved through engaging in forms of tourism that offer 'voluntary simplicity' (Hall, 2011) rather than in more hedonistic, consumptive touristic experiences. Irrespective of these debates, however, the evidence suggests that promoting the growth of tourism through encouraging more frequent participation in tourism, as opposed to more people becoming tourists, may in fact have negative consequences on prosperity in both destinations and tourism-generating regions.

Conclusion

What, then, can be inferred from this discussion of the theoretical relationship between tourism and development? Immediately, it must be stressed that it has not simply been the purpose of this chapter to repeat the frequent (yet justifiable) argument that tourism must be better planned and controlled to reduce its undoubted environmental impacts. Rather, the intention has been to reassess tourism's potential to contribute to development based upon contemporary understandings of development which, inevitably, embrace an environmental/sustainability dimension. However, the conclusion that must be drawn is that more effective planning and control is necessary. Re-visioning development as prosperity, as opposed to economic growth, as well as recognition of the 'de-developmental' consequences of the growth imperative, points to the need for policies that allow for only targeted growth in tourism within an overall slowing down (or even de-growth) of tourism globally – continuing growth is not sustainable in the longer term. While this means that established destinations, particularly in the developed world, will need to 'downsize' their international tourism sectors, opportunities exist to develop domestic tourism utilizing more sustainable forms of transport and, perhaps, less resource-intensive forms of tourism. Moreover, the link between increasing wealth/consumption and 'social recession' points to a paradox; more frequent consumption of tourism may be seen as a means of escape from such recession, but may actually contribute to it. Hence, Krippendorf's (1987) suggestion that the solution to the 'problem' of tourism lies in improving tourists' home society – in the context of this chapter, enhancing their prosperity – thereby diminishing their perceived need to consume tourism, is perhaps of greater relevance now than 30 years ago. What is certain, however, is that it can no longer be business as usual.

References

Brida, J.G. and Zapata, S. (2010) Cruise tourism: economic, socio-cultural and environmental impacts. *International Journal of Leisure and Tourism Marketing* 1(3), 205–226. DOI: 10.1504/IJLTM.2010.029585.

Britton, S.G. (1982) The political economy of tourism in the Third World. *Annals of Tourism Research* 9(3), 331–358. DOI: 10.1016/0160-7383(82)90018-4.

Cadarso, M.Á.., Gómez, N., López, L.A. and Tobarra, M.Á.. (2016) Calculating tourism's carbon footprint: measuring the impact of investments. *Journal of Cleaner Production* 111, 529–537. DOI: 10.1016/j.jclepro.2014.09.019.

CNN (2018) Teen tells climate negotiators they aren't mature enough. Available at: https://edition.cnn.com/2018/12/16/world/greta-thunberg-cop24/index.html?utm_term=image&utm_source=twCNN&utm_content=2018-12-16T23%3A30%3A07&utm_medium=social (accessed 1 January 2019).

Croall, J. (1995) *Preserve or Destroy: Tourism and the Environment*. Calouste Gulbenkian Foundation, London.

Dag Hammarskjöld Institute (1975) What now? Dag Hammarskjöld report on development and international cooperation. Available at: http://www.daghammarskjold.se/wp-content/uploads/2016/07/What-Now-1975.pdf (accessed 2 January 2019).

Daly, H. (2014) *From Uneconomic Growth to a Steady State Economy*. Edward Elgar, Cheltenham, UK.

de Kadt, E. (1979) *Tourism: Passport to Development?* Oxford University Press, New York.

DeLeire, T. and Kalil, A. (2010) Does consumption buy happiness? Evidence from the United States. *International Review of Economics* 57(2), 163–176. DOI: 10.1007/s12232-010-0093-6.

Dobruszkes, F. (2009) New Europe, new low-cost air services. *Journal of Transport Geography* 17(6), 423–432. DOI: 10.1016/j.jtrangeo.2009.05.005.

Dubois, G., Peeters, P., Ceron, J.-P. and Gössling, S. (2011) The future tourism mobility of the world population: emission growth versus climate policy. *Transportation Research Part A: Policy and Practice* 45(10), 1031–1042. DOI: 10.1016/j.tra.2009.11.004.

Gilovich, T. and Kumar, A. (2015) We'll always have Paris: the hedonic payoff from experiential and material investments. In: Zanna, M. and Olson, J. (eds) *Advances in Experimental Social Psychology*. Vol. 51. Elsevier, Oxford, UK, pp. 147–187.

Godfrey, K. (1996) Towards sustainability? Tourism in the Republic of Cyprus. In: Harrison, L. and Husbands, W. (eds) *Practising Responsible Tourism: International Case Studies in Tourism Planning, Policy and Development*. John Wiley & Sons, Chichester, UK, pp. 58–79.

Goulet, D. (1989) Participation in development: new avenues. *World Development* 17(2), 165–178. DOI: 10.1016/0305-750X(89)90242-8.

Hall, C.M. (2011) Consumerism, tourism and voluntary simplicity: we all have to consume, but do we really have to travel so much to be happy? *Tourism Recreation Research* 36(3), 298–303. DOI: 10.1080/02508281.2011.11081675.

Harrigan, J. and Mosley, P. (1991) Evaluating the impact of World Bank structural adjustment lending: 1980–87. *Journal of Development Studies* 27(3), 63–94. DOI: 10.1080/00220389108422204.

Harrison, D. and Sharpley, R. (eds) (2017) *Mass Tourism in a Small World*. CAB International, Wallingford, UK.

Hunter, C.J. (1995) On the need to re-conceptualise sustainable tourism development. *Journal of Sustainable Tourism* 3(3), 155–165. DOI: 10.1080/09669589509510720.

IATA (2017) 2036 forecast reveals air passengers will nearly double to 7.8 billion. International Air Transport Association. Available at: https://www.iata.org/pressroom/pr/Pages/2017-10-24-01.aspx (accessed 4 January 2019).

Jackson, T. (2016) *Prosperity Without Growth: Foundations for the Economy of Tomorrow*, 2nd edn. Routledge, Abingdon, UK.

Jenkins, C. (1991) Development strategies. In: Lickorish, L., Jefferson, A., Bodlender, J. and Jenkins, C. (eds) *Developing Tourism Destinations. Longman*. London, UK, pp. 59–118.

Jorgenson, A.K. and Clark, B. (2012) Are the economy and the environment decoupling? A comparative international study, 1960–2005. *American Journal of Sociology* 118(1), 1–44. DOI: 10.1086/665990.

Knutsson, B. (2009) The intellectual history of development: towards a potential widening repertoire. *Perspectives* No. 13, Göteborgs Universitet: School of Global Studies. Available at: https://www.globalstudies.gu.se/digitalAssets/1272/1272997_Perspectives_13.pdf (accessed 2 January 2019).

Krippendorf, J. (1987) *The Holiday Makers*. Heinemann, Oxford, UK.

Lazarus, J. (2008) Participation in poverty reduction strategy papers: reviewing the past, assessing the present and predicting the future. *Third World Quarterly* 29(6), 1205–1221. DOI: 10.1080/01436590802201188.

Leys, C. (1995) *The Rise and Fall of Development Theory*. James Currey, Oxford, UK.

Mabogunje, A. (1980) *The Development Process: A Spatial Perspective*. Hutchinson, London.

McKercher, B. (1993) Some fundamental truths about tourism: understanding tourism's social and environmental impacts. *Journal of Sustainable Tourism* 1(1), 6–16. DOI: 10.1080/09669589309450697.

Misturelli, F. and Heffernan, C. (2008) What is poverty? A diachronic exploration of the discourse on poverty from the 1970s to the 2000s. *The European Journal of Development Research* 20(4), 666–684. DOI: 10.1080/09578810802464888.

Nawijn, J. (2011) Happiness through vacationing: just a temporary boost or long-term benefits? *Journal of Happiness Studies* 12(4), 651–665. DOI: 10.1007/s10902-010-9221-y.

Payne, A. and Phillips, N. (2010) *Development*. Polity Press, Cambridge, UK.

Pigram, J. (1990) Sustainable tourism: policy options. *Journal of Tourism Studies* 1(2), 2–9.

Potter, R. (2014) The nature of development studies. In: Desai, V. and Potter, R. (eds) *The Companion to Development Studies*, 3rd edn. Routledge, Abingdon, UK, pp. 16–20.

Rahnema, M. (1997) Towards post-development: searching for signposts, a new language and new paradigms. In: Rahnema, M. and Bawtree, V. (eds) *The Post-Development Reader*. Zed Books, London, pp. 377–403.

Redclift, M. (1987) *Sustainable Development: Exploring the Contradictions*. Routledge, London.

Rutherford, J. (2008) Well-being, economic growth and social recession. Middlesex University e-prints. Available at: http://eprints.mdx.ac.uk/4058/1/Rutherford-Well_being.pdf (accessed 2 January 2019).

Schumpeter, J. (1976) *Capitalism, Socialism and Democracy, 5th edn*. George Allen & Unwin, London.

Seers, D. (1969) The meaning of development. *International Development Review* 11(4), 2–6.

Seers, D. (1977) The new meaning of development. *International Development Review* 19(3), 2–7.

Sen, A. (1999) *Development as Freedom*. Anchor Books, New York.

Sharpley, R. (2000) Tourism and sustainable development: exploring the theoretical divide. *Journal of Sustainable Tourism* 8(1), 1–19. DOI: 10.1080/09669580008667346.

Sharpley, R. (2015) Tourism: a vehicle for development? In: Sharpley, R. and Telfer, D. (eds) *Tourism and Development: Concepts and Issues*, 2nd edn. Channel View Publications, Bristol, pp. 3–30.

Sharpley, R. and McGrath, P. (2016) Tourism in Cambodia: opportunities and challenges. In: Brickell, K. and Springer, S. (eds) *Handbook of Contemporary Cambodia*. Routledge, Abingdon, UK, pp. 87–98.

Sharpley, R. and Telfer, D. (eds) (2015) *Tourism and Development: Concepts and Issues*, 2nd edn. Channel View Publications, Bristol, UK.

Sidaway, J. (2007) Spaces of post-development. *Progress in Human Geography* 3(3), 345–361.

Sklair, L. (ed.) (1994) *Capitalism and Development*. Routledge, London.

Telfer, D. (2015) The evolution of development theory and tourism. In: Sharpley, R. and Telfer, D. (eds) *Tourism and Development: Concepts and Issues*, 2nd edn. Channel View Publications, Bristol, UK, pp. 31–73.

Torkington, K., Stanford, D. and Guiver, J. (2020) Discourse(s) of growth and sustainability in national tourism policy. *Journal of Sustainable Tourism* 28(7), 1041–1062.

Turner, G. (2008) A comparison of *The Limits to Growth* with 30 years of reality. *Global Environmental Change* 18(3), 397–411. DOI: 10.1016/j.gloenvcha.2008.05.001.

Turner, L. and Ash, J. (1975) *The Golden Hordes: International Tourism and the Pleasure Periphery*. Constable, London.

UN (2018) SDG Indicators. United Nations Sustainable Development Goals. Available at: https://unstats.un.org/sdgs/metadata/ (accessed 31 December 2018).

UNDP (2010) *Human Development Report 2010*. OUP, New York.

UNEP (2011) Decoupling natural resource use and environmental impacts from economic growth. United Nations Environment Programme. International Resource Panel. Available at: http://www.gci.org.uk/Documents/Decoupling_Report_English.pdf (accessed 5 January 2019).

UNWTO (2018a) *UNWTO Tourism Highlights, 2018 edition*. UN World Tourism Organization, Madrid.

UNWTO (2018b) About UNWTO. UN World Tourism Organization. Available at: https://www.unwto.org/who-we-are (accessed 6 January 2019).

van der Voet, E., van Oers, L., Moll, S., Schütz, H., Bringezu, S. *et al*. (2005) Policy Review on Decoupling: Development of Indicators to Assess Decoupling of Economic Development and Environmental

Pressure in the EU-25 and AC-3 Countries. Leiden University: CML report 166. Available at: https://openaccess.leidenuniv.nl/bitstream/handle/1887/11934/11_587_047.pdf?sequence=1 (accessed 5 January 2019).
WCED (1987) *Our Common Future*. Oxford University Press, Oxford, UK.
Wood, K. and House, S. (1991) *The Good Tourist: A Worldwide Guide for the Green Traveller*. Mandarin, London.
World Data Atlas (2018) Contribution of travel and tourism to total GDP. Available at: https://knoema.com/atlas/topics/Tourism/Travel-and-Tourism-Total-Contribution-to-GDP/Contribution-of-travel-and-tourism-to-GDP-percent-of-GDP (accessed 30 December 2018).
World Inequality Report (2018) Global income inequality dynamics. Available at: https://wir2018.wid.world/part-2.html (accessed 6 January 2019).
WTO (1980) *Manila Declaration on World Tourism*. World Tourism Organization, Madrid.
WTTC (2018) *Travel and Tourism World Economic Impact 2018*. World Travel and Tourism Council. Available at: https://www.wttc.org/-/media/files/reports/economic-impact-research/regions-2018/world2018.pdf (accessed 30 December 2018).
Young, G. (1977) *Tourism: Blessing or Blight?* Penguin, Harmondsworth, UK.

3 Sustainability and Tourism Growth

Larry Dwyer[1]* and Alison Gill[2]
[1]University of Technology, Sydney and Griffith Institute for Tourism (GIFT), Queensland, Australia; [2]Simon Fraser University, Vancouver, Canada

Introduction

An implicit assumption in much of the tourism planning literature is that the market requires increasing economic growth. According to the prevailing pro-growth mindset, more is better, despite the fact that tourism's obsessive drive for expansion is destroying the very natural and socio-cultural environments that attract visitation. Indeed, a focus on economic issues has led to the relative neglect of the costs of tourism growth to the socio-cultural fabric of destinations, as well as to the quality of environmental resources that are important attractors of visitation. Tourism planning typically consists of a SWOT analysis, followed by strategies to build on strengths to promote growth, address weaknesses to growth, counter threats to growth, and capitalize on opportunities for growth. Moreover, in the World Travel and Tourism Council's *Blueprint for New Tourism* (WTTC, 2003), which purports to address sustainability issues, tourism growth is lauded as an important goal for all destinations. Similarly, an explicit growth ethic also underpins numerous UN World Tourism Organization publications that promote tourism as a driver of economic growth, inclusive development and environmental sustainability (see: http://www2.unwto.org/), as well as the bulk of the destination competitiveness literature in which better attractions and better management imply greater demand for tourist industry products and services.

Despite this pervasiveness of the 'growth ethic' in policy and practice, its appropriateness remains a largely unexamined issue in the tourism research literature (Dwyer, 2018). This is arguably surprising given that markets are social institutions that typically do not produce optimal economic social or environmental outcomes. More specifically, market failures of various kinds mean that actual market outcomes are not efficient. Hence, while acknowledging that tourism growth is often associated with economic benefits to a destination, it is important to note that it can also generate income inequalities, destroy local industries and create greater dependency of developing economies on developed ones, with adverse effects on small business. Local-level negative effects include increased prices of consumer goods and services, increased price of land and housing beyond local affordability, increasing inequality between rich and poor and increasing demands on public services and facilities. At the same time, however, the mounting evidence that economic growth, measured by growth of GDP, does not measure quality of life, social progress, human development or happiness, continues largely to be ignored (Sharpley, 2009; Kubiszewski *et al.*, 2013).

*Corresponding author: larry.dwyer@uts.edu.au

 Tourism in Development: Reflective Essays (eds P.U.C. Dieke *et al.*)
DOI: 10.1079/9781789242812.0003

The question is, then: Can these two contrasting dynamics be reconciled? Or, more precisely, can tourism be developed in a way that contributes to economic development, employment and poverty reduction without destroying culture and damaging the environment? Given that significant increases in global tourism growth in the future are forecast, there is considerable urgency to exploring and implementing new sustainability approaches, not least because there is a growing consensus that *sustainable* tourism is possible only if stakeholders take serious account of the interactions between economic, environmental, social and cultural factors.

The Problem with GDP

Studies of economic growth in tourism have traditionally focused on its ability to increase a destination's GDP, levels of employment, foreign exchange earnings and so on. GDP in particular has long been the world's yardstick of economic progress; it is widely employed by policy makers, economists, international agencies and the media as the primary scorecard of a nation's economic health and well-being. The assumption underpinning the use of GDP to assess well-being is that the higher the level of economic production, the better people are able to satisfy their needs. However, GDP was never designed to measure social progress or social welfare (Kubiszewski *et al.*, 2013; Kubiszewski and Costanza, 2015). Several of its limitations may be listed as follows:

- GDP is merely a gross tally of products and services bought and sold, with no distinction between transactions that enhance well-being and those that diminish it.
- GDP has a narrow focus on economic production. The more a nation depletes or degrades its natural resources or culture to produce goods and services, the more the GDP increases. In the GDP measure, every transaction is positive as long as money changes hands.
- Higher GDP is associated with increased carbon emissions, environmental problems and congestion, leaving people with lower economic welfare and lower levels of happiness.
- GDP ignores a number of factors that are important for well-being because they take place outside the market. It thus ignores the crucial economic functions performed in civic engagement and in the household and volunteer sectors, as well as transactions in the informal (non-cash, barter) economy. Many of these omitted activities are of essential importance in tourism worldwide (for example, when households host VFR travellers or when locals volunteer to organize special events or when locals act as informal guides or escorts).
- GDP makes no adjustment for leisure time that is widely regarded as essential to human well-being and an important objective of productivity increases in any tourism destination.
- GDP measures the quantitative output and consumption of goods and services, ignoring quality attributes that can satisfy and enhance consumer well-being.
- GDP measures ignore income distribution, despite the fact that a dollar's worth of increased income to a poor person produces more additional welfare than a dollar's increased income to a rich person.
- A destination's GDP ignores intrinsic development factors, such as lifestyle choice, environmental quality, life expectancy, standards of healthcare or freedom from oppression.

Consequently, at a time when new political narratives are necessary to identify how our societies should develop, prominent economists now advocate a shift of emphasis from a 'production-oriented' measurement system to one focused on the well-being of current and future generations; that is, toward broader measures of social progress (Stiglitz *et al.*, 2009). In this effort, new measures of economic, social and natural wealth are required that go beyond the narrow GDP measure.

Sustainability

'Sustainable development' is that which is environmentally, socially and economically sustainable, meeting the needs of the present without compromising the welfare of future generations (UNEP/UNWTO, 2005). For sustainable tourism development, a suitable balance must be established between these three dimensions (environmental, social and economic) to guarantee its long-term sustainability. Yet, while tourism's economic impacts have long been studied intensively, the economic pillar of sustainability has been relatively neglected. This chapter aims to remedy this gap, at least partially.

Economic approaches to sustainability frame the issue in terms of human well-being (utility). Historically, economists have argued that what determines the ability of a given set of humans to improve their well-being (utility) is the quantity and quality of capital assets (wealth) available at the time. Hence, a simple intergenerational rule is that development is sustainable if it does not decrease the capacity to provide non-declining per capita utility into the indefinite future. The capacity to provide utility is conceptually embodied in four forms of capital: built, human, social and natural.

- *Built capital (Kb)*: the tools, machines, buildings, technologies and infrastructure that enhance productivity.
- *Human capital (Kh)*: the skills and knowledge of the workforce, including ability to develop and innovate and employ new technologies.
- *Social capital (Ks)*: concerns the relationships between individuals, between institutions (including government), and between individuals and institutions (languages, institutions, educational system, laws).
- *Natural capital (Kn)*: the value of the goods and services provided by nature, including renewable components such as clean air and water, ecosystems and solar energy, and non-renewable stocks such as mineral deposits, biodiversity and fossil fuels.

Wealth creation is the process of using these four types of capital in combination to produce the flows of goods and services that people want or need. The condition for sustainable development is that each generation leaves the next generation a stock of productive capacity, in the form of capital assets and technology, that is capable of sustaining utility or well-being per capita at a level no less than that enjoyed by the current generation. Notationally, $dK/dt \geq 0$, where $K = Kb + Kh + Ks + Kn$. This implies that the change in the real value of aggregate assets at a point in time, dK/dt, must be at least zero in the aggregate. However, an underlying assumption is that changes in the constituent components of the overall stock, K, can be traded off for each other. In other words, if some part of natural capital, Kn, is depleted, then this is consistent with sustainability as long as adequate investment is made in other forms of wealth, such as human, social or physical capital.

Since sustainability depends on the maintenance of the capital stock, an important issue is whether it is the total stock of capital that must be maintained, with substitution allowed between the different capital forms, or whether certain components of capital, in particular natural capital, are non-substitutable, in that they contribute to welfare in a unique way that cannot be replicated by another capital stock (Ekins *et al.*, 2003; Dietz and Neumayer, 2007). This has led to the discussion of different degrees of sustainability, with two major positions being taken up by researchers, namely: 'weak' and 'strong' sustainability.

Weak sustainability

Weak sustainability is about maintaining total capital stock ($K = Kb + Kh + Ks + Kn$) where different types of capital are assumed to be perfectly substitutable. According to proponents of weak sustainability, any one form of capital can be depleted provided 'proceeds' are reinvested in other forms of capital. Growth occurs when $dK/dt > 0$. This assumes that any type of capital is perfectly substitutable for natural or social capital as an input to production. From the 'adjusted net savings' standpoint, for example, a nation which reinvested all of its profits from the exploitation of non-renewable natural resources in the formation of physical capital (for example, telecommunications or road systems) or human capital through its educational system would impose no net opportunity cost on the country's future citizens. By implication, therefore, if $dK/dt > 0$, then any degradation (-Kn) of natural assets or social capital (-Ks) can be compensated for by investments in built or physical capital making the investment 'sustainable'. In other words, Kn or Ks can be destroyed as a result of growth provided one or both of Kb and Kh are increased to compensate.

The weak sustainability thesis, which is broadly associated with neo-liberalism, is limited because, quite evidently, there are some unique assets crucial to well-being and, perhaps, the survival of all life forms on earth that have no substitutes. These include, for example, the basic life-support functions associated with various ecosystem services provided by the atmosphere, land and the oceans. Thus, some components of Kn must not be allowed to decline given the potentially catastrophic consequences that could follow.

Strong sustainability

In contrast, strong sustainability does not regard economic, social and environmental capital as perfectly substitutable. Rather, strong sustainability assumes a very limited degree of substitution between human and natural capital stocks, its proponents arguing that much of the stock of natural resources and ecological functions is irreplaceable. This fact implies that there are inherent limits to growth faced by human economies. Consequently, strong sustainability means treating natural capital (Kn) separately since it cannot ultimately be substituted for physical or human capital. Putting it another way, physical capital cannot, regardless of price, replace the services and amenities provided by nature, particularly basic life-support services such as the ozone layer, climate regulation, the food chain, the balance between alkalinity and acidity, clean water and so on. By implication, therefore, investments in built infrastructure (Kb) cannot adequately compensate future generations for losses of complex ecological communities and other 'critical components' of natural capital. That is, if Kn is degraded beyond threshold levels, the unique services of ecological systems have no substitute, resulting in irreversible harm to global biodiversity and the long-term collapse of essential ecosystems. Hence, many proponents of strong sustainability require that natural capital remains constant in aggregation, but allow one form of natural capital to be substituted for another, subject to certain ecological constraints (Dietz and Neumayer, 2007).

Researchers have developed various operational rules to ensure the sustainable management of natural capital stocks (Ekins *et al.*, 2003). More specifically, several driving principles underpin the strong sustainability principle, limiting the reduction in natural capital Kn. These include the *Responsibility Principle* and the *Precautionary Principle*, both of which drive the stewardship perspective in respect of tourism development. The Responsibility Principle affirms that access to environmental resources carries attendant responsibilities to use them in an ecologically sustainable, economically efficient and socially fair manner (Lenzen *et al.*, 2007). This principle supports the notion that, in all cases, resident values should be important in determining the type and scale of tourism development. Similarly, the Precautionary Principle affirms that the combination of irreversibility and scientific uncertainty should make us cautious about depleting natural and social capital (Fennell and Ebert, 2004). In a tourism context, the Precautionary Principle would preclude development actions that impose potential irreversible effects on natural and sociocultural environments, both in the present and future.

It is likely that the strong sustainability thesis will evolve over time to distinguish between types of natural and social capital that could be destroyed or degraded without large costs and those that should not be changed in any respect given the risks involved. Thus, Kn may comprise types of natural capital Kn1, Kn2 and so on (Kn1 may, for example, be a complex ecosystem while Kn2 is a small swamp). Similarly, social capital Ks may comprise types Ks1, Ks2 and so on (Ks1 may be the degradation of a unique culture, while Ks2 relates to a small job loss). This would give stakeholders the opportunity to develop and apply threshold levels essential to maintain the integrity of different types of natural and social capital stocks, according to their importance to both present and future generations. The strong sustainability rule applied to different types of economic, natural and social capital can then be written as follows:

Strong economic (built plus human) sustainability: $dK/dt \geq 0$; $dKb1, dKh1 \geq 0$ - - -
Strong environmental sustainability: $dK/dt \geq 0$; $dKn1 \geq 0$ - - -
Strong social sustainability: $dK/dt \geq 0$; $dKs1 \geq 0$ - - -

Economic, social and natural capital are strongly interactive. As the stock of natural assets declines, social unrest is more likely to occur as individuals and groups seek access to fresh water, unpolluted environments, more productive farmlands and more stable weather systems. This will affect economic growth. In turn, social capital can lead to environmental improvements by way of volunteerism, participation in the sharing economy, common property regimes and other social behaviour to reduce resource exploitation and levels of negative externalities.

Tourism and the Sustainable Development Goals (SDG)

In 2016, the United Nations General Assembly adopted the 2030 Agenda for Sustainable Development with the 17 Global Sustainable Development Goals (SDGs) as the plan of action for global sustainable development. The 2030 Agenda calls for all stakeholders to act in collaborative partnerships to implement the SDGs, while explicitly encouraging businesses to apply their creativity and innovation to solving sustainable development challenges. The UNWTO has argued forcefully that the tourism industry shares with local residents, governments and communities the obligation to protect and maintain the natural and cultural heritage resources of our planet, both to sustain economies and to be passed on unimpaired to future generations. Tourism is explicitly mentioned in three SDGs: 8.9, 12.b and 14.7, while it is also considered to be very relevant to the achievement of the other goals. The implications for tourism meeting the SDGs are listed in Box 3.1.

Can Tourism Succeed in Its Sustainability Agenda?

Achieving the SDGs will require significant and deep transformations in education systems, healthcare, energy use, land use, urban planning and the deployment of information technologies. Moreover, these transformations will require strong government leadership working in partnership with business and civil society, not least because integrating the SDGs by 2030 into national strategies, budgets, audits, procurement policies, regulatory and human resource management and other dimensions of public policy poses major challenges for both developed and developing countries. While some destinations are better at long-term planning and implementation, all need to improve their public policies and governance processes if they are to achieve all the SDGs by 2030. Success will require massive innovation, learning and sharing of best practices of governance within and among destinations. Some countries have established dedicated co-ordination units, action plans and accountability systems, while others lag behind on some or all of these dimensions.

The United Nations Environment Programme (UNEP/UNWTO, 2012) has identified tourism as one of ten economic sectors leading the transformation to the global green economy. Despite the optimism expressed by UNEP, however, we note that substantial adverse social and environmental impacts continue to be associated with tourism growth globally and, hence, current practice or 'business as usual' in tourism planning and management seems impossible to reconcile with sustainability. We would argue that the underlying causes of the unsustainable direction of tourism development are embedded in the type of mindset possessed by key tourism stakeholders – not only destination managers, industry operators and host communities but also tourists themselves (Pollock, 2015; Dwyer, 2018). To effect the serious change in practice required to achieve the SDGs, new ways of thinking and behaviour must replace the mindset that underpins the current destructive practices associated with tourism growth. There are three elements that are essential to a new mindset, namely: the necessity for new business models; an ecological perspective; and good governance.

The need for new business models

The prevailing neo-liberal paradigm regards business firms as principally responsible to their shareholders (profitability) and not to society as a whole, except regarding conformance to state laws (Friedman, 1970). This view that the firm has largely optional social obligations provides

Box 3.1. Sustainable Development Goals and tourism.

SDG 1 – Eliminate poverty worldwide.

Tourism creates income and jobs at local and community levels. It plays an important role in national poverty reduction strategies and entrepreneurship development. Low skills requirement and local recruitment can empower relatively disadvantaged persons, particularly youth and women.

SDG 2 – Eliminate hunger, achieve food security and nutrition, and promote sustainable agriculture.

Tourism can spur sustainable agriculture via production and supply to hotels, and sales of local produce to tourists. Agro-tourism can generate additional income while enhancing the value of the tourism experience.

SDG 3 – Ensure healthy lives and promote well-being for all.

Tax revenue generated from tourism can be reinvested in healthcare and services, improving maternal health, reducing child mortality and preventing disease. Visitor fees collected in protected areas can contribute to health services.

SDG 4 – Ensure inclusive and equitable quality education and promote lifelong learning.

Tourism can promote inclusiveness. A skilful workforce is crucial for tourism growth. Improved education provides opportunities for direct and indirect jobs for youth and women, and benefits persons with special needs.

SDG 5 – Promote gender equality and female empowerment.

Tourism can empower women, particularly through the provision of direct jobs and income-generation in tourism and hospitality related enterprises. Tourism can be a tool for women to fully engage and lead in every aspect of social development.

SDG 6 - Ensure availability and sustainable management of water and sanitation.

Tourism investment in utilities can play a critical role in achieving water access and security, as well as hygiene and sanitation for all. The efficient use of water in tourism, pollution control and technology efficiency conserves scarce resources.

SDG 7 – Ensure access to affordable, reliable, sustainable energy.

As an energy-intensive sector, tourism can accelerate the shift towards increased renewable energy shares in the global energy mix. By promoting investments in clean energy sources, tourism can help to reduce greenhouse gases, mitigate climate change and contribute to energy efficiency.

SDG 8 – Promote sustained, inclusive and sustainable economic growth and employment.

Tourism is an important export earner globally, currently providing one in ten jobs worldwide. Quality work opportunities in tourism, particularly for youth and women, and diversification policies through tourism value chains, can enhance tourism's positive socio-economic impacts.

SDG 9 – Build resilient infrastructure, promote inclusive and sustainable industrialization and foster innovation.

Tourism development relies on good public and private infrastructure. The sector can influence public policy for infrastructure development that is sustainable, innovative and resource-efficient, thus attracting tourists and other sources of foreign investment.

SDG 10 – Reduce inequality within and among countries.

By engaging local populations and a range of stakeholders in its development, tourism can be a powerful tool for reducing inequalities. Tourism is an effective means for economic integration and diversification, giving people the opportunity to prosper in their place of origin and contributing to urban renewal and rural development.

SDG 11 – Make cities and human settlements inclusive, safe, resilient and sustainable.

Tourism can advance urban infrastructure and accessibility, promote regeneration and preserve cultural and natural heritage, assets on which it depends. Investment in green infrastructure (more efficient transport, reduced air pollution) will result in smarter and greener cities for residents and tourists.

Continued

SDG 12 – Ensure sustainable consumption and production patterns.
The tourism sector needs to adopt sustainable consumption and production modes. Tools to monitor sustainable development impacts for tourism including for energy, water, waste, biodiversity and job creation, will result in enhanced economic, social and environmental outcomes.

SDG 13 – Take urgent action to mitigate and adapt to climate change.
Tourism contributes to and is affected by climate change. By reducing their carbon footprint, tourism stakeholders can benefit from low carbon growth, and lead the global response to climate change.

SDG 14 – Conserve and sustainably use the oceans, seas and marine resources for sustainable development.
Coastal and marine tourism rely on healthy marine ecosystems. Tourism development must be a part of Integrated Coastal Zone Management in order to help conserve and preserve fragile marine ecosystems, promote the blue economy, and contribute to the sustainable use of marine resources.

SDG 15 – Protect, restore and promote sustainable use of terrestrial ecosystems and address biodiversity loss.
Rich biodiversity and natural heritage of destinations are strong motivations for tourist visitation. Tourism can play a major role in fragile zones, if sustainably managed, not only in conserving and preserving biodiversity, but also in generating revenue as an alternative livelihood to local communities.

SDG 16 – Promote peaceful and inclusive societies, with justice for all.
Involving billions of encounters between people of diverse cultural backgrounds, tourism can foster multi-cultural and inter-faith tolerance and understanding, laying the foundation for more inclusive institutions and peaceful societies. Tourism can engage and benefit local communities and foster peace following crises.

SDG 17 – Enable robust global partnership for sustainable development.
Due to its cross-sectoral nature, tourism has the ability to strengthen private/public partnerships and engage multiple stakeholders – international, national, regional and local – to work together to achieve the SDGs and other common goals. Sound public policy and innovative financing are important for achieving the 2030 Agenda.

Adapted from UNWTO/UNDP (2017), Table 1.

the rationale to ignore market failures, such as pollution, congestion and social problems, associated with production, and an incentive to 'externalize' as many social and environmental considerations as possible.

An alternative perspective rejects the conceptual validity and long-term performance implications of the traditional profit-maximizing, shareholder-centric view of business (Porter and Kramer, 2011; Mackay and Sisodia, 2014; Markman *et al.*, 2016). Although continuing to seek financial returns, firms (so-called 'hybrid', 'conscious' or 'sustainable, ethical, entrepreneurial enterprises') should strive to create long-term value for all their stakeholders through the generation of various other kinds of societal wealth. In other words, this alternative approach to business, with roots in the social entrepreneurship literature, proposes that firms and organizations should seek social and environmental goals alongside financial goals; specifically, businesses must think beyond immediate, instrumental, self-interest objectives and consider a broader context of benefits to customers, employees, suppliers and communities as well as the social and natural environments of which they are a part. In short, the social and environmental impacts of firm operations should be built into business models, rather than being addressed as 'optional extras'.

In support of this approach, there is growing evidence that responsible businesses enjoy higher brand equity and profitability than those focused primarily on profit. In addition, other benefits include efficiencies and cost savings; improved market positioning; better stakeholder relationships; improved 'strategic' decision making; and 'wider' community benefits relating to destination competitiveness (Mackay and Sisodia, 2014). At the same time, improved environmental management can enhance the social reputation of the firm's products for which customers are willing to pay higher prices, with positive influence on financial performance (Dwyer, 2005).

There is a large research effort under way to explore the conceptual foundations of emerging alternative business models and to address their implications for industry practice. The social entrepreneurship approach recognizes the importance of each of the four types of capital that must be balanced under the strong sustainability perspective. By adopting a social entrepreneurship approach, tourism can act as a social force that promotes the SDGs through a wider vision of tourism's role in global communities.

The ecological perspective

Whatever the particular business model adopted, we would argue that it is essential also to adopt an ecological perspective to drive sustainable development. Indeed, there is evidence that such a mindset is gaining favour among planning and development. This reflects a broader perspective than that of conventional economics, which views the economy as a sub-system embedded in the larger systems of society and the biosphere with biophysical limits to growth. Biodiversity concerns must play an important role in destination planning and investment in respect of built capital formation and, from the ecological perspective which reflects the strong sustainability position, sustainability is achieved through those modes of production and consumption that fit within the constraints posed by ecological limits.

With respect to the economic lens of sustainability analysis, the emerging field of ecological economics has much to offer tourism research and practice with guidance as to how tourism can change current practice to achieve the various SDGs. The ecological perspective, characterized by a strong applied focus, encourages critical work that draws on and integrates elements of ecological science, economics and the analysis of values, behaviours, cultural practices, institutional structures and societal dynamics. From the premises of strong sustainability, it follows that economic policy has a responsibility to the greater ecological world and that sustainable development must, therefore, take a broader approach to valuing natural resources and ecological functions. Economic analysis should seek to stimulate provisioning processes that enhance the broad quality of life, standard of living, happiness, growth and development for the common good, while recognizing the conflicting and co-operative interaction within and between classes, genders, ethnic groups, species, institutions, nations and regions (Kubiszewski and Costanza, 2015).

The importance of good governance

Governance best suited to addressing sustainable futures is located at the local or regional scale where the significance of impacts on places with respect to their social, cultural and environmental specificities is unique. However, governance at the local level needs to be understood within the global framework of neo-liberalism that continues to promote a 'growth first' agenda (Peck and Tickell, 2002). In the specific context of tourism, the concept of sustainability is now deeply embedded in tourism policy documents and governance models across all scales from the global to the local (McCool and Bosak, 2016; Dwyer, 2018); the challenge lies in how to operationalize these ideas – and at what scale.

Challenges regarding governance for sustainability arise because political and economic systems are not readily conducive to the necessary collaborative decision making across the three pillars of sustainability (Westley *et al.*, 2011). As Meadowcroft (2007, p. 313) reminds us, '[t]he sort of radical decoupling of economic activity from environmental burdens that sustainable development implies will require iterative processes of reform stretching over many decades'. More pessimistically, perhaps, the frustrations that emerge from three decades of research and discussion on sustainable tourism development have caused some scholars (for example, Sharpley, 2009) to conclude that the idea and applicability of sustainable tourism is at a dead end.

Recent tourism research has engaged with complex adaptive systems approaches (Baggio *et al.*, 2010; Scuttari *et al.*, 2016) and, in so doing, concepts such as 'resilience' have gained popularity, directing attention to shorter, more definable and achievable goals (Calgaro *et al.*, 2014). Some researchers argue that attention to the concept of 'resilience' and its concern with responses to shocks, stressors and adaption, offers an alternative focus for destination-level governance compared to the fuzzier concept of 'sustainability' (Lew, 2014). However, Saarinen and Gill (2019) support the idea that, 'resilience' should be perceived as an integral part of the sustainability paradigm and a key element of pathway creation in the process of moving towards sustainable futures - a position supported by a high-level United Nations Report (UN, 2012).

The wider societal spread of disenchantment with the term 'sustainability' may suggest a lack of awareness that sustainability is a long-term process, not an end state (Smith and Stirling, 2010). This is evident in the lack of progress in implementing the variously interpreted, 'grand' concept of sustainability, amid increasing calls for the adoption of more transitional approaches (Van Assche *et al.*, 2014). The field of transition management has emerged over the past two decades in response to the challenges of developing large-scale societal transformation for sustainable futures. Grounded in the concept of complex adaptive systems, this field embraces a broad multi-disciplinary arena that addresses socio-technical, socio-institutional and socio-ecological domains (Loorbach *et al.*, 2017). The core ideas of transition management approaches include: multi-actor dynamics; reframing the problem; the importance of experimentation; and the importance of continuous learning and evaluating (Loorbach *et al.*, 2017). The aim is to create disruptive systemic changes within regimes, countering lock-in and creating new pathways towards sustainability.

Issues of tourism sustainability fall primarily within the socio-institutional approach to transition management, typically with a focus on the role of agency and governance, addressing such issues as power, politics and agency to support advocacy of strong sustainability. While much policy is place-based and enacted at a community and regional scale, understanding the governance of destination resilience through a socio-ecological systems (SES) lens requires consideration of multiple scales of complexity with interaction between system components that are subject to regulation and policies occurring through networks that transcend regional, national and international scales (Calgaro *et al.*, 2014).

While the term 'transition' occurs frequently in the sustainable tourism literature, there has been relatively little attention to the extensive body of theory and practice of 'transition management' that has developed across many disciplines. The earliest work by tourism researchers is that by Gössling *et al.* (2012) who examine transition management as a tool for sustainable tourism involving stakeholder involvement and policy implementation processes. They conclude that transition management 'provides a valuable theoretical framework, while the dialetics of stakeholder involvement and policy implementation are an essential precondition for successful governance' (Gössling *et al.*, 2012, p. 899). Utilizing Gössling *et al.* (2012)s framework, Scuttari *et al.* (2016) apply a complexity-based approach to examine sustainable mobility in the tourism context through an examination of stakeholder involvement and policy implementation processes. They highlight the public sector's special role, the time demands of transition management and the value of both formal and informal partnerships. In a recent study of sustainability governance in the mountain resort of Whistler, Gill and Williams (2018) revealed that, although not overtly identified as such, transition management concepts and practices were embedded into the policy and practice of developing a new pathway to sustainability. In Whistler's case, the immediate success of the sustainability planning process was subsequently impeded by changes in local politics. However, it is important to remember that the path to sustainability is not a linear one but subject to loops, stalls and readjustments of direction, with no steady state envisaged.

As researchers have highlighted, transitioning towards sustainability is volatile as it is grounded in power and politics (Dredge, 2019; Hall, 2019). This, perhaps, represents one of the biggest challenges to the achievement of sustainability, requiring closer examination of the political dimensions of sustainability transitions (Meadowcroft (2011). Since politicians' jobs depend upon meeting the immediate needs of voters, dominant forms of government, favouring short-term gains over long-term responsibility may not be especially conducive to sustainability planning (Bosselmann *et al.*, 2008). Nevertheless, slow progress towards sustainability governance is advancing on many fronts, with the transition management approach offering a useful model for tourism researchers and planners to embrace more fully a strong sustainability approach to tourism development.

Measuring Sustainability: The MST Index

The UN World Tourism Organization (UNTWO) is exhorting the industry to adopt more sustainable practices. As part of this process, one critical challenge is to create destination-based reporting standards that integrate economic, social and environmental dimensions of local tourism impacts to assess whether sustainable development principles are being effectively applied in practice. Developments in tourism statistics, tourism satellite accounts, environmental-economic accounting and sustainability indicators have each served to advance the understanding of the measurement challenges and opportunities. With the support of the United Nations Statistics Division, UNWTO (2017, 2018a, b) has launched the initiative *Towards a Statistical Framework for Measuring Sustainable Tourism* (MST). The aim is to extend the current statistical frameworks used in tourism beyond their economic focus, to incorporate environmental and social dimensions at relevant spatial levels: global, national and sub-national. The MST acknowledges the dependency of tourism activity on 'social' infrastructure, including a suitably qualified workforce, transport infrastructure and public facilities. The anticipated outcome is an international statistical framework for measuring tourism's role in sustainable development.

The MST project builds upon earlier work to achieve three primary objectives:

- Integrating the established measurement frameworks of the Tourism Satellite Accounts (TSA: Recommended Methodological Framework) and the System of Environmental-Economic Accounting (SEEA Central Framework) to provide a platform for the measurement of sustainable tourism. The MST will provide a strong statistical base to support monitoring tourism progress towards the UN Sustainable Development Goals (SDGs) as part of the 2030 Development Agenda.
- Developing indicators for global monitoring of sustainable tourism and promoting tourism within the SDG indicator framework. This will facilitate comparison of the performance of the tourism sector and the impacts of different policies on a consistent basis with other sectors in different destinations and countries.
- Advancing the development of sub-national tourism statistics, recognizing the importance of location-specific information in decision making on tourism. The MST will foster participation and growing commitment from industry and government stakeholders at all levels resulting in more co-ordinated and equitable commitments to achieve sustainable development goals.

The proposed MST will provide a basis for improving co-ordination in data collection and organization (including identification of data sources and data gaps), as well as for improving institutional arrangements for the governance and management of statistics on sustainable tourism. It will also provide a cross-functional platform for discussing the interactions and dependencies between economic and social development, environmental protection and mitigation of potential risks for the sustainable growth of tourism.

Development of the MST is in its early stages. Four extensions or adaptations of the index relevant to tourism are envisaged:

- extensions within the TSA framework for describing tourism industries;
- connections between SEEA based accounts for individual environmental flows (for example, water, energy, waste) and tourism activity;
- accounting for produced and environmental assets relevant to tourism; and
- spatial accounting for tourism activity applying the logic of SEEA-based land and ecosystem accounting.

While the current choice of indicators, based on TSA and SEEA measures, represents a good approximation of the important sustainability-related concepts, the selection will be improved in the future as better statistics become available. An ideal set of well-being indicators should come from an internationally harmonized data collection based on common definitions and survey practices, and collected as part of the official statistical system of countries. This will improve measures of society's well-being, prosperity and progress as well as environmental sustainability, appropriate to informing destination performance in achieving the SDGs.

In particular, development of the MST can be further informed by recent work on the development of social indicators as part of the Genuine Progress Indicator (GPI). Given the limitations of GDP as a measure of economic well-being summarized earlier in this chapter, more comprehensive indicators are required to consolidate economic, environmental and social elements into a common framework to show net progress in development processes. The GPI consolidates critical economic, social and environmental factors into a single framework in order to give a more accurate measure picture of the progress (or lack of progress) towards enhancing the well-being of destination residents resulting from economic growth, including tourism-led growth (Stiglitz *et al.*, 2009; Kubiszewski and Costanza, 2015). At both national and international levels, the development of statistics must respond to the needs of users. In the case of tourism, this will include both government and private sectors. Good governance requires that the various types of analysis and specific areas of focus within the broad coverage of sustainable tourism are undertaken with full participation of relevant stakeholders.

GPI accounts and applications are increasing worldwide (Bagstad and Shammin, 2012); If appropriately used and understood, the GPI can help chart and guide the course of development The component measures are monetized and, therefore, offer a substantial degree of objectivity. While not a tourism index *per se*, it has clear application to measuring the sustainable development of any destination tourism industry if used in association with TSA.

The Challenges Ahead

Achievement of a sustainable and desirable future for tourism development requires a shift away from 'business as usual' with its focus on maximizing production and consumption (GDP) towards improving genuine human well-being (GPI or something similar). This shift will require governance processes that give much more attention to environmental protection, full employment, social equity, better product quality and durability, ecological sustainability and greater efficiency in resource use. A strong sustainability perspective should be maintained in project and investment assessment, recognising pervasive uncertainties, responsible and cautious behaviour, and incorporating safe minimum standards (moving from egocentric to ecocentric behaviour). Alternative measures of progress, such as GPI, are continually evolving and being revised and refined (Stiglitz *et al.*, 2009). A comprehensive suite of measures with particular relevance to tourism is under development within the MST framework.

The mission statements of firms and DMO should explicitly state socio-economic prosperity rather than tourist numbers or expenditure as key performance indicators. This will involve targeting 'ideal tourists' who are characterized by attributes such as, 'responsible', 'caring', 'involved', 'environmentally aware', 'culturally sensitive', 'seeking meaningful social encounters' and so on, in addition to being high spenders (Dwyer, 2016).The role of social norms, including fairness, inequity aversion and reciprocal altruism, in determining purchasing behaviour is now a growing research area in behavioural economics which links into behavioural science and cognitive and social psychology. This research has substantial implications for studies of tourist behaviour and sustainability practices (Dwyer, 2018).

For the new business models to be successful in tourism policy and planning, a deeper scientific understanding of the complex interactions between the four types of capital – built. natural, social and human – is needed. It is likely that a 'dashboard' approach with multiple indicators and comparing and interpreting them simultaneously will provide the best data to promote sustainable development. Only by better understanding the full range of impacts of our economy can we protect and maintain the planet's ecosystems. This does not of course preclude the use of standard economic performance measures, such as GDP, to determine (narrow) economic performance. However, the additional measures should serve to integrate current knowledge of how ecology, economics, psychology and sociology collectively contribute to establishing and measuring sustainable well-being. This will enhance their usefulness in charting and guiding the course of industry development including tourism.

There is a need to improve methods to value unpriced items that are not traded in any market, valuing costs and benefits of projects that occur well into the future and valuing the effects on species other than humans (Fennell and Ebert, 2004). As noted above, the challenges of steering the direction of sustainable governance lie in integrating conflicting values and dealing with risk, uncertainties and ambivalence. As a consequence, governance issues confront a wicked set of barriers to achieving sustainable tourism development.

Conclusions

The strong approach to measurement of sustainability requires analysis of the evolution over time of the different stocks of capital (economic, environmental, human and social) that sustain the various dimensions of well-being, particularly in respect of how decisions taken today affect these stocks. Over time, the types of variables that comprise progressive indexes, such as the MST and GPI, should be informed by each of the 17 SDGs. Developing integrated measures of progress attached to these goals offers the global community generally, and tourism stakeholders in particular, the opportunity to better understand the nature of sustainable well-being, how to measure it and how to achieve it.

In the context of tourism development, any new metrics must have the broad support of stakeholders. Good governance requires that any 'top-down' process of developing measures be complemented with a 'bottom-up' engagement of civil society that includes city and regional governments, non-governmental organizations, business and industry associations, environmental groups and other parties, including destination residents.

References

Baggio, R., Scott, N. and Cooper, C. (2010) Improving tourism destination governance: a complexity science approach. *Tourism Review* 65(4), 51–60. DOI: 10.1108/16605371011093863.

Bagstad, K.J. and Shammin, M.R. (2012) Can the Genuine Progress Indicator better inform sustainable regional progress? A case study for Northeast Ohio. *Ecological Indicators* 18, 330–341. DOI: 10.1016/j.ecolind.2011.11.026.

Bosselmann, K., Engel, R. and Taylor, P. (2008) *Governance for Sustainability: Issues, Challenges, Successes*. IUCN, Gland, Switzerland.

Calgaro, E., Lloyd, K. and Dominey-Howes, D. (2014) From vulnerability to transformation: a framework for assessing the vulnerability and resilience of tourism destinations. *Journal of Sustainable Tourism* 22(3), 341–360. DOI: 10.1080/09669582.2013.826229.

Dietz, S. and Neumayer, E. (2007) Weak and strong sustainability in the SEEA: concepts and measurement. *Ecological Economics* 61(4), 617–626. DOI: 10.1016/j.ecolecon.2006.09.007.

Dredge, D. (2019) Governance, tourism and resilience: a long way to go? In: Saarinen, J. and Gill, A.M. (eds) *Resilient Destinations and Tourism*. Routledge, Abingdon, UK, pp. 48–66.

Dwyer, L. (2005) Relevance of triple bottom line reporting to achievement of sustainable tourism: a scoping study. *Tourism Review International* 9(1), 79–938. DOI: 10.3727/154427205774791726.

Dwyer, L. (2016) Who is the ideal tourist? *Acta Turistica* 28(2), 151–183.

Dwyer, L. (2018) Saluting while the ship sinks: the necessity for tourism paradigm change. *Journal of Sustainable Tourism* 26(1), 29–48. DOI: 10.1080/09669582.2017.1308372.

Ekins, P., Simon, S., Deutsch, L., Folke, C. and De Groot, R. (2003) A framework for the practical application of the concepts of critical natural capital and strong sustainability. *Ecological Economics* 44(2-3), 165–185. DOI: 10.1016/S0921-8009(02)00272-0.

Fennell, D.A. and Ebert, K. (2004) Tourism and the precautionary principle. *Journal of Sustainable Tourism* 12(6), 461–479. DOI: 10.1080/09669580408667249.

Friedman, M. (1970) The social responsibility of business is to increase its profits. *The New York Times Magazine,* 13 September, pp. 32–33, 122–126.

Gill, A.M. and Williams, P.W. (2018) Transitioning towards sustainability in the mountain resort community of Whistler, British Columbia. *Tourism Recreation Research* 43(4), 528–539. DOI: 10.1080/02508281.2018.1509828.

Gössling, S., Hall, C.M., Ekström, F., Engeset, A.B. and Aall, C. (2012) Transition management: a tool for implementing sustainable tourism scenarios? *Journal of Sustainable Tourism* 20(6), 899–916. DOI: 10.1080/09669582.2012.699062.

Hall, C.M. (2019) Resilience theory and tourism. In: Saarinen, J. and Gill, A.M. (eds) *Resilient Destinations and Tourism*. Routledge, Abingdon, UK, pp. 34–47.

Kubiszewski, I. and Costanza, R. (2015) Measuring genuine social progress. In: Murray, J., McBain, D. and Wiedmann, T. (eds) *The Sustainability Practitioner's Guide to Social Analysis and Assessment*. Common Ground Publishing, Champaign, Illinois.

Kubiszewski, I., Costanza, R., Franco, C., Lawn, P., Talberth, J. *et al.* (2013) Beyond GDP: measuring and achieving global genuine progress. *Ecological Economics* 93, 57–68. DOI: 10.1016/j.ecolecon.2013.04.019.

Lenzen, M., Murray, J., Sack, F. and Wiedmann, T. (2007) Shared producer and consumer responsibility – theory and practice. *Ecological Economics* 61(1), 27–42. DOI: 10.1016/j.ecolecon.2006.05.018.

Lew, A.A. (2014) Scale, change and resilience in community tourism planning. *Tourism Geographies* 16(1), 14–22. DOI: 10.1080/14616688.2013.864325.

Loorbach, D., Frantzeskaki, N. and Avelino, F. (2017) Sustainability transitions research: transforming science and practice for societal change. *Annual Review of Environment and Resources* 42(1), 599–626. DOI: 10.1146/annurev-environ-102014-021340.

Mackay, J. and Sisodia, R. (2014) *Conscious Capitalism: Liberating the Heroic Spirit of Business*. Harvard Business Review Press, Cambridge, Massachusetts.

Markman, G.D., Russo, M., Lumpkin, G.T., Jennings, P.D.D. and Mair, J. (2016) Entrepreneurship as a platform for pursuing multiple goals: a special issue on sustainability, ethics, and entrepreneurship. *Journal of Management Studies* 53(5), 673–694. DOI: 10.1111/joms.12214.

McCool, S. and Bosak, K. (2016) *Reframing Sustainable Tourism*. Springer, Berlin.

Meadowcroft, J. (2007) Who is in charge here? Governance for sustainable development in a complex world. *Journal of Environmental Policy & Planning* 9(3-4), 299–314. DOI: 10.1080/15239080701631544.

Meadowcroft, J. (2011) Engaging with the politics of sustainability transitions. *Environmental Innovation and Societal Transitions* 1, 70–75. DOI: 10.1016/j.eist.2011.02.003.

Peck, J. and Tickell, A. (2002) Neoliberalizing space. *Antipode* 34(3), 380–404. DOI: 10.1111/1467-8330.00247.

Pollock, A. (2015) *Social Entrepreneurship in Tourism: The Conscious Travel Approach*. Tourism, Innovation Partnership for Social Entrepreneurship. Available at: http://www.tipse.org/conscious-tourism-pdf-download/ (accessed 20 June 2019).

Porter, M.E. and Kramer, M.R. (2011) Creating shared value. *Harvard Business Review* 89(1/2), 62–77.

Saarinen, J. and Gill, A.M. (2019) Introduction: placing resilience in the sustainability frame. In: Saarinen, J. and Gill, A.M. (eds) *Resilient Destinations and Tourism*. Routledge, Abingdon, UK, pp. 3–12.

Scuttari, A., Volgger, M. and Pechlaner, H. (2016) Transition management towards sustainable mobility in alpine destinations: realities and realpolitik in Italy's South Tyrol region. *Journal of Sustainable Tourism* 24(3), 463–483. DOI: 10.1080/09669582.2015.1136634.

Sharpley, R. (2009) *Tourism Development and the Environment: Beyond Sustainability?* Earthscan, London.

Smith, A. and Stirling, A. (2010) The politics of social-ecological resilience and sustainable socio-technical transitions. *Ecology and Society* 15(1), 1–13. DOI: 10.5751/ES-03218-150111.

Stiglitz, J., Sen, A. and Fitoussi, J.P. (2009) *The Measurement of Economic Performance and Social Progress Revisited. Reflections and Overview*. Commission on the Measurement of Economic Performance and Social Progress, Paris.

UN (2012) *Resilient People, Resilient Planet: A Future Worth Choosing: Overview. United Nations Secretary-General's High-Level Panel on Global Sustainability*. United Nations, New York.

UNEP/UNWTO (2005) Making Tourism More Sustainable: A Guide for Policy Makers. United Nations Environment Programme / UN World Tourism Organization. Available at: http://www.unep.fr/shared/publications/pdf/dtix0592xpa-tourismpolicyen.pdf

UNEP/UNWTO (2012) *Tourism in the Green Economy: Background Report*. UN World Tourism Organization, Madrid.

UNWTO (2017) *Statistical Framework for Measuring Sustainable Tourism Statistics and Tourism Satellite Account.* Seventeenth Meeting, UNWTO Headquarters, Madrid, Spain, 24-25 January 2017. Available at: http://cf.cdn.unwto.org/sites/all/files/pdf/item_3.2_1.pdf

UNWTO (2018a) *Measuring Sustainable Tourism*. World Tourism Organization, Madrid.

UNWTO (2018b) *Statistical Framework for Measuring Sustainable Tourism. Consultation Draft prepared for initial round of consultation with the UNWTO Committee on Tourism Statistics and TSA and the Working Group of Experts on Measuring Sustainable Tourism. February*. UN World Tourism Organization, Madrid.

UNWTO/UNDP (2017) *Tourism and the Sustainable Development Goals – Journey to 2030, Highlights*. UN World Tourism Organisation, Madrid.

Van Assche, K., Beunen, R. and Duineveld, M. (2014) *Evolutionary Governance Theory: An Introduction*. Springer, Cham, Switzerland.

Westley, F., Olsson, P., Folke, C., Homer-Dixon, T., Vredenburg, H. *et al*. (2011) Tipping toward sustainability: emerging pathways of transformation. *AMBIO* 40(7), 762–780. DOI: 10.1007/s13280-011-0186-9.

WTTC (2003) *Blueprint for New Tourism*. World Travel and Tourism Council, London.

4 Tourism Development: From 'Please Come' to 'Please Go' to 'Please Come Back' (After the Virus)

Richard Butler*
Strathclyde University, Glasgow, Scotland

Introduction

A rather long academic lifetime of working in tourism has led this writer to believe that much of the material written about tourism has been done so from a variety of positions, including self-interest, ignorance, selective recall, ideological intent, nationalistic ideals and wishful thinking along with, occasionally, insight and intellectual rigour. In my own case, unfortunately, the first three, but hopefully also the last two, probably apply. In this chapter, I will probably reveal elements of all these five positions in attempting to comment on the way that tourism has developed itself and the environments in which it is found over the past six decades during which my interest in the subject has been focused. That timespan is longer than some writers have claimed modern mass tourism has existed (Leigh, 2013), and it encompasses the demise of many older resorts, the continual emergence of a vast number of new destinations, drastic changes in transportation and access, changes in attitudes towards tourism itself, generally from positive to negative, and, inevitably, a seemingly never-ending increase in tourist numbers in every part of the world (UNWTO, 2018). However, because of the effect of the Covid-19 virus, 2020 may be one of very few years on record that this trend is broken (for example, declines were experienced in 2001 (9/11) and in 2009 following the global financial crisis of 2008), a pattern previously consistent despite regional wars, economic depression, fuel shortages, other epidemics and terrorism. Despite all the criticisms, the revelation of its negative, as well as positive, impacts on destinations and, most recently, the appearance of the media-generated phenomenon of overtourism (Dodds and Butler, 2019a, b), tourism has continued to grow and to spread globally, often, it would appear, without formal direction or control, and to diversify into an increasing number of sub-forms, the latter often misidentified and mislabelled and, as noted by McKercher (2007), inevitably miscounted.

So much for context. This chapter will discuss this pattern of development and some of the enabling forces that have influenced and shaped tourism and its effects and comment on what this writer sees as key issues that warrant further examination. These include the pattern of development in destinations, the determinants of the spatial spread of tourism, the fallacies that have grown up around tourism and the lack of influence of academia on the phenomenon despite decades of study. In conclusion, some thought is given to potential future issues and directions for the development of tourism.

*richard.butler@strath.ac.uk

 Tourism in Development: Reflective Essays (eds P.U.C. Dieke *et al.*)
DOI: 10.1079/9781789242812.0004

Development Patterns

Change in tourism

The way tourism has developed reveals much about the way it is regarded, primarily as a source of income and employment by the suppliers and of pleasure for the consumers. This sometimes appears to be forgotten when a whole host of other objectives are ascribed to tourism, particularly in the context of its oxymoronic sustainable form. While tourism is capable of supporting other interests, such as heritage conservation, indigenous traditions, ecosystems and whole communities, that is rarely its primary purpose, and it is equally capable of harming those other interests and changing them irrevocably, sometimes accidentally (Butler, 2018).

A fundamental paradox at the heart of tourism is that the overall patterns and nature of development of tourism are both evolutionary and revolutionary at the same time. Tourism is also constantly vulnerable to change (Aramberri and Butler, 2005), as the Covid-19 virus has made abundantly clear. Change in tourism, in both supply and demand, is evolutionary in that it builds on what has gone before and what is existing at present, using those attributes and facilities in a generally similar pattern (e.g. a focus on beach use, cultural heritage visitation and general sightseeing). It is, therefore, somewhat predictable. The Tourism Area Life Cycle (Butler, 1980) is based on this supposition, arguing for a consistency in the pattern of destination development in stages that are recognizable throughout resorts globally, albeit with many local variations in scale and rate of change, reflecting differing conditions in each location. At the same time, tourism development is revolutionary, with shocks to the system that quickly and fundamentally alter all aspects of development (McKercher, 1999; Russell and Faulkner, 1999). Some of these shocks are positive, such as the intervention of entrepreneurs and innovators (Russell, 2006, 2010) at both the resort and regional scale, and some have been positive at the global scale through the actions of individuals (Butler and Russell, 2010) such as Thomas Cook and Walt Disney. Other shocks can be negative, including economic recessions, conflict (Isaac *et al.*, 2019), physical disasters, (tsunamis, earthquakes, fire and flood), and health-related problems. A failure to appreciate that both forms of change shape tourism simultaneously, although on different timescales, can result in inappropriate development and failure to anticipate future problems.

Spatial patterns

The spatial pattern of tourism development is also both evolutionary and revolutionary. It is evolutionary in the sense that existing destinations soon face competition from similar and often closely located newer developments (Christaller, 1964), and revolutionary as illustrated by Hall (2006), when the effects of innovations in transportation or changes in travel restrictions suddenly come into effect. Holiday destinations become weekend retreats (Wolfe, 1952) and often, eventually, sites for day visits, while at the global scale, long-haul destinations, once the privilege of the elite, become available to many, often to the chagrin of that elite who relocate to even more remote locations (see, for example, thebrando.com).

Whereas many older tourist destinations began, and continued for many years, as communities whose economic focus was not tourism but gradually experienced their economies becoming more and more dependent on tourism for their viability, the pattern over recent decades has been much more varied. The early 'instant' resorts such as Cancún (established in 1970) represented a new phase in tourism development, with destinations being purpose-built (in the case of Cancún with major federal government investment) and communities created to meet tourism demand. Indeed, some of them were established to create such demand, catering to a market that did not initially exist. For example, the Disney theme parks developed a model ('I believe in being an innovator' – Walt Disney, cited in Shani and Logan, 2010) that has been copied in various forms of theme parks throughout the world, incorporating not only attractions and amusements but also integrated accommodation and other services. In cases such as Disney World in Florida, these theme parks operate as private municipalities with virtual total control over the scale, nature and rate of development, and such an arrangement may partly explain why successful theme parks have maintained their market and appeal for over six decades (as in the case of Disneyland California).

The Nature of Tourism Development and Attitudes Towards Tourism

The success of specific purpose-built destinations has prompted further developments on a variety of scales. The desire to be, or appear to be, sustainable has led to the creation of exclusive resorts aimed at a specific tourism segment, which is often assumed to be not only responsible and sustainably inclined but also affluent and, therefore, doubly desirable (Butler, 2018). The sustainable (so-called) development of Marlon Brando's Pacific Island as a luxury (rates from $1200 a night) hotel destination is one example:

> The Brando is a unique luxury resort on French Polynesia's breathtakingly beautiful private island of Tetiaroa....the Brando offers carefree luxury in the midst of pristine nature. With access to the island by private plane, the resort features 35 villas on white-sand beaches frequented by sea turtles, manta rays and exotic birds. (thebrando.com, n.d.)

In the same vein are eco-lodges in Amazonia and other developments of private islands and remote retreats where the elite traveller can escape the masses for a little longer and feel good about it, despite the inevitable long-haul travel to reach such places (Wheeller, 1993). The motivations for this pattern of development are both economic and spiritual; economic, as entry is limited by high cost, and spiritual, as the benefit is in the sense of seclusion and escapism in an exclusive environment. It symbolizes a form of escape from a problem that many travel writers and some academics complain about, namely how to escape the majority of other tourists who 'ruin' experiences by their presence, let alone their behaviour. Such elitism is hardly new; John Ruskin complained of the horrors of tourism in Venice in the mid-19th century, writing his guide to Venetian architectural structures of 500,000 words 'for the few travellers who still care for her monuments' (cited in Hewison, 2014). Likewise, Thomas Cook was criticized around the same time (1865) by the British consul at Spezia for bringing middle-class tourists to destinations that previously had been exclusive to those 'with the effortless gentlemanly cultural capital to appreciate what they were seeing' (Walton, 2010). Such views are still common in both what Walton describes as 'the influential sectors of the London press' and perhaps among academics who believe that tourists seek authenticity (MacCannell, 1976) and generally fail to find it (perhaps with the unwritten message that such tourists would not recognize it if they did find it).

Why such disdain exists towards the great majority of participants in the subject that we study is puzzling. One should surely regard the fact that so many (albeit still a small minority) of the world's population are able to engage in holidays and travel to destinations, gaining pleasure and excitement and spreading money where they go, as a great achievement, although at the same time recognizing that the resulting impacts are not always positive and that there are real costs and problems involved with such mass travel. It should be appreciated that mass tourism, on a per capita basis, is almost certainly less harmful environmentally and possibly socially than much of the misnamed 'responsible' or 'alternative' forms of tourism which may place heavy individual demands on the environment in terms of energy use, water consumption, wildlife disturbance and local cultures and ways of life. To suggest that relocating large numbers of tourists to alternative destinations in order to relieve pressure on 'hotspots' is unrealistic, not least because the most heavily visited sites are generally unique and iconic (Weidenfeld, 2010) and cannot be duplicated and because the alternatives do not have the capacity to take the large numbers that would be involved. Long-haul tourists in particular (e.g. Chinese in Europe or vice versa) will have travelled to see the sights, and visiting the outskirts of Paris or London instead of the Eiffel Tower and the Louvre or Buckingham Palace and Big Ben will hardly be acceptable, no more than visiting China and not seeing the Great Wall or the Forbidden Palace. The more likely result is that such tourists would still flock to see these iconic sights as well as going to other locations, thus spreading and not resolving the problem of overcrowding.

Tourist Use of, and in, Destinations

Part of the problem of negative impacts of, and negative attitudes towards, tourism stem from the changing ways in which tourists (and short-term recreational visitors) utilize the destinations that they visit. For many decades, even millennia, visitors to specific sites (sights) went to see those features and landscapes, generally in what they perceived to be their 'natural' or 'unspoilt' state. These visits were mostly of a passive nature; that is, viewing the sights (painting, writing about them, more lately photographing them) and then departing. Increasingly, today, however, an emphasis is on *using* these locations for activities rather than visiting them for their inherent attributes. Thus, viewpoints become launching points for hang-gliding and jumping, designated trails become race tracks for mountain bikes and off-road vehicles, lakes the setting for boating activities, and mountains for 'summit bagging' by coach parties, as predicted for Mount Everest (Whipple, 2016).

Increasingly, such activities are also becoming mechanized and organized, creating a greater likelihood of disturbance for other users and residents of these locations by involving large parties of visitors. This is not a diatribe against modern forms of recreation and tourism, nor against mechanization. Such changes have mostly been evolutionary, for forms of leisure consistently have been created or have evolved from utilitarian activities, such as snowmobiling, often with unexpected and undesired results as well as economic and user benefits (Butler, 1974). It is not surprising that such a trend would develop in tourism, but it is a recognition that modern tourism is changing the way in which tourists behave, now increasingly using and doing rather than visiting and viewing locations. Thus, many places become visited for what the visitor can do while there, not because of the inherent attraction of the place itself. This is not new; Beau Nash drew large numbers of visitors to Bath in the 18th century because, through his promotion, the town became the place to be seen and to engage socially (Towner, 2010), not because Bath was particularly worth visiting *per se* at that time. The impacts of such changes are both extensive and intensive on the communities and environments involved and, in the case of Bath, created a reputation and a positive image as a result of the architectural developments of Nash's time (and later associations with writings such as those of Jane Austen) that still draw many tourists today.

Emerging Problems of Tourism Development

Many attempts to develop destinations and regions for tourism are successful in their initial stages but run into problems, including negative reactions of residents and visitors as tourism increases in scale and changes in nature (resorts such as Benidorm and Alicante in Spain, encouraged to develop tourism by Franco for prestige reasons (Aramberri, 2005), are examples). Another example has been the creation of the 'North Coast 500'. This is a newly promoted driving route, so named because it generally parallels the north-west, north and north-east coasts of Scotland for a distance of approximately 500 miles. It was a local initiative created in 2015 with the support of a number of regional bodies, including the Tourism Project Board of the North Highland Initiative, and was widely endorsed as a means of providing employment and income to the scattered small communities in that part of northern Scotland. Such has been the success of the project and its promotion that it has become known as 'Scotland's Route 66' and is claimed to be one of the 'top five coastal routes in the world' (Kerr, 2015). It is a deliberately created attraction, based on using existing roads and impressive wild scenery along with cultural and natural heritage features, to draw tourists, but there appears to have been little thought given to potential problems that could emerge should it prove successful in attracting those visitors. The roads, for a large part of the route, are single-track in width with passing places, requiring drivers to follow locally established patterns relating to oncoming traffic, overtaking, parking in the passing places, and being able to reverse considerable distances on narrow roads with blind summits and many bends. Those drivers with caravans (trailers) find this extremely difficult. Additional vehicles on such roads also can hold up local traffic, including livestock movements, and, more seriously, delay emergency vehicles. As the route has become more popular, it has been used as a race track, with car drivers, motorcyclists and cyclists competing to complete the route in the shortest possible time (24 hours) (heavywhalley, 2017). These problems have been exacerbated by social media users both in popularizing the route regardless of problems and in the images and contents being shown on various outlets. The website for the route (northcoast500.com) and related messages on Facebook show illustrations mostly of individuals rather than the impressive scenery, while links to the website, including sponsorship by luxury brands such as Aston Martin, generally have little in common with the region itself. Accommodation along the route is limited and, as a result, wild camping and other illegal overnight stops by mobile homes and caravans along the route have increased along with problems of rubbish and pollution because of the lack of services at stopping points. This development illustrates the point made earlier about the way visitors distort the intended or traditional use of attractions, in this case converting a slow scenic drive through a thinly populated area for more active, sometimes competitive and certainly inappropriate uses, prompting police action and charges being laid (Dickson, 2019).

Filling (or Killing) Time

Tourist behaviour is a sensitive and difficult issue that is an inherent part of tourism development. One of the greatest problems facing tourists in their holiday environment is what to do. Most humans are creatures of habit and subject to constraints on their time because of work and other obligations. When on holiday, most of these constraints are removed temporarily although some obligations, such as those involving social and family considerations, remain, thus leaving holidaymakers with the universal problem of how to fill their time when on vacation. The tourist industry has responded from earliest times with the provision of entertainment, amusements, guides and attractions (Weidenfeld *et al.*, 2016) and it is clear that modern tourists are no less demanding of time-filling pastimes than those in earlier times.

The provision of additional attractions can be a cause of opposition to further development by permanent residents of destinations as such development may result in significant and permanent physical change in their home environments. While in some cases these additions may be welcomed as also providing entertainment and services for local residents, they can then become another area of disagreement over the nature and scale of tourism development among residents, sometimes revealing differences of opinion between long-term and recent residents of destinations (Brougham and Butler, 1981). The implicit freedom from constraints that being on holiday implies can also result in what is almost universally regarded as inappropriate behaviour by tourists, from trying to open coffins at funerals (Horne, 2016) to drunken excessiveness in places such as Magaluf and Ayia Napa that has resulted in bans on specific hedonistic activities (Euronews.com, 2020). Such issues reveal clearly the problems which can arise when physical development in, and improved access to, destinations takes place without appropriate consideration being given to the implications of additional tourists and their behaviour and impacts on the communities involved. As noted by Plog (1973), changes in the image and nature of a destination may occur over time and reduce the appeal of a destination to its traditional market segment. Attempts to respond to changing tastes and accessibility bring the risk of alienating an existing market and altering the image of a destination in ways not always intended or welcomed. How much change, and at what rate and scale, is significant in impacting on a destination's image and appeal has not been determined except in hindsight and reinforces the significance of anticipating tipping points and perhaps using chaos theory (McKercher, 1999; Russell and Faulkner, 1999) in anticipating the occurrence and impact of change in destinations.

The (Un)sharing Economy and Development Issues

Tourism is subject to many myths and misperceptions. A relatively recent one relates to its suggested links with what has been called 'the sharing economy (or society)' (OECD, 2016), an illustration of which is the spectacular rise in popularity of Airbnb, the renting of residential properties on a short-term basis for tourist accommodation, mostly through social media and the internet in particular. In reality, there is nothing sharing about this recent phenomenon. As Timothy (2008) has noted, it is generated primarily, as is much of tourism and other development, by greed, although often with claims to be part of the 'moral economy' through connecting with strangers (Germann Molz, 2013). By claiming such a linkage, it is imbuing involvement in such practices with the false sense of self-righteousness and superiority also found in eco-ego tourism (Wheeller, 1993) and 'good tourism' (Wood and House, 1991). It manifests itself in individuals renting out their properties for financial gain, in many cases at the expense of the comfort, privacy and general quality of life of neighbouring permanent residents. It is clear that at least a part of the resentment being shown at what has been termed 'overtourism' in numerous locations stems from such actions (Dodds and Butler, 2019a). Local residents are forced to 'share' their space and often their amenities with tourists, as many communities have been slow or reluctant to pass and then enforce laws limiting or prohibiting such rentals. The result of this failure can be seen in protests about such developments occurring in places including Edinburgh, Palma, Barcelona, Venice and Prague, and, as Rickly (2019) notes, in many cases it introduces tourism and tourists into residential parts of communities where it might not otherwise have occurred. In this situation, the 'sharing' is one-sided and neither voluntary nor welcomed by many local residents in affected areas, as they experience direct disbenefits such as disturbance, but few, if any, direct benefits.

This issue typifies a common feature of many forms of development, including tourism, which has been clearly seen in the furore of overtourism: the common existence of contrasting viewpoints between local residents in favour and those opposed to further development of tourism. It is rarely as simple as those benefitting economically from tourism being in favour of development and those not benefitting directly being opposed, although such a scenario is not unusual. Tourism proponents, such as Destination Management Organizations (DMOs), are prone to promote the 'welcoming' nature and natural hospitality of local communities to visitors, and, as such, a positive scenario is generally common, at least in the early stages of development of any kind in most places. Indeed, while Doxey's (1975) model of changing attitudes may be overly simplistic and unidirectional, it cannot be ignored.

In most cases, it is primarily the economic argument for development which is promoted and it has been known for a long time (Brougham and Butler, 1981; Mathieson and Wall, 1982) that many residents of tourist destinations feel that tourism is economically beneficial to their community even if they personally do not receive any direct economic benefit from it. Although improvements in infrastructure, in access, in services relating, at least in part, to the growth of, and because of, tourism can be welcomed and appreciated, if such improvements are seasonal, then eventually complants may arise that they have only been provided for tourists and not for local residents. Community involvement in tourism planning and development was proposed decades ago by Murphy (1985), but such participation requires knowledge and information and these are often in short supply and incomplete for residents when development is initially proposed. When some of the less desirable effects of tourism are experienced, communities are prone to protest and even take action, sometimes political (Martin, 2006), to counter further development. Effective resident involvement in community development can be seen in the case of Whistler in British Columbia (Sheppard, 2017), although changes in external circumstances can make achievement of goals increasingly difficult or unattainable (Gill and Williams, 2018) and demonstrate the vulnerability of tourism to exogenous forces noted earlier.

Implications for Future Tourism Development

In terms of the future for the development of tourism, one is reminded that almost all soothsayers, forecasters and scenario writers claim that the present is always a time of change, the start of something different that they are able to foresee. In reality, change is always occurring in tourism as in most things, and tourism, as shown over the past decades, causes and responds to change in other phenomena. One of these phenomena is political change (Butler and Suntikul, 2017), something which has been relatively little studied in tourism and yet exerts major influence on tourism growth and development. Political change means more than simply different relationships between nation states, although these are of critical importance for the removal or lessening of barriers to tourist travel, such as visas. In recent years, the rapid growth of both Chinese visitors abroad and foreign visits to China has demonstrated what changes in political ideology can bring about (Sofield, 2017). Similar patterns of growth have been seen in Hong Kong (reflecting changes in visas for visitors from mainland China), in Vietnam after *doi moi* (Suntikul *et al.*, 2010), and the island of Ireland after the end of conflict, while political changes brought about both the closure of Cuba to tourism and its reopening, although visitation is still subject to restrictions by the USA on political grounds (Butler and Suntikul, 2017). It would be naïve to imagine that political relationships will remain static in the future. However, from the point of view of tourism, the hope would be that tensions would be reduced, for example between Iran and its potential markets, as well as in other parts of the Middle East including Israel and Palestine, and Tunisia.

Where such tensions have erupted into conflict, the impact on tourism is normally rapid, although often temporary if violence ends and is not repeated. War, while generally catastrophic for tourism and its development during periods of conflict, has proven to be of considerable benefit to tourism once permanent peace or long-term stability has been assured (Butler and Suntikul, 2013). The artefacts left after conflict (including battlefields, castles and cemeteries) become tourist attractions and attract millions of visitors to sites such as the Demilitarized Zone in Korea, Civil War battlefields in the USA, the Normandy beaches of D-Day, sites of the First World War in Flanders and death camps in eastern Europe. Some, but certainly not all, of this visitation is dark tourism, as many well-visited former military sites long predate any personal links with the visitors, such as the battlefield of Waterloo or Thermopylae, and, regretfully but almost inevitably, new 'attractions' in this vein will emerge and become part of the new tourist routes. What is perhaps clear about tourism to these sites is that often the conflicts are unexpected or certainly unplanned, as is their role in future tourism. Checkpoint Charlie and the Berlin Wall were not anticipated in their creation and certainly not expected to become tourist attractions once their military use was ended (so much so that the checkpoint was demolished, along with most of the wall, but a re-creation has been established in response to tourist demand). It does behove those responsible for conservation to appreciate the importance of military- and conflict-related artefacts and sites to future tourists and to take steps to preserve them for future generations of travellers, even if the origins of the features are viewed as inappropriate at the current time.

To date, conventional tourism has grown steadily since records began, although technological advances and changes in social-cultural mores (Gretzel, 2017) are bringing about significant changes in the choice of destinations to be visited and activities to be engaged in. Whether these forces will replace physical travel with virtual travel is unlikely in my opinion, but then I am a person brought up before the advent of much of modern technology. To me, travel, even if it is at times uncomfortable and even unreliable, is an exciting experience, particularly when the destination is new to me, although I am not sure that I agree with Robert Louis Stevenson that 'To travel hopefully is a better thing than to arrive' (Stevenson, 1874). To assume that future generations will be desirous or even willing to forsake visiting 'real' sites and sights in exchange for a virtual visit is unrealistic to this writer but, as noted below, circumstances might force such a change in behaviour. However, at a time when space travel is envisaged as being close (as supposedly it has been for the last few years according to promotions by would-be providers of such an experience such as Virgin Galactic.com), an inability to get to current destinations seems unlikely.

The current time (2020), however, is little different in many respects, except one, and that is the ongoing reaction to and effects of the Covid-19 virus pandemic. It might be reasonable to assume that once this health crisis is over, tourism will return to its traditional path of continuous global growth with perhaps an even more rapid rate of growth in the immediate aftermath, once confidence is regained and restrictions on travel removed. However, there are other elements which might modify or change such expectations. One is the fact that the economic slowdown has revealed an immediate improvement in environmental conditions because of behavioural restrictions; a fact that will be seized on by all concerned about the quality of the environment and, in particular, climate change. This may result in arguments that the temporary severe reductions in air travel be made permanent, at least to some degree, with opponents of tourism in general and of tourist travel by air in particular citing an opportunity to reduce what may be claimed as 'unnecessary' travel, which might well include most tourist travel (Wood and House, 1991; Croall, 1995; Hall, 2011). This would present a challenging situation and doubtless a controversial discussion about the merits and necessity of leisure in its entirety, including tourism (Nawijn, 2011), and its role in alleviating poverty and supporting economies in developing countries (Lazarus, 2008). On the one hand, the green lobby would likely unite to show the very real merits of reduced travel from an environmental point of view (Elliott, 2020). On the other hand, one would expect bodies such as UNWTO and WTTC, along with associated tourist industries, to demonstrate the need to restore economies, particularly those of developing countries highly dependent on tourism, and to allow millions of workers to return to their jobs and enable the world to avoid continued recession. The UNWTO has already produced a tag line 'staying home today means travelling tomorrow' (UNWTO.org), which indicates what its policy is likely to be once the current travel restrictions are lifted.

If one is forced to predict, it may be that air travel will be restored quickly to a level fairly close to that which existed before the health crisis but with fewer players, as some airline companies will have been forced out of operation. However, cruise shipping may take much longer to recover given its history of infections before Covid-19 and possibly a lingering fear among potential customers of such contamination remaining. Some ports that have already restricted the number of cruise vessels allowed to dock may continue with these or even impose stricter controls as they would be operating from a position of relative strength before previous levels of visitation were resumed. Otherwise one might expect tourism to resume much as before, and pent-up desire might result in even more record-breaking levels of visitation to some locations, at least in the short term. One would hope that lessons would have been learned on how to improve the management and control of tourism during the period of reduced demand and that these could be instituted before high levels are regained. The end of the great perturbation that is occurring will provide an opportunity, if not to start again, then at least to restart operations in a more appropriate manner and at more appropriate levels of visitation in order to avoid not just the probably temporary problems of overtourism as currently seen, but also the more serious long-term over-taxing of human and physical environments and communities. In the long run, tourism is likely to survive and continue, if only because evolution normally beats revolution, despite temporary shocks and short-term chaos.

References

Aramberri, J. (2005) How global is tourism? In: Aramberri, J. and Butler, R.W. (eds) *Tourism Development: Issues for a Vulnerable Industry*. Channel View Publications, Clevedon, UK, pp. 127–152.

Aramberri, J. and Butler, R.W. (2005) Tourism development: vulnerability and resilience. In: Aramberri, J. and Butler, R.W. (eds) *Tourism Development: Issues for a Vulnerable Industry*. Channel View Publications, Clevedon, UK, pp. 293–308.

Brougham, J.E. and Butler, R.W. (1981) The application of segregation analysis to explain resident attitudes to social impacts of tourism. *Annals of Tourism Research* 8(4), 569–590.

Butler, R.W. (1974) Snowmobiles - a blessing and a curse. *Journal of Travel Research* 13(1), 13–16.

Butler, R.W. (1980) The concept of a tourist area cycle of evolution: implications for management of resources. *The Canadian Geographer/Le Géographe canadien* 24(1), 5–12. DOI: 10.1111/j.1541-0064.1980.tb00970.x.

Butler, R.W. (2018) Sustainable tourism in sensitive environments: a wolf in sheep's clothing? *Sustainability* 10(6), 1789. DOI: 10.3390/su10061789.

Butler, R.W. and Russell, R. (2010) *Giants of Tourism: Key Individuals in the Development of Tourism*. CAB International, Wallingford, UK.

Butler, R.W. and Suntikul, W. (2013) *Tourism and War*. Channel View Publications, Clevedon, UK.

Butler, R.W. and Suntikul, W. (2017) *Tourism and Political Change*, 2nd edn. Goodfellow Publishers, Oxford, UK.

Christaller, W. (1964) Some considerations of tourism location in Europe: the peripheral regions-underdeveloped countries-recreation areas. *Papers of the Regional Science Association* 12(1), 95–105. DOI: 10.1007/BF01941243.

Croall, J. (1995) *Preserve or Destroy: Tourism and the Environment*. Calouste Gulbenkian Foundation, London.

Dickson, B. (2019) Bad driving crackdown by police on the North Coast 500 trail. *The Sunday Times*, 31 August, p. 31.

Dodds, R. and Butler, R.W. (2019a) *Overtourism: Issues, Realities and Solutions*. De Gruyter, Berlin.

Dodds, R. and Butler, R. (2019b) The phenomena of overtourism: a review. *International Journal of Tourism Cities* 5(4), 519–528. DOI: 10.1108/IJTC-06-2019-0090.

Doxey, G.V. (1975) A causation theory of visitor-resident irritants: methodology and research inferences. *Proceedings of the Travel Research Association 6th Annual Conference Travel Research Association: San Diego*, pp. 195–198.

Elliott, F.A. (2020) Blue skies, dolphins and clear canals in Venice – the beautiful side effects of the coronavirus pandemic. *The Telegraph*. Available at: www.telegraph.co.uk/travel/news/coronavirus-nature-environment-swans-venice-clear-skies-china/ (accessed 27 March 2020).

Euronews.com (2020) Spanish islands ban pub crawls in crackdown on drunk tourists. Available at: www.euronews.com/2020/01/17/spanish-islands-ban-pub crawls-in crack down-on drunk tourists (accessed 5 January 2020).

Germann Molz, J. (2013) Social networking technologies and the moral economy of alternative tourism: the case of couchsurfing.org. *Annals of Tourism Research* 43, 210–230. DOI: 10.1016/j.annals.2013.08.001.

Gill, A.M. and Williams, P.W. (2018) Transitioning towards sustainability in the mountain resort community of Whistler, British Columbia. *Tourism Recreation Research* 43(4), 528–539. DOI: 10.1080/02508281.2018.1509828.

Gretzel, U. (2017) Social media activism in tourism. *Journal of Hospitality and Tourism* 15(2), 1–14.

Hall, C.M. (2006) Space-time accessibility and the TALC: the role of geographies of spatial interaction and mobility in contributing to an improved understanding of tourism. In: Butler, R.W. (ed.) *The Tourism Area Life Cycle Conceptual and Theoretical Issues*. Channel View Publications, Clevedon, UK, pp. 83–100.

Hall, C.M. (2011) Consumerism, tourism and voluntary simplicity: we all have to consume, but do we really have to travel so much to be happy? *Tourism Recreation Research* 36(3), 298–303. DOI: 10.1080/02508281.2011.11081675.

heavywhalley (2017) Race track or tourist magnet? The North Coast 500. Any Thoughts? heavywhalley. Available at: worldpress.com/2017/05/24/race-track-or-tourist-magnet-the north-coast-500-any-thoughts (accessed 30 October 2018).

Hewison, R. (2014) A film of real artistry. *Culture: The Sunday Times*, 5 October.

Horne, M. (2016) Snap-happy tourists disrupt funerals at island cathedral. *The Times*, 6 August, p. 21.

Isaac, R.K., Çakmak, E. and Butler, R. (eds) (2019) *Tourism and Hospitality in Conflict-ridden Destinations*. Routledge, Abingdon, UK.

Kerr, D. (2015) North Coast 500 named in top five coastal routes in the world. *The Press and Journal*, 6 August. Available at: www.pressandjournal.co.uk/fp/news/highlands/657962/ north-coast-500-named-in-top-five-coastal- routes-in-the world (accessed 21 August 2018).

Lazarus, J. (2008) Participation in poverty reduction strategy papers: reviewing the past, assessing the present and predicting the future. *Third World Quarterly* 29(6), 1205–1221. DOI: 10.1080/01436590802201188.

Leigh, J. (2013) 'Peak Oil' confronts society and tourism: a futuristic view. In: Leigh, J., Webster, C. and Ivanov, S. (eds) *Future Tourism: Political, Social and Economic Challenges*. Routledge, Abingdon, UK, pp. 7–20.

MacCannell, D. (1976) *The Tourist: A New Theory of the Leisure Class*. Schocken Books, New York.

Martin, B. (2006) The TALC and politics. In: Butler, R.W. (ed.) *The Tourism Area Life Cycle: Applications and Modifications*. Channel View Publications, Clevedon, UK, pp. 227–249.

Mathieson, A. and Wall, G. (1982) *Tourism: Economic, Physical and Social Impacts*. Longman, Harlow, UK.

McKercher, B. (1999) A chaos approach to tourism. *Tourism Management* 20(4), 425–434. DOI: 10.1016/S0261-5177(99)00008-4.

McKercher, B. (2007) Phantom demand: how some research 'proves' demand when none really exists. *Culture, Tourism and the Media, Proceedings of the 5th DeHaan Tourism Management Conference*.

Murphy, P. (1985) *Tourism: A Community Approach*. Methuen, London.

Nawijn, J. (2011) Happiness through vacationing: just a temporary boost or long-term benefits? *Journal of Happiness Studies* 12(4), 651–665. DOI: 10.1007/s10902-010-9221-y.

OECD (2016) *Policies for the tourism sharing economy. Tourism Trends and Policies 2016*. OECD, Paris, pp. 89–120.

Plog, S. (1973) Why destinations rise and fall in popularity. *Cornell Hotel and Restaurant Quarterly* 13, 6–13.

Rickly, J.M. (2019) Overtourism and authenticity. In: Dodds, R. and Butler, R.W. (eds) *Overtourism Issues, Realities and Solutions*. De Gruyter, Berlin, pp. 46–71.

Russell, R. (2006) Chaos theory and managerial approaches. In: Buhalis, D. and Costa, C. (eds) *Tourism Management Dynamics, Trends, Management and Tools*. Butterworth-Heinemann, Oxford, pp. 108–115.

Russell, R. (2010) Keith Williams: chaos maker on the Gold Coast. In: Butler, R.W. and Russell, R. (eds) *Giants of Tourism: Key Individuals in the Development of Tourism*. CAB International, Wallingford, UK, pp. 243–254.

Russell, R. and Faulkner, B. (1999) Movers and shakers: chaos makers in tourism development. *Tourism Management* 20(4), 411–423. DOI: 10.1016/S0261-5177(99)00014-X.

Shani, A. and Logan, R. (2010) Walt Disney's World of entertainment attractions. In: Butler, R.W. and Russell, R. (eds) *Giants of Tourism: Key Individuals in the Development of Tourism*. CAB International, Wallingford, pp. 155–169.

Sheppard, V.A. (2017) Resilience and destination governance: Whistler. B.C. In: Butler, R.W. (ed.) *Tourism and Resilience*. CAB International, Wallingford, pp. 69–80.

Sofield, T. (2017) China: government policy and tourism development. In: Butler, R.W. and Suntikul, W. (eds) *Tourism and Political Change*, 2nd edn. Goodfellow Publishers, Oxford, UK, pp. 38–54.

Stevenson, R.L.S. (1874) *El Dorado. Essay printed in London (also in Virginibus Puerique (1904))*. Scribner, New York.

Suntikul, W., Butler, R. and Airey, D. (2010) The influence of foreign direct investment on accommodation patterns in Vietnam as a result of the open-door policy. *Current Issues in Tourism* 13(3), 261–277. DOI: 10.1080/13683500902937628.

thebrando.com (n.d.) Resort. Available at: https://thebrando.com/resort/ (accessed 27 March 2020).

Timothy, D.J. (2008) Tourism and political boundaries: places of friendship, spaces of antagonism. Keynote address. *Bridging the North-South Divide through Sustainable Tourism Development*, Stenden University, Leeuwarden, The Netherlands.

Towner, J. (2010) The master of ceremonies: Beau Nash and the rise of Bath, 1700-1750. In: Butler, R.W. and Russell, R. (eds) *Giants of Tourism: Key Individuals in the Development of Tourism*. CAB International, Wallingford, pp. 18–31.

UNWTO (2018) *UNWTO Tourism Highlights, 2018 Edition*. UN World Tourism Organization, Madrid.

Walton, J.K. (2010) Thomas Cook: image and reality. In: Butler, R.W. and Russell, R. (eds) *Giants of Tourism: Key Individuals in the Development of Tourism*. CAB International, Wallingford, UK, pp. 81–92.

Weidenfeld, A. (2010) Iconicity and 'flagshipness' of tourist attractions. *Annals of Tourism Research* 37(3), 851–854. DOI: 10.1016/j.annals.2010.02.007.

Weidenfeld, A., Butler, R.W. and Williams, A. (2016) *Visitor Attractions and Events: Locations and Linkages*. Routledge, Abingdon, UK.

Wheeller, B. (1993) Sustaining the ego. *Journal of Sustainable Tourism* 1(2), 121–129. DOI: 10.1080/09669589309450710.

Whipple, T. (2016) Coach parties set to conquer Everest. *The Times*, 12 December.

Wolfe, R.I. (1952) Wasaga Beach: the divorce from the geographic environment. *The Canadian Geographer/Le Géographe canadien* 1(2), 57–66. DOI: 10.1111/j.1541-0064.1952.tb01711.x.

Wood, K. and House, S. (1991) *The Good Tourist: A Worldwide Guide for the Green Traveller*. Mandarin, London.

5 Tourism as a Development Option: Perspectives from the World Bank

Shaun Mann*
World Bank, Washington, DC, USA

Context: Lessons from the First 50 Years

Arguably one of the largest, and still-growing, global economic sectors, tourism is quite evidently part of the development conversation. Tourism affords developing countries the opportunity to use innate geography, history, nature and intangible assets such as culture, cuisine and customs to attract visitors and investors and, through them, to support a range of employment- and tax-generating local goods and service providers. In 2018, tourism sector activities generated over US$10 trillion in direct contributions to global GDP (UNWTO, 2019), representing more than 10% of the global economy. In that same year, developing countries accounted for about 38% of all tourism-related transactions, a figure that has more than doubled in the past 30 years. Moreover, almost every country in the world and every region of every country is dealing with, or benefitting from, an influx of tourism, and if they are not, they want to be.

This constantly evolving social and economic phenomenon moves money and people to every corner of the planet (and beyond (Northon, 2019)). However, as has long been recognized, this swelling tide of tourists also brings a range of social, environmental and infrastructure issues to the host destinations, often straining local capacity to supply and manage the increased needs for land, water, energy, waste disposal, housing, safety and security, to say nothing of the challenge of managing increased traffic volumes. This inherent – inherent because the supply of tourism services (production) and their consumption (not necessarily their sale[1]) happen simultaneously – collision between supply and demand is at the heart of the challenge for tourism as a viable, and ultimately sustainable, development option.

Context is everything. Many developing countries are highly indebted, geographically remote and connected by poor infrastructure, less economically diversified, challenged with consistently low productivity and often affected by conflict and political instability. As such, this is a complex and fragile environment to successfully develop and sustain investments – public and private – in tourism. Countries with a youth bulge (a common characteristic of developing countries) have a potential asset but also face heightened demand for jobs while contending with low education and skills levels. In contrast, in other countries where the population is ageing, there are increasing concerns about fiscal sustainability. In addition, several middle-income countries – often with very poor regions – are in the midst of stalled economic transformations. Moreover, these challenges are compounded by

*shaunhmann@yahoo.com

DOI: 10.1079/9781789242812.0005

externalities such as ebbing global growth prospects, trade policy uncertainties, the changing nature of work, climate change impacts and geopolitics in constant flux.

Further complicating the supply-side challenges by the sheer weight of their market influence are global value chains that dominate and drive demand in source markets (World Bank, 2020). Unlike manufacturers of goods, like computers or cars, the tourism sector offers a bundle of local services (accommodation, transport, activities, etc.) that are bought and packaged by an international buyer (tour operator) and then tailored and sold on into multiple different markets and market segments as a finished product; in this case a package holiday. For instance, TUI, the largest tour operator in the world, markets and sells a ten-day beach holiday to Mombasa in 26 different countries; this holiday is tailored, according to the services offered, into eight different market segments. Mombasa is just one of more than 50 options for ten-day beach holiday destinations offered by TUI. At the same time, these traditional market distribution channels are also changing. Today's disruptors – low-cost airlines, online travel agencies, Google, YouTube, Airbnb, Uber, social media and so on – connect products and destinations to markets in completely different ways. Consequently, travel is becoming cheaper and information more accessible to a much larger proportion of an ever-growing world population; the pie has grown exponentially along with tourism's piece of it. In particular, growing middle classes – with more disposable income – in huge population clusters in India, China and Africa are new sources of travellers.

Within this context, the purpose of this essay is to reflect primarily on the supply-side issues and on the reality of successfully harnessing tourism and achieving desired impacts in developing countries: good-quality jobs, natural and cultural heritage conservation, improved standards of living, steady foreign exchange earnings, stronger linkages with local goods and services suppliers and more positive interactions between guests and destinations and their hosts. Specifically, drawing on the experiences of the past three decades as well as a cohort of over 80 World Bank Group tourism-related lending and advisory projects between 2000 and 2019 in 62 countries, the essay highlights lessons and ex-post evaluations and observations drawn from these projects and the consequential dialogues with governments and the private sector. It also suggests some guiding principles for tourism development activities going forward. In addition, the demand-side is examined from the perspective of the influencers of changing demand patterns and distribution channels, such as global value chains and disruptive technologies, and the effects they are having on destinations and their ability to respond.

Supply-side Issues: The Legacy of Government Involvement

Tourism 1.0: the first phase of capital investment in tourism was opportunistic and poorly planned

From the supply-side perspective, the fundamental appeal of tourism to developing country governments has always been its foreign exchange earning and job creation potential. Unsurprisingly, therefore, this has tended to underpin government-led investment strategies. Starting in the 1950s and 1960s – when most developing countries became independent sovereign nations – governments played the lead role in investing in the sector though two distinct approaches. On the one hand, some governments invested directly in hotels and tour operations and ran them as state-owned enterprises (e.g. India, Bulgaria, Indonesia, Ethiopia, Nigeria, Tanzania, Kenya, Uganda and Zimbabwe). On the other hand, others took a spatial approach, providing land and infrastructure in defined regions, usually coastal, and then fiscal incentives to encourage and de-risk private sector investment (e.g. Tunisia, Egypt, Mexico, Dominican Republic and Thailand). However, as tourism grew and the private sector became increasingly interested in the business opportunities it offered, a perfect storm was brewing. The 1970s saw a surge in private sector investment in hotel development on the supply side that was, in turn, both fuelling and fuelled by a rapid demand-side expansion of travel agencies, tour operations and package or mass tourism. Notably, the commencement of Boeing 747 (jumbo jet) commercial flights in the 1970s added to the viability

Box 5.1. Senegal's tourism development in the mid-1970s (extracts from World Bank Project Completion Report in 1990 of the Petite Cote Tourism Project, implemented between 1977 and 1987).

By the early 1970s, Senegal had begun to realize the possibilities for growth in the tourism sector. However, tourism potential remained largely untapped. By 1975, the Government had taken a number of measures to expand tourism and, as a result, development had been rapid but haphazard. A co-ordinated effort was envisioned under the project to strengthen and promote the sector and to develop an integrated tourist resort in the Petite Cote. The project also included the restoration of Goree Island, an important historic site. On the basis of a feasibility study, a resort site was selected at Sali. The project comprised the provision of infrastructure and common facilities for a resort of about 2500 beds on about 12 lots. A second phase was to be developed later but only after the first phase had proven the market. An implementing organization (SAPCO) was to be established to develop and manage the site. The project also included technical assistance to provide better co-ordination and planning in the tourism sector, to establish an Economic Analysis and Planning Unit in the Government Tourist Agency, and to help prepare and implement a marketing programme for the entire country. Implementation of the project took several years longer than originally foreseen, primarily because of indecisiveness on the part of the Government. The infrastructure at Sali was completed and eight hotels were built, providing 1762 beds, which is about 30% below the number originally foreseen.

of long-haul destinations because these aircraft carried 150% more passengers than any other aircraft. Consequently, tourism became big business across the globe.

This first phase of investment in tourism development required significant amounts of capital to establish an adequate infrastructure platform as well as commercial financing for hotel development. Hence, International Finance Institutions (IFIs), such as the World Bank and Inter-American Development Bank, were heavily invested in supplying both financing for public goods, such as infrastructure, as well as providing concessional finance to local intermediary banks to on-lend to private sector hotel developers. This rapid growth and investment happened mainly in countries with a viable beach product, the demand driver of the time. The example of Senegal was a typical approach (see Box 5.1 below)

In land-locked countries or those without an obvious or heavily promoted beach product, tourism took root at a slower pace, but was still typically state-led. In Uganda, for instance, a wholly government-owned enterprise, Uganda Hotels, borrowed money from the government-owned Uganda Development Bank to develop 22 hotels and lodges around Uganda between 1962 and 1970. While there was no real commercial or broader tourism development strategy driving these investments, they were intended to open up areas of Uganda to commerce and tourism – lodges and hotels were built in national parks and each major district capital.

In those countries with an identified and viable beach tourism product – then defined as those destinations having easy access to European and North American markets – tourism development was generally rushed and poorly planned with governments and investors being greedy for growth. Where governments opted to 'leave it to the private sector', they effectively ceded control of growth and market forces took over. In the most popular destinations, excessive hotel capacity led to cut-throat competition and a race to the bottom. The two examples below, both with about 900 hotels and similar numbers of tourists, highlight two obvious but critical ingredients for sustainable tourism development: planning and governance. In Cancún, the intention was the economic development of the neglected Yucatan Peninsula by the federal government. The Mexican government created a dedicated tourism planning and development agency to facilitate a specific strategy of tourism development. In the second example of Tunisia, a *de facto* one-party state between 1956 and 2011, the government plan was simply to make concessional finance available to the domestic private sector and allowing those investors – often with no experience in tourism but with deep political connections – to develop the plant and product.

Example 1: Cancún – when benefits and costs collide *(see also McLean, 2018)*

The Cancún project originated from Mexico's central bank's search for greater export earnings, job creation and regional development in the face of slowing post-revolutionary economic growth, global competition and internal demographic growth. The resort city of Cancún stands as a monument to the ground-breaking strategy formulated by the Mexican state in the 1960s. However, while Cancún has consistently surpassed all expectations in terms of its overall earnings, the impacts of the project, particularly social and environmental, have not been so successful. Currently, Cancún accounts for approximately one third of Mexico's tourism earnings and boasts over 650,000 residents, with migrants still attracted to the city and the surrounding region in search of work in the tourism economy. Repeated construction booms initiated by state-sponsored phases of infrastructure and hotel expansion created thousands of jobs, as did the actual tourism industry once hotels and supporting businesses began operating. In 2018, roughly 900 hotels were in operation, providing services to nearly 7 million tourists who account for over US$7 billion in tourism expenditure.

With Cancún, the Banco de México hoped to create a self-sustaining and globally competitive resort area, the first of five planned sites that would form the backbone of the new policy of state-sponsored tourism development in the country. Yet, selling the project proved difficult, as officials confronted the scepticism of international lenders and the vicissitudes of the Mexican political system. At the same time, Mexico's six-year presidential system and the lack of institutional continuity that accompanied it also threatened to undermine the project. Nevertheless, the federal government financed the first nine hotels and, by emphasizing the benefits of job creation in a region with few prospects for economic development and promising to ensure national control of the industry, secured political approval and funding for the project.

However, once construction started, problems arising from the city's rapid growth quickly surfaced. As early as the mid-1970s, residents of Cancún faced land shortages and the virtual absence of any housing stock, which pushed them into illegal shanty towns, the *colonias populares*, located outside the parameters of the planned city. More specifically, while the federal tourism development fund and tourism planning body (INFRATUR, later known as FONATUR) managed to keep the hotel zone isolated on a barrier island, the high rate of migration resulted in the growth of squatter settlements without water, sewage or electricity on the adjacent mainland. The involvement of the Inter-American Development Bank (IADB) in the resort's funding subjected the Cancún project to considerable scrutiny while regional elites, feeling marginalized by federal control of the project, used the shanty towns' neglect to lobby for greater local autonomy. FONATUR subsequently responded by setting up an agency, Fideicomiso Puerto Juárez, which worked to regularize the illegal settlements from the mid-1970s.

Cancún's performance in its first decade therefore demonstrated the potential of state-led tourism to generate economic growth and create jobs, but the city's growing pains also suggested that a broader, whole-of-government approach to tourism development might have been more beneficial. For instance, there was a clear lack of a policy and legal framework regarding public environmental management and, over the past 30 years, this has contributed to the environmental deterioration of Cancún. The migration of thousands of largely rural Mexicans to a fast-paced and pressured urban environment also had unforeseen social consequences, but perhaps the most significant oversight was the absence of a defined strategy focused on building local capacity to provide goods and services to the burgeoning tourism volumes. Indeed, the all-inclusive business model pursued by the hotel investors provided few opportunities for local businesses and, meanwhile, the government was content to count growing numbers. Thus, the experience of Cancún demonstrates that the private sector will always be motivated by maximizing profits and that, without mechanisms to regulate and respond to changing social or environmental circumstances, governments risk ceding development control at the cost of sustainability.

Example 2: Tunisia's race to the bottom 1960-2010

In 2010, the World Bank conducted a forensic tourism value chain analysis in Tunisia as part of a sector diagnostic. During interviews with British tourists staying in four-star hotels in one of the main beach resort areas, it emerged that many tourists were on package tours that amounted to a half-board daily rate of just 8 Euros. Several were quoted as saying that they were 'on the dole' (on state benefits) and that it was cheaper for them to holiday in Tunisia than it was to live in the UK. They bought the package before going back home to 'sign on', then bought another package, continuing to do this for three to six months of the year. How did this happen?

Tunisian tourism developed rapidly, from just 4000 beds in 1962 to around 250,000 in 2010 shared across 900 or so hotels and, with nearly 7 million arrivals in 2010, Tunisia was the third most visited destination in Africa (after Egypt and South Africa). While initially state-led through the 'societe nationale hoteliere et touristique' (National Society of Hotels and Tourism, SHTT), incentives provided by the government through cheap loans facilitated by the banking sector and in the provision of land and tax relief resulted in a surge of private sector investment between 1968 and 2000. The resultant abundance of hotel beds, referred to above, and the consequential pressure on their owners to sell this perishable product placed large tour operators with access to markets in a powerful buying position in which they effectively were able to dictate room rates and market positioning to hotel owners and, by default, dictate the perception of Tunisia as a destination. This, in turn, led to a business model based on high volumes and low profit margins and prevented the hotel owners from investing in maintenance and training and from developing their own market access strategy, thereby leading to a decline in the quality of services offered. This narrow source of supply resulted in a very short value chain where only two cost elements captured nearly 95% of the total value of the package: (i) European tour operators who buy rooms for the upcoming season from (ii) hotel owners, and package these for sale through retail outlets in the high streets and on the internet in source market countries. In the case of Tunisia in 2010, 90% of the holiday tourism market was captured by only five tour operators.

In a more normal tourism system, sources of supply (distribution channels) are much more diversified, markets are more diversified, there is healthier competition among hotels which are more differentiated, travellers can reach the destination independently at a competitive cost (i.e. the air ticket is decoupled from a tour operator package), and there is more out-of-pocket expenditure (the average in Tunisia for package tourists was around 5 Euros per day) outside the hotel.

In the case of Tunisia, the narrow focus of the government on the expansion of hotels succeeded for a couple of decades, up to the early 2000s, when the sector was hit by a series of shocks. In particular, the availability of cheap credit and the 'beds only' strategy led to a stock of undifferentiated hotel rooms just at a time when new and more sophisticated competitors were entering the market, while the events of 11 September 2001 in New York and April 2002 in Djerba (the terrorist bombing of a synagogue in which a number of tourists lost their lives) led to revenue shocks that exposed underlying structural weaknesses.[2] The credit quality of the tourism sector had deteriorated and accommodation assets were in severe financial distress. The tourists mentioned at the start of this example who were interviewed were staying in hotels leased by tour operators from their owners for the UK winter season. The tour operators were in complete control of every cost driver along the value chain and the owners had abandoned their hotels and left the outstanding debts with the central bank.

Given the level of state support and assistance, tourism does not operate in many countries in conditions of perfect competition – in the late 1990s, for example, Cyprus found itself in a similar situation to that in Tunisia (Sharpley, 2003). Tourism is unique in that supply and demand occur simultaneously and in the same place. The pressures on emerging or existing tourism destinations are only going to increase and the need for a more rigorous process of designing, planning, implementing, managing and monitoring tourism development is the first lesson taken from destinations that have gone through a life cycle (Butler, 1980). The collision of benefits and costs is a complex dynamic to manage and requires much more focus on tourism to be well conceived (why tourism and what kind of tourism?), well planned (what investments, where and when?), implemented according to a plan and constantly monitored. It is probably fair to conclude that this process of conceiving, planning, implementing, managing and monitoring has not been common practice in the past 50 years of tourism development in the context of developing countries.

Tourism 2.0: the second phase was characterized by master planning, little public investment and excessive regulation and taxation

Following an initial period of tourism development – depending on the country, this period roughly covered the 1950s to 1980s – based on capital investment where governments took the lead in either investing or facilitating investment, an element of control was lost as the private sector was the main driver of tourism growth. This resulted in rapid, opportunistic and unplanned growth. In reaction to these opportunities and risks, many countries consequently developed national tourism master plans as a potential means of optimizing tourism benefits. This too, however, had a legacy. Tourism master plans, while comprehensive and well-meaning tomes, were simply impossible to implement. In particular, such plans highlighted the cross-cutting nature of the ecosystem required to generate tourism development, typically covering interventions across almost every part of government. The cost of implementing these plans and the impracticality of co-ordinating the required inputs across the different parts of government deterred most governments from embarking on any capital investments. Hence, the lowest-hanging fruit, least-cost and most visible actions that most governments took were to focus on milking the cash cow that tourism had become through regulation and taxation. As a host of new taxes and licensing regimes were imposed, tourism industries became over-regulated and many governments simply adopted a *laissez-faire* approach to tourism development. Coincidentally, and as exemplified in Box 5.2, this was also a period when institutions like the World Bank stopped funding tourism projects, viewing them as a private sector-led activity (Hawkins and Mann, 2007).

This marginalization and isolation (from the rest of government) of tourism led to a series of systemic market, institutional, resource and co-ordination failures that typify tourism development challenges facing many destinations today. In particular, poor co-ordination and communication between different government ministries, institutions and stakeholders with respect to tourism planning, tourism asset management and tourism destination management are probably an inevitable consequence of tourism-related responsibilities being spread across multiple agencies and government levels. Additionally, the absence of clear investment planning has resulted in tourism-related institutions being focused on institution building rather than skilled functions; arguably, they have no other choice. This has also contributed to a gradual disconnect between the public sector line ministry and other stakeholder groups, such

Box 5.2. Withdrawal of the World Bank from 'tourism' investments in 1980 (extracts from World Bank Project Completion Report in 1990 of the Petite Cote Tourism Project implemented between 1977 and 1987).

'...It is difficult not to conclude that following the abolition of the Bank's Tourism Department, project supervision (in Senegal) began to suffer from frequent changes in management at headquarters, high turnover of supervision staff and much longer intervals between visits. The project was supervised by four different Bank divisions involving approximately 12 staff members and several consultants. The project dragged on for years for lack of Government decisions concerning relatively simple matters.'

'Undoubtedly, by withdrawing from the tourism sector, the Bank lost some leverage with the Government's tourism agencies. It must also have been evident to the staff that there was no future in the sector.'

Table 5.1. Typical systemic failures in tourism destinations.

Institutional failures	• At central government level, there is no recognition that tourism requires a whole-of-government approach. • Tourism ministry staff are under-skilled and under-motivated. • Few tourism ministries are also responsible for nature and culture resource bases. • Ministries are often mandated to develop plans but not to implement them. • There exists poor co-ordination and communication between different government ministries, institutions and stakeholders with respect to tourism planning, tourism asset management and tourism destination management. This is further complicated with tourism-related responsibilities being spread across multiple agencies and government levels as a devolved subject area. Similarly, there is poor communication and co-ordination with other tourism stakeholders (e.g. with respect to conservation, education, SMEs). This contributes to conflict in land use (e.g. environment vs infrastructure projects, fishing vs tourism, energy vs tourism, mining vs tourism), and causes administration and decision-making processes to become complex and slow. • There is a reliance on regulation to manage issues such as quality, safety and standards, but limited enforcement of the regulations. In addition, rent-seeking practices subvert the regulatory process.
Market failures	• Limited attention is given to product diversification and destination development. This restricts competitiveness, recreation and tourism choices, the development of niche markets and the ability to value-add to the visitor experience. • There exists a lack of effective mechanisms to manage, standardize and enforce best practices. This has contributed to poor interpretation, poor visitor experiences and poor conservation of environmental and cultural values. • Market distortions emerge because of government policies, rent seeking and subsidized government operations (e.g. tariff rates, room rates, domestic aviation etc.).
Resource failures	• Limited emphasis is placed by both the public and private sector on human capital development and an inclusive employment agenda across the tourism and hospitality value chains. • A lack of comprehensive visitor research and data, ongoing research into products and markets and market intelligence limits the effectiveness of product development decisions and marketing strategies as well as the ability to address seasonality and to make the most of regional attributes. • Access to investment and financing is often lacking, especially for SMEs.

as the private sector and communities involved in delivering tourism goods and services. A review of over 60 World Bank lending projects in the tourism sector highlights the main features of typical systemic failures encountered; these are summarized in Table 5.1.

The fragmented nature of tourism requires multiple central government ministries as well as different levels of local government to undertake various roles in response to political, social, environmental and infrastructural issues that affect tourism development. However, the fact that the full responsibility for tourism does not lie with one line ministry makes this a challenging co-ordination effort that depends greatly on the political economy around the sector. Achieving effective outcomes from tourism policy is inevitably dependent on a country's economic, environment, industry, public-private partnership, foreign direct investment and other related policies. Thus, without a supportive political economy, tourism policy has little chance of having any real influence on tourism development.

Tourism 3.0: stronger planning and integrated investments

Since the early 2000s, there has been a sea change in the recognition governments accord to tourism as an economic driver. In part, this is a result of the more explicit presence of sustainable tourism in global development objectives, the most recent iteration of which are the Sustainable Development Goals (SDGs) in which tourism targets have been included in three of 17 globally agreed goals (UN, n.d.).[3] Also, the re-engagement of lenders such as the World Bank, Inter-American Development Bank, Asian Development Bank, and the African Development Bank in tourism-related projects, as well as the willingness of bilateral funders such as USAID (USA Agency for International Development), the UK's DfID (Department for International Development), Germany's KfW and GiZ (Kreditanstalt für Wiederaufbau and Deutsche Gesellschaft für Internationale Zusammenarbeit GmbH), Japan's JICA (Japan International Cooperation Agency), France's AFD (Agence Française de Développement), Sweden's SIDA (Sweden International Development Cooperation Agency), Australia's AusAID (Australian Agency for International Development) to support tourism-led growth opportunities has further encouraged governments to harness the economic benefits of a competitive tourism sector. Within the World Bank, this growth in client–country demand for tourism-related support, lending and advice has strengthened approaches to analytics (World Bank, 2019a), theories of change (World Bank, 2019b), impact measurement and project implementation. In particular, it has led to the recognition of an ecosystem of interlinked spheres of influence that affect the achievement of competitive and sustainable tourism sectors (Fig. 5.1).

This ecosystem perspective in project interventions has shifted the delivery model within the World Bank Group as well as changing the dialogue with governments in a number of ways. First, the recognition that delivering positive tourism-related economic impacts cuts across these different touchpoints has engaged government stakeholders beyond the ministries of tourism. Second, there has been a spatial focus on specific destinations within the national context that is more aligned with decentralization trends in most developing countries. This allows for longer-term thinking where a country can plan for more bite-sized destination- or region-specific investments rather than trying to implement a national master plan (without the resources to see it through). Also, this has highlighted the need for destination management and co-ordination, two serious systemic failures in earlier phases of tourism development. Third, driven by a tourism economy, there is growing opportunity for private capital investments in infrastructure (e.g. roads, energy, water, airports, cruise terminals, solid waste management) that lowers the public cost of opening new destination or new markets. The approach of leveraging public investments to crowd in private capital has also strengthened the rigour of demand assessment and the feasibility for financial and economic returns. Finally, the scale of tourism-related projects has increased; throughout the 1990s and early 2000s, the average World Bank tourism project size was US$27 million but, since 2012, this has risen to US$88 million.

These perspective shifts have triggered a more internally collaborative World Bank Group response. For instance, the Indonesia Tourism Development Project, launched in 2018 (World Bank, 2018a), aims to develop the first three of ten new destinations featuring collaboration across multiple sectors of the World Bank Group: Education, Environment and Natural Resources, Transport, Digital Development, Social, Urban, Rural and Resilience Macroeconomics, Trade and Investment, Water, and Governance, as well as IFC (International Finance Corporation) Advisory Services. The project's support to the development of selected tourism destinations is seen as the start of a longer-term engagement with the government on tourism destination development. The delivery platform and good practices established during the implementation

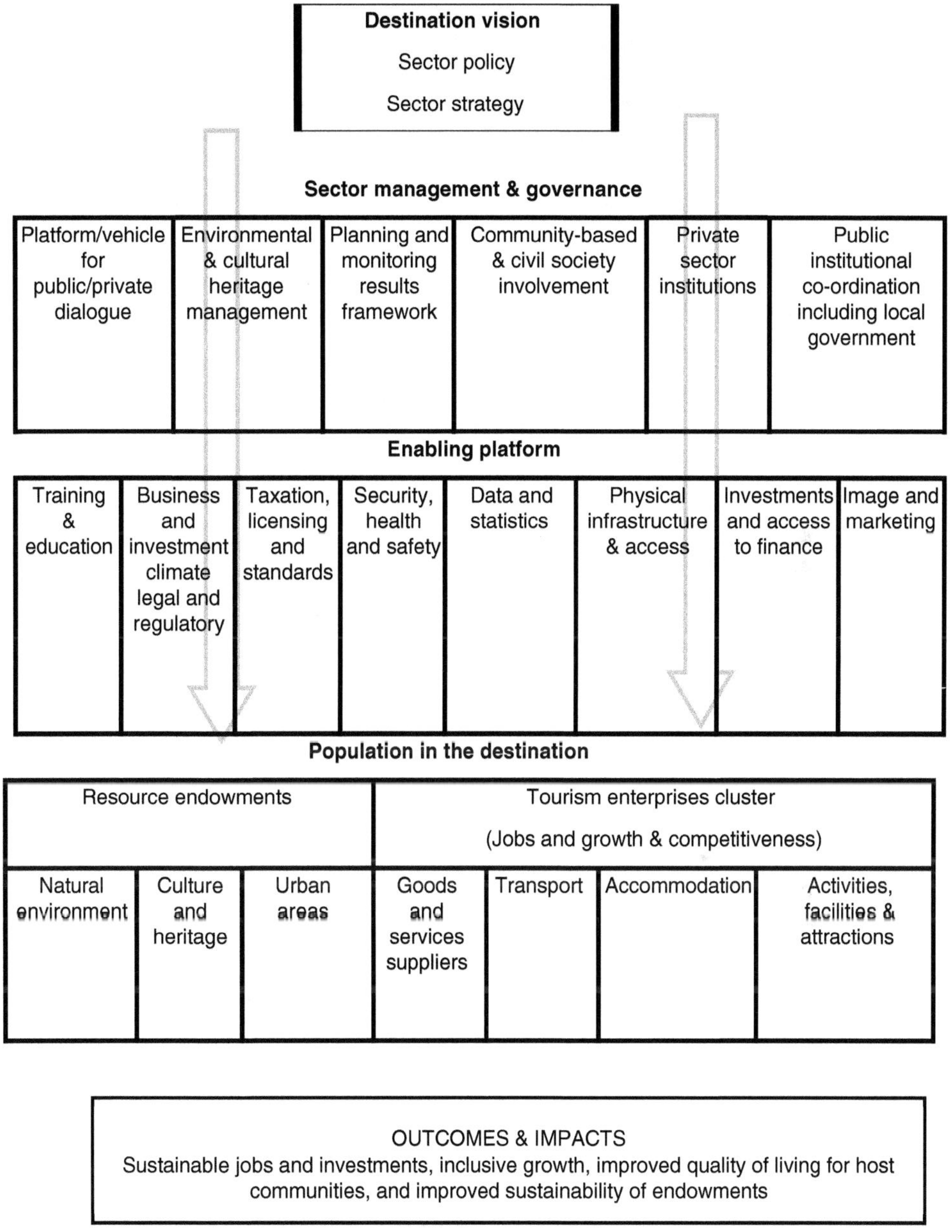

Fig. 5.1. The Tourism Ecosystem.

of this initial project will serve as a collaborative platform for central ministries, as well as provincial and local governments, for developing tourism destinations across Indonesia. It will also serve as the foundation for further collaboration (both advisory and financing) between the World Bank Group and the government in developing other tourism destinations.

In another example, in Georgia, the government, together with the World Bank (World Bank, 2015) invested about US$600 million between 2005 and 2018 for revitalization of municipal infrastructure and restoration of historical buildings (most of which have a cultural heritage value) throughout Georgia. However, the government recognized that restoring buildings and municipal infrastructure alone would not be enough to trigger and sustain local economic transformation; therefore, it initiated a series of projects focused on using tourism as a lever to generate economic transformation. These projects (now on the fourth Regional Development Project) took the approach of using tourism potential as a means of prioritizing investments. Destination management and planning capacity was built during the design and implementation process.

Tourism 4.0: the future: building local ecosystems to deliver local benefits

What seems apparent from the tomes of unimplemented tourism master plans, the clash of costs and benefits occurring in highly concentrated tourist-rich areas and the complexity and political economy of the tourism ecosystem is that achieving genuinely sustainable tourism economies is something that requires a lot more cost and maintenance overhead than expected or estimated. Realistically, these are just too many dots to connect at a national level. In so many instances across developing countries, the systemic reforms required to change the way institutions work in order to facilitate more functional tourism ecosystems are very difficult to achieve. However, the principle that the tourism product or experience is produced and consumed in the same place does offer an opportunity for connecting the dots in that place, and there are a growing number of tourism projects working with ambitious and reform-minded mayors or regional administrations. In the example of a project in Albania, summarized in Box 5.3, newly elected mayors in four municipalities have taken the initiative of establishing tourism working groups (in effect, destination management organizations) and offering budget and office space to support nascent destination management partnerships that are, in turn, supported by project activities linked to market and product development, data and statistics, locally based tourism business licensing and registration, co-ordination and planning, and leveraging other development partners to provide local training opportunities. In short, a local tourism ecosystem has been established.

If there is a trend in the current pipeline of tourism projects in the World Bank, it is that the clients are increasingly district-, municipal- and village-level administrations. Project development objectives are more localized and aimed at targeted places and populations, and combinations of lending, policy and grant instruments are being used more creatively. For instance, Development Policy Loans (DPLs[4]) that directly support a country's budget and are used to strengthen public financial management, improve the investment climate and address bottlenecks to improve service delivery

Box 5.3: Building Albania's local tourism ecosystems (extract from 2016 Project Appraisal Document; see World Bank, 2016)

...The Project Development Objective is *to improve urban infrastructure, enhance cultural assets, and strengthen institutional capacity to support tourism related local economic development in selected municipalities in the south of Albania*.

...In support of the overall development objective of the project, the activities in this component are focused on stimulating and sustaining growth in the tourism economy in selected municipalities in the south of Albania by complementing and leveraging the capital investments supported under Component 1. This component will support the following activities: (i) a market research and market development strategy for the south of Albania, (ii) a destination management capacity assessment, (iii) piloting the establishment of destination management entities in selected urban areas, (iv) implementing a product and market development strategy, and, (v) establishing a system for tracking sustainable tourism indicators. This component will also include activities that strengthen systems and the capacity for tourism data collection between central government and municipalities.

can tackle macroeconomic issues while targeted investment loans (Investment Project Financing) are focused capital-intensive investments, rural development, credit and grant delivery, community-based development and institution building. In Morocco, for instance, an Inclusive Green Growth DPL developed with the World Bank (World Bank, 2018b) was articulated around three key objectives that were also reflected in a national programme and in the sector strategies for agriculture, water, energy, tourism, and fisheries. Policy actions within each of these sectors were defined as prior actions that would release budget capital to finance specific projects. For example, Morocco's coastal zones are home to 56% of Morocco's population and provide important contributions to its economic growth and prosperity, and sectors such as tourism, agriculture and fisheries depend on coastal ecosystems. However, demographic and economic pressures have led to a progressive degradation of these coastal ecosystems. Activities linked to tourism, such as construction, sand-mining, insufficient sewerage and solid-waste collection and disposal have come at a cost. One objective in the DPL was, therefore, 'improving the management of natural capital', linked to indicators such as 'the development of integrated coastal zone management plans'. In this way, a strong policy lever was used to incentivize practical planning instruments that, in turn, triggered local spending to implement the coastal plans that mitigate the negative costs.

Conclusion

As mentioned earlier in this essay, context is everything. There is no standard tourism destination success template. In recognition of the fact that so much depends on local champions, destination life cycle factors, the macro environment and the demand position, better questions are now being asked: Who are the local champions? Which are the specific life cycle factors and how can they be affected? What aspects of the macro environment influence tourism supply capability? And what demand factors specifically affect tourists' length of stay or expenditure? Building a sustainable tourism economy in any destination is a continuous process of making optimal use of environmental and cultural resources, involving host communities and ensuring fairly distributed benefits along tourism value chains, providing value-for-money experiences to visitors and constant monitoring of environmental and social well-being. Again, the questions being asked are more nuanced: How can environmental resources such as coastlines be used sustainably for communities, tourists and businesses? How can policies and legislation ensure equitable value chains? And how can this be measured and monitored? Along with better diagnosis, theory of change, more innovative financing instruments being applied, improved results monitoring and broader engagement with governments, tourism is evolving as a development option.

The pace of this evolution is accelerating with the ubiquity of digital tools, big data and social media (World Bank, 2018c). On the demand side, digital platforms that help travellers plan, book and share their trips online have disrupted conventional distribution systems. For instance, by 2030, Airbnb – an accommodation booking platform – expects hosts in developing countries to have received 400 million guests (Elevate, 2019). The people who visit destinations and buy travel services can now share their opinions across various platforms. They can share these opinions with a global audience before, during and after their trip, influencing the choices that future consumers will make. The growth of user-generated content (UGC) is unprecedented and destinations and travel companies no longer have a choice about whether to engage with online audiences. Managing online reputation is now central to running a thriving tourism business. For emerging destinations, this is new and unknown territory.

Big data refers to electronically generated and collected information. In contrast to regular structured data sources for the tourism sector, such as arrivals data, this customer-based data can be provided knowingly through guest surveys and reviews on websites as well as collected unknowingly through web navigation data, booking and payment data and cell phone location data. Tourism organizations have started to use big data for many different purposes and in different ways. In Italy, for example, policy makers collected 21 million messages on Twitter and Instagram during a period of six weeks in 2016. Using specialized software, they were able to identify key words and concepts and break down the information into 20 fields of analysis, including location, accommodation, facilities, sports, food and wine and so on (CrowdRiff, 2017). Similarly, the city of Barcelona is using data collected through mobile phone providers to track visitor volume in and around Sagrada Familia Cathedral, one of the city's most visited attractions. The data, used to monitor visitor flows, provides insight into the length of stay of the visitors, the busiest times and the route visitors take before and after their visit (Roberts, 2017).

On the supply side, e-government systems are becoming integrated and licensing, employment, investment, tax and accounting data can be collected, collated and analysed across different databases in different ministries. This is improving the government's ability to regulate and measure the economic impact of businesses.

Overall, then, the future of tourism as a development option looks positive. Developing country governments are recognizing its potential and the effort and investments required to unleash it, as well as understanding their role in this process better. Institutions such as the World Bank are also supporting this in a more integrated and impactful way through better upfront diagnostics, more structured theories of change built into project interventions, better co-ordination across the institution and with other development partners, and a push to use public investments to crowd in the private sector. Older, and perhaps less inclusive business models are changing as stakeholders in destinations gain more control through better planning, improved co-ordination, more targeted investments and marketing, easier access to markets, and more awareness of the environmental and social pitfalls of too much tourism.

Notes

[1] An important distinction because in some cases services are presold, such as with package tourism, and there is often a huge mark-up on the presale price that doesn't materialize at the destination.

[2] It must also be noted the terrorist attacks on tourists on the beach at Sousse in 2015 and the subsequent decline in tourism revealed the dependency of the Tunisian economy on the tourism sector.

[3] Goal 8: Promote sustained, inclusive and sustainable economic growth, full and productive employment and decent work for all. *'By 2030, devise and implement policies to promote sustainable tourism that creates jobs and promotes local culture and products.'*

[4] Goal 10: Ensure sustainable consumption and production patterns. *'Develop and implement tools to monitor sustainable development impacts for sustainable tourism which creates jobs, promotes local culture and products.'*

[5] Goal 14: Conserve and sustainably use the oceans, seas and marine resources for sustainable development. *By 2030 increase the economic benefits of SIDS and LCDs from the sustainable use of marine resources, including through sustainable management of fisheries, aquaculture and tourism.'*

[6] [4]Investment Project Financing provides IBRD loan, IDA credit/grant and guarantee financing to governments for activities that create the physical/social infrastructure necessary to reduce poverty and create sustainable development. Development Policy Financing provides IBRD loan, IDA credit/grant and guarantee budget support to governments or a political subdivision for a programme of policy and institutional actions to help achieve sustainable, shared growth and poverty reduction. Program-for-Results links disbursement of funds directly to the delivery of defined results, helping countries improve the design and implementation of their own development programmes and achieve lasting results by strengthening institutions and building capacity. Trust funds and grants allow scaling up of activities, notably in fragile and crisis-affected situations; enable the Bank Group to provide support when our ability to lend is limited; provide immediate assistance in response to natural disasters and other emergencies; and pilot innovations that are later mainstreamed into our operations. Private sector options for financing, direct investment and guarantees are provided by MIGA and IFC. Guarantees can also be provided through World Bank (IBRD/IDA) for private sector projects. Multiphase Programmatic Approach (MPA) allows countries the flexibility to implement an approach to achieve development objectives in stages when: the development challenge is complex; it would take a longer time to achieve the objectives; it would take a longer time to prepare one large project; the solution needs a broader and comprehensive approach; or when a stop-and-go approach is not feasible. Projects under an MPA programme may be financed by Investment Project Financing or Program-for-Results Financing, or their combination.

References

Butler, R.W. (1980) The concept of a tourist area cycle of evolution: implications for management of resources. *The Canadian Geographer/Le Géographe canadien* 24(1), 5–12. DOI: 10.1111/j.1541-0064.1980.tb00970.x.

CrowdRiff (2017) The DMO's Guide to Attracting More Travelers with Instagram. Available at: https://cdn2.hubspot.net/hubfs/2719325/eBooks/SKIFT%20INSTAGRAM%20EBOOK%20(v2).pdf (accessed 5 December 2019).

Elevate (2019) Advancing sustainable tourism through home sharing. *Airbnb Citizen*. Available at: https://s3.amazonaws.com/airbnbcitizenmedia/2017/11/Advancing_Sustainable_Tourism_Through_Home_Sharing_171128.pdf (accessed 9 January 2020).

Hawkins, D.E. and Mann, S. (2007) The World Bank's role in tourism development. *Annals of Tourism Research* 34(2), 348–363. DOI: 10.1016/j.annals.2006.10.004.

McLean, M. (2018) Squatters in paradise: state-led tourism development in Cancún 1970-1995 (unpublished report).

Northon, K. (2019) NASA Opens International Space Station to New Commercial Opportunities, Private Astronauts. NASA Press Release, 7 June. Available at: https://www.nasa.gov/press-release/nasa-opens-international-space-station-to-new-commercial-opportunities-private (accessed 1 August 2019).

Roberts, F. (2017) Barcelona IoT, big data projects help manage tourists at popular attractions. Internet of Business. Available at: https://internetofbusiness.com/barcelona-iot-big-data-tourist/ (accessed 12 January 2020).

Sharpley, R. (2003) Tourism, modernisation and development on the island of Cyprus: challenges and policy responses. *Journal of Sustainable Tourism* 11(2-3), 246–265. DOI: 10.1080/09669580308667205.

UN (n.d.) Sustainable development goals. United Nations. Available at: https://sustainabledevelopment.un.org/topics/sustainabledevelopmentgoals (accessed 12 January 2020).

UNWTO (2019) *UNWTO Tourism Highlights, 2018 Edition*. United Nations World Tourism Organization, Madrid.

World Bank (2015) *Project Appraisal Document: Georgia Third Regional Development Project*. World Bank Group. Available at: http://documents.worldbank.org/curated/en/571991468000614527/pdf/PAD1263-PAD-P150696-R2015-0115-1-Box391466B-OUO-9.pdf (accessed 12 January 2020).

World Bank (2016) *Project Appraisal Document: Albania Project for Integrated Urban and Tourism Development*. World Bank Group. Available at: http://documents.worldbank.org/curated/en/330301479438041183/pdf/Albania-PAD-10272016.pdf (accessed 12 January 2020).

World Bank (2018a) *Project Appraisal Document: Indonesia Tourism Development Project*. World Bank Group. Available at: http://documents.worldbank.org/curated/en/839781527910281861/pdf/Indonesia-Tourism-PAD-05102018.pdf (accessed 12 January 2020).

World Bank (2018b) Implementation and Completion Results Report for Inclusive Green Growth Development Policy Loans (1 and 2). World Bank Group. Available at: documents.worldbank.org › curated › ICR-Main-Document-04302018

World Bank (2018c) *The Wisdom of the Crowds: Harnessing the Power of User-Generated Content for Sustainable Tourism Development*. World Bank Group, Unpublished report.

World Bank (2019a) Tourism or Development: Tourism Diagnostic Toolkit. World Bank Group. Available at: http://documents.worldbank.org/curated/en/240451562621614728/Tourism-Diagnostic-Toolkit (accessed 12 January 2020).

World Bank (2019b) Tourism for Development: Tourism Theory of Change. World Bank Group.

World Bank (2020) World Development Report: Trading for Development in the Age of Global Value Chains. Washington DC: World Bank Group. Available at: https://www.worldbank.org/en/publication/wdr2020 (accessed 10 January 2020).

6 Tourism in Nigeria: New Policy and Planning Directions Needed for a Neglected Sector

Peter U.C. Dieke[1]*, Nneoma G. Ololo[2] and Afamefuna P. Eyisi[3]
[1]Makerere University Business School, Kampala, Uganda; [2]University of Nigeria, Nsukka, Nigeria; [3]Murdoch University, Perth, Western Australia

Introduction

This chapter reflects on a neglected export sector in the Nigerian economy, that of tourism, in order to identify whether some refinements to the existing principles of developing tourism are merited. It then goes on to hint at both the issues that deserve consideration for future tourism development in Nigeria and also the general implications that the continued neglect of the sector portends.

In considering these reflections, the logical question must be: by what criteria can one describe tourism in Nigeria as being neglected? First, tourism is essentially a service rather than a merchandise business sector; hence, it is relatively consumer- , rather than producer- , oriented. It dominates many, if not most, global economies through its share in job creation, investment and fiscal revenues, and by its overall contribution to development through its spillover effects on other areas of economic activity. However, while service is common to tourism everywhere, in Nigeria it is viewed as almost servitude, characterized by poor–rich societal relations and reinforced by the special situation in the country where the tourism sector, by world standards, is still at a relatively early stage of growth. Therefore, the blitheness and perhaps even the disdain with which tourism services are treated in Nigeria can be thought of as strangely inappropriate.

Second, as tourism is recognized as both a demand- and supply-driven activity, the suggestion here is that Nigeria has, by any comparative standard, done poorly on both counts. From the demand perspective, the explanation lies in aspects of the culture of the society and of certain economic realities (Adejuwon, 1986; Okorafor, 1995; Bankole, 2002). In line with their family life traditions and extended family commitments, most Nigerians do not view tourism as an integral part of their lifestyle. This implies that the values of the society profoundly influence their holiday-taking behaviour, which is further compounded by the low income levels that reflect their economic conditions. In relation to aspects of supply, as we shall see, despite the richness and variety of tourism products in the country, they are very much underutilized; both the relatively low volume of visitors to the country and their typically short length of stay point to areas for concern. And third, the tourism sector in Nigeria, commensurate with its relative economic significance, has not

*Corresponding author: pucdieke@gmail.com

DOI: 10.1079/9781789242812.0006

received the attention it deserves in national development policies *vis-à-vis* other sectors. This is not, however, surprising given that tourism has played little or no significant part in the economic and social development activity of the country (Kendell, 1990).

Though there are no simple answers to these dilemmas, they must be resolved one way or the other, prompting some fundamental questions. For example, is it possible to speak of tourism in Nigeria without considering the requirement that policy and development planners for tourism should be accountable? Or should the priority be the quest for clear political leadership without which not only does tourism lose much of its significance, but also the requirement that tourism activity, its nature and scope should be relevant to the needs of the society whose resources nourish it goes unrecognized? These questions and others are central to our examination in this chapter on a topic which is both timely and important. Here, the old maxim of Mao Tse-Tung is instructive: 'No investigation, no right to speak' (see Emery, 1995, p. 8). Thus, in the next section, we consider the case of Nigeria, setting out the background to the country as a prelude to a consideration of tourism which examines the situation in the country's tourism sector as a whole.

Background to Nigeria

The country

Nigeria, located on the western coast of Africa and occupying a land area of 923,769 km^2 (or 356,669 m^2), is Africa's most populous country. The population in 2019 was estimated at 200,788,000, with an estimated projection of 263,839,000 by 2030 and a population density of 221 people per km^2 (or 572 per m^2), reportedly one of the highest in mainland Africa (Falola *et al.*, 2019). The climate is tropical but variable with a rainy season from April to October and a dry season from November to March. Nigeria became a British territory in 1914 following the joining of the Southern and Northern Protectorates, subsequently achieving independence on 1 October 1960. In 1963, Nigeria became a republic and is currently a democratic state in which multi-party politics is permitted.

There are over 250 ethnic groups with different languages and diverse religious backgrounds in Nigeria. Each ethnic group inhabits a territory which it calls its ancestral home. The English language is the official language of business and government, although the indigenous languages of the ethnic groups are used as well. The largest ethnic groups are the Hausa-Fulani in the north (36%), the Yoruba in the south-west (15.5%) and the Igbo in the south-east (15.2%). About 50% of Nigerians are Muslims (mainly in the north) and the other 50% are Christians (in the south). Nigeria is divided into 37 administrative divisions: 36 states and one territory, the latter being the Federal Capital Territory in Abuja, situated in the centre of the country, which was created by decree in 1976, replacing Lagos, the former capital, in 1991. Lagos has, however, retained its standing as the country's leading commercial and industrial city. There are three tiers of government – federal, state and local levels. The country has been further divided into six geopolitical zones, grouped largely by ethnic background: North-Central, North-East, North-West, South-East, South-South and South-West.

Although Nigeria is now a full democracy, this situation is relatively recent and can be traced back to 1999 when the country returned to civilian rule. Previously, for about 30 years, the country had been plagued by the successive replacement of civilian governments by military regimes in the following order: 1960–1963 (civilian government); 1963–1966 (civilian government – 1st Republic); 1966–1979 (military government); 1979–1983 (civilian government – 2nd Republic); 1983–1999 (military government); 1999 to present (civilian government – 3rd Republic) (Falola *et al.*, 2019). Special note must be made that the hiatus of these was the civil war of 1967–1970 which nearly ended the country's existence as a unified political entity. These dysfunctional political systems, reflected through continuous insecurity, fractural social relationships, pervasive corruption and mismanagement (Achebe, 1983; Ogbeidi, 2012), had implications at several levels, not least of which were their impacts on the economy and society which frustrated the achievement of the wider goals of development.

The economy

The Nigerian economy has had a truncated history. Space limitations here do not allow for a detailed consideration of the country's economic history; suffice to say that the following analysis discusses, in order of importance, the performance of the key five sectors of the Nigerian economy as published by the country's National Bureau of Statistics (2019). Typical of most African countries, the economy of Nigeria is undiversified with mining, essentially petroleum, the dominant sector, this being the main source of foreign exchange earnings and government revenues. In the third quarter (Q3) of 2019, oil accounted for 97.11% of the mining sector output which contributed 9.8% to overall real gross domestic product (GDP). The agricultural sector comprises crop production, livestock, forestry and fishing activities. Crop production is the largest driver of the sector, accounting for 91.6% of sector in Q3 of 2019. Agriculture contributed 29.25% to overall real GDP during the quarter. The manufacturing sector contributed 8.74% to real GDP, lower than the 8.84% recorded in Q3 of 2018 and the 9.08% recorded in the second quarter (Q2) of 2019. The construction sector is small and its contribution to total real GDP was 3.01% in Q3, the same as its contribution in the same quarter of the previous year, but lower than in the preceding quarter where it contributed 4.45%. Finally, the trade sector, mainly accommodation and food service, is similarly small. In Q3 of 2019, the sector accounted for 0.87% of real GDP, the same as its contribution in the same period in 2018 but higher than its Q2 of 2019 contribution of 0.64%.

The description above merits brief reflections with respect to the key economic sectors – petroleum and agriculture – and their evident implications for the economy. Petroleum is the 'powerhouse' of the country's economy. However, questions are now being asked about Nigeria's excessive reliance on oil as a dependable revenue source. In a recent public lecture, the core messages of which are instructive, Chief John Nnia Nwodo, the current President General, Ohanaeze Ndigbo Worldwide -- a socio-cultural organization of the Igbo ethnic group in Nigeria -- was unequivocal: 'Our economy is failing and the slide is unstoppable . . . oil is becoming a non-dependable commodity for getting revenue' (2020, p.1). He argued that the threat will arise from supply–demand imbalance. First, America's discovery of shale-oil technology will increase its daily production of oil to 12 billion barrels of crude oil. When compared to Nigeria's daily production of 1.8 billion barrels and also the fact that the world production of crude oil is growing at a phenomenal rate, the implication for Nigeria is obvious: 'so in the next 3 years our oil will be worthless in the international market owing to increase in supply and decrease in demand'. Second, and relatedly, Nwodo (2020) points to the emergence of electric – as opposed to fossil fuel – powered engines:

> Right now, in the last one and a half years, the number of engines in terms of vehicles and machinery demanded in China, Europe and America, there have been more electric engines demanded in these countries [and regions] than fossil oil-based engines. What does this say? It also says that, apart from the increase in the production of crude oil, fossil oil and shale oil which will reduce the demand for our oil, the demands for engines that consume our oil are reducing globally.

Since crude oil is the mainstay in Nigeria's economy, the implications for the local economy are self-evident; the scenario described above will have a disastrous effect on export earnings and employment in the country.

The implications for the agricultural sector are two-fold. First, vagaries of the weather, such as droughts, might decrease harvests and lead to low foreign currency accrual for investment purposes. A low harvest also means insufficient food production and so increased food imports are needed to feed the growing population. In addition, with the terms of trade moving against it, Nigeria has, over the years, suffered a debt crisis. Second, and related, there are potential consequences for this sector arising from the emergence of drone technology. Again, Nwodo (2020) weighs in, sharing his perspective on the remarkable growth in the number of agricultural companies that use digital knowhow for production. Citing Israel as an example of a country at the forefront of the use of such technology, he observed that: 'Israel can now determine the chemical deficiency of a plant in a farm without a foot on the farm, just by using drone technology. They can scan through the farm and know which plant has chemical deficiency and they can cure it by drone technology.' Relating this situation to Nigeria, he laments: 'Our problem is leadership; what are we doing to determine the right policies?'

Given the economic challenges discussed above, this chapter suggests that significant potential lies in the development of the tourism sector. Hence, it is to this that we now turn, considering three strands of tourism: the historical dimension, the current situation, and development issues.

Tourism Development in Nigeria

A short historical perspective

Steps to develop tourism in Nigeria, at least from the perspective of policy for and planning of the sector, can be traced back to six historical landmarks: 1959, 1962, 1976, 1991, 1992 and 2006. The first was the 1959 report (Government of Nigeria, 1959) of an ad hoc committee involving a group of American travel agents, set up by the federal government to advise on tourism development possibilities. The government accepted the report of this advisory committee and this resulted in the setting up of the Nigerian Tourist Association (NTA) in 1962 (Government of Nigeria, 1962). However, the NTA was found to be ineffective, poorly organized and poorly funded and was consequently replaced in 1976 by the Nigerian Tourist Board (NTB) (Government of Nigeria, 1976).

The establishment of the NTB ushered in positive measures to provide a stimulus to tourism, including the inclusion of tourism in the Nigerian Development Plan, the formulation of tourism policy and an institutional framework, the establishment of a Tourism Committee in each State of the Federation, and the renaming of the Federal Ministry of Trade to the Federal Ministry of Trade and Tourism (Okorafor, 1995; Yusuff and Akinde, 2015; Bassey and Egon, 2016). To crystallize these efforts, a three-day national tourism development conference was held in Lagos on 10 July 1990 (see Kendell, 1990). The outcome of the conference was seen in the publication of a consultancy report, the *Federal Republic of Nigeria Tourism Development Strategy*, published in 1991 (see TPL Associates and Amhurst Tourism Services, 1991). Essentially, this was a *strategy* document rather than a *plan*, setting out a number of development options or development scenarios and suggesting ways in which these might be realized. To progress matters, in 1992 the government published, in line with the recommendations of the strategy document, a new Decree 81 which abolished the NTB and replaced it with the Nigerian Tourism Development Corporation (NTDC) (Government of Nigeria, 1992). The belief was, as always, that the defunct NTB had been unable to develop the tourism sector, in part because it enjoyed limited authority to enable it to stimulate development and in part because it was, at the federal level, at an inferior level relative to the main economic ministries. Even at the state level, tourism committees had limited impact. In comparison, the newly established NTDC was conceived as an aggressive and commercially oriented parastatal corporation with the responsibility to promote, market and disseminate information to the industry. Subsequently, the Federal Ministry of Culture, Tourism and National Orientation was established in 2000.

The final positive effort to develop tourism was the publication of a Tourism Development Master Plan in 2006 (Government of Nigeria, UNWTO and UNDP, 2006). This Master Plan is a comprehensive, up-to-date reference point document against which we can assess Nigeria's tourism policy planning objectives. But the desire of tourism leaders in Nigeria to pursue tourism must be accompanied by a similar desire and commitment from the Nigerian populace. However, these two-fold desires, no matter how well-intentioned, do not necessarily work as these intentions must be accompanied by situational factors which transform desire into actuality. In other words, words must be matched with action.

Current trends

This section describes, in brief, the tourism product and characterizes the product with respect to Nigeria, considers the extent and impact of tourism in the country, and discusses its administrative structure for tourism.

The tourism product

A tourism product refers to the sum total of all those facilities, amenities and services, including the natural environment, which attract visitors to a destination. In Nigeria, this product has three main features: (i) culture- and heritage-based attractions (songs, dress pattern and style, traditional music, belief system, cuisine, indigenous technology, folklores, festivals, monument and heritage sites); (ii) nature-based attractions (national parks and game reserves, lakes, waterfalls, mountains, hills, plateaus, springs, beaches, rocks and scenic beauty); and (iii) built attractions – zoos, museums, airports and hotels. These features prompt two general reflections.

First, the Nigerian tourism product is largely based on cultural and heritage attractions. This has to do with the country's more than 250 ethnicities, with diverse cultural activities. Each of these ethnic groups has its unique cuisines, cultural and traditional events and festivals, belief systems, religious activities, dress style, languages, traditional music, unique way of life, monuments, deities, shrines and sacred groves (Government of Nigeria, UNWTO and UNDP, 2006). Most of these tourism assets are not yet harnessed, although they offer great development potential. At the same time, however, there are in fact numerous examples of 'successful' cultural tourism events and festivals held annually which must be noted, including: Abuja carnival (November), Calabar carnival in Cross River State (December), the Argungu fishing festival in Kebbi State (February) and festivals of arts and culture celebrated in various states in Nigeria. Second, neither the nature-based nor the built attractions in Nigeria are in themselves unique; they are available to a greater or lesser extent in several sub-Saharan African countries. Thus, in marketing terms, these are, for Nigeria, secondary rather than primary attractions. In this sense, no one activity or site is sufficiently attractive to motivate visitors but, collectively, they offer a variety of tourism options. Arguably, this is not an unusual situation, with very few countries having attractions which constitute unique selling propositions (USPs).

Extent and impact of tourism

The volume and value of international tourism in Nigeria can be ascertained from the most recent data available from the United Nations World Tourism Organization (UNWTO), which are based on estimates by the Nigerian Tourism Development Corporation (NTDC). The volume data refer to tourist arrivals disaggregated on main source markets and principal purpose of visit, while the value information reflects gross earnings from such tourism in the form of tourism expenditure accruing to the country (UNWTO, 2018). In addition, statistics on the capacity and utilization of commercial accommodation facilities provide valuable planning information on the volume of both international and domestic tourism in the country as a whole and by geographical area (UNWTO, 2019).

In 2016, an estimated 5.3 million international tourists visited Nigeria, of whom only 2.1 million (approximately 40%) were leisure tourists (UNWTO, 2018). The major motivations for visitor arrivals were visiting friends and relatives – the so-called VFR category of visitors (31%) and business (29%) (UNWTO, 2019). It was further noted that 45% of all visitors came from three continents – Africa, 1.8 million (33%), Europe, 367,425 (6.98%) and the Americas, 310,000 (5.88%). At a country level, visitors from six countries comprised 24% of the total arrivals in the following order of importance: USA, 246,562 (4.68%), UK, 213,368 (4.05%), China, 207,927 (3.95%), and three African countries, namely, Benin, 164,828 (3.13%), Ghana, 149,974 (2.8%) and Niger, 393,670 (4.68%).

The patterns of visitor spending in 2016 were similar (UNWTO, 2019). As a consequence, tourism expenditure contributed just 0.3% to the GDP and net foreign exchange earnings of US$1070 million (excluding international transport, which is a major source of tourism-induced revenue). Average length of stay was seven nights. Tourism employed about 1,429,400 Nigerians directly and indirectly in a wide spectrum of activities. Direct jobs, estimated at around 1,208,000, were, unsurprisingly, in hotels and restaurants that supplied and served tourists, and provided earnings for workers and investors in the form of wages and profits. Indirect employment estimates gave an approximate figure of about 221,400 individuals whose activities, such as handicrafts and taxi driving, were tourism-related.

In analysing the above statistics, a number of general observations can be made. First, the figures mask more than they reveal, considering that there is a dearth of reliable tourism statistics in Nigeria and these should be interpreted with caution. Although the data have to be interpreted cautiously, they are likely to be based on either over- or underestimates, both reflecting the well-known problems of tourism statistics (Jenkins, 2015) and on grounds that Nigeria has not so far implemented Tourism Satellite Accounts (TSA) as a means of improving the collection of tourism financial and economic data and to facilitate comparisons between countries of the impact of tourism activity. None the less, the figures do provide an indication of the nature and scope of tourism and its place in the economy of Nigeria. As this role increases, so too does the need for more reliable time series data as a basis for policy formulation and planning parameters.

Second, it is clear that international tourism in Nigeria is a low-volume industry that relies on regional African markets. This has been consistent over the years. In addition, the sector has a very small core of genuine holidaymakers and a heavy dependency on three non-African markets, and on the USA, the UK and China. As the sector is still in a fairly embryonic state in Nigeria and with an annual number of visitor arrivals over the four successive years (2012–2016) fluctuating between 4.6 million and 5.3 million (UNWTO, 2018), there are no grounds for believing that visitor arrivals and tourism receipts will increase in the foreseeable future, cognisant of the country's low priority given to the sector to date, the security and other related challenges facing Nigeria.

Administrative structure for tourism

Earlier, we provided a brief historical account of tourism in Nigeria. Here, special note must be taken of the 1990, 1992 and 2006 trends with respect to tourism administration. First, the policy objectives for tourism development in 1990 specifically sought to: protect and promote Nigeria's cultural heritage as a resource for home-grown socio-economic development; encourage community and public partnerships in tourism development; generate foreign exchange, enhance income redistribution, alleviate poverty and create employment; promote Nigeria as a desirable tourism destination within the context of Africa's cultural renaissance; promote geopolitical integration, healthy international co-operation and understanding; and ensure environmental sustainability in the development of tourism resources (Government of Nigeria, 1990). Second, the Nigerian Tourism Development Corporation, an apex tourism body with the responsibility of steering tourism activities in the country, was, as previously discussed, established in 1992 to replace the defunct National Tourism Board (Nigerian Tourism Development Corporation, 1992). Finally, the 2006 Master Plan document sought to strengthen the institutional and capacity support for tourism (Government of Nigeria, UNWTO and UNDP, 2006).

Unfortunately, the present federal political leadership of the country made drastic changes to the administrative structure of tourism on taking office in 2015, enhancing the impression of benign neglect of the sector in the following ways. At the federal level, the tourism ministry was dissolved, its status becoming subsumed under the Federal Ministry of Information and Culture (FMIC). The FMIC comprises 18 parastatals with their respective director-generals to oversee each parastatal. Nine of these parastatals are directly involved in tourism development and cultural resource management/preservation, while the other nine are indirectly involved in marketing Nigerian tourism (see Table 6.1).

As can be seen in Table 6.1, NTDC, as Nigeria's apex tourism body, is one of the parastatals subsumed under the current Federal Ministry of Information and Culture. While NTDC operates at the national level to oversee tourism development and formulate tourism policies, at the state level the NTDC Act empowered each of the 36 states in the country to have a state tourism board (STB) to champion tourism development in the state. The Act also made provisions for the establishment of a Local Government Tourism Committee (LGTC) in the 774 local governments in the country.

At the level of the federating units of the country, notably the Federal Capital Territory

Table 6.1. Parastatals - Federal Ministry of Information and Culture.

S/N	Direct tourism parastatals	Indirect tourism parastatals
1	Centre for Blacks and African Arts and Culture (CBAAC)	Advertising Practitioners Council of Nigeria
2	National Commission for Museum and Monuments (NCMM)	Federal Radio Corporation
3	National Council for Arts and Culture (NCAC)	National Broadcasting Commission
4	National Gallery of Art (NGA)	National Film and Video Censors Board
5	Nigerian Institute for Cultural Orientation (NICO)	News Agency of Nigeria
6	National Orientation Agency (NOA)	Nigerian Film Corporation
7	Nigerian Institute for Hospitality and Tourism (NIHOTOUR)	Nigerian Press Council
8	National Theatre/National Troupe of Nigeria (NT/NTN)	Nigerian Television Authority
9	Nigeria Tourism Development Corporation (NTDC)	Voice of Nigeria

(From https://fmic.gov.ng)

(FCT), each of the 36 states have now formulated tourism slogans to market their respective areas (Table 6.2).

The pertinent reflective questions must be: Do these complex arrangements advance the cause of the tourism sector in Nigeria? Do the policies and plans work? If so, what is the evidence? And if not, why not? What has been done to arrest the situation? What else needs to be done?

Tourism development issues

The questions above are striking in light of the following insightful observations:

> It is the measure of our self-delusion that we can talk about developing tourism in Nigeria. Only a masochist with an exuberant taste for self-violence will pick Nigeria for a holiday, only a character out of Tutuola seeking to know punishment and poverty at first hand! No, Nigeria may be a paradise for adventurers and pirates, but not tourists. (Achebe, 1983, p. 11)
>
> The tourism sector in Nigeria is a good example of benign neglect. Although government has created a legislative structure for tourism, it is ineffective. There is a lack of specific policy and adequate fund, with little or no coordination between the federal and state levels . . . The Nigeria experience clearly demonstrates that political instability and tourism are not good bed-mates. A policy in itself would do little to overcome image problems associated with political upheaval. (Jenkins, 1991)

The issues implicit in the characterizations above are legion and there are a number of general development issues that can be associated with tourism in Nigeria. These stem from several factors, including: first, general disappointment with the economic returns from the tourism sector; second, insufficient knowledge of the market mix of international tourism; third, the social and political discontent with tourism and, in particular, the market-driven nature of the sector; fourth, the inability of the Government of Nigeria, because of their bureaucratic structure, to react to market changes; and, finally, the general level of development of the country (see also Jenkins, 1994). In subsequent sections we pursue the nature of the 'benign neglect' of the tourism sector in Nigeria.

Neglect of the Tourism Sector

The previous discussion examined the development of tourism in Nigeria. Now is the time to ask why the potential for tourism in this country has been ignored. To the extent that the answers represent new methods in the development of tourism, some general prognostications can also be made.

Table 6.2. Official (country) tourism slogans and states in Nigeria.

S/N	State	State capital	Descriptive slogan/title
1	Abia	Umuahia	God's Own State
2	Adamawa	Yola	Land of Beauty
3	Akwa Ibom	Uyo	Land of Promise
4	Anambra	Awka	Light of the Nation
5	Bauchi	Bauchi	Pearl of Tourism
6	Bayelsa	Yenegoa	Glory of all Lands
7	Benue	Makurdi	Food Basket of the Nation
8	Borno	Maiduguri	Home of Peace
9	Cross River	Calabar	The People's Paradise
10	Delta	Asaba	The Big Heart
11	Ebonyi	Abakaliki	Salt of the Nation
12	Edo	Benin City	Heartbeat of the Nation
13	Ekiti	Ado-Ekiti	Land of Honour and Integrity
14	Enugu	Enugu	Coal City State
15	Gombe	Gombe	Jewel of the Savannah
16	Imo	Owerri	Eastern Heartland
17	Jigawa	Dutse	The New World
18	Kaduna	Kaduna	Centre of Learning
19	Kano	Kano	Centre of Commerce
20	Katsina	Katsina	Home of Hospitality
21	Kebbi	Benin-Kebbi	Land of Equity
22	Kogi	Lokoja	The Confluence State
23	Kwara	Ilorin	State of Harmony
24	Lagos	Ikeja	Centre of Excellence
25	Nasarawa	Lafia	Home of Solid Minerals
26	Niger	Minna	The Power State
27	Ogun	Abeokuta	Gateway State
28	Ondo	Akure	Sunshine State
29	Osun	Oshogbo	Land of Virtue
30	Oyo	Ibadan	Pace Setter State
31	Plateau	Jos	Home of Peace and Tourism
32	Rivers	Port Harcourt	Treasure Base of the Nation
33	Sokoto	Sokoto	Seat of the Caliphate
34	Taraba	Jalingo	Nature's Gift to the Nation
35	Yobe	Damaturu	Pride of the Sahel
36	Zamfara	Gusau	Farming is Our Pride
37	Federal Capital Territory (F.C.T.)	Abuja	Centre of Unity

(From https://en.m.wikipedia.org/wiki/List_of_Nigerian_states%27_nicknames)

The evidence

The 'Nigeria criticism' narrative decrying the neglect of the tourism sector is not hard to find. Some of the criticisms are perhaps subjective, whereas others are persuasive arguments; as a recent article in the *Guardian* pointed out (Abuh, 2017), the current federal government policy objective to diversify the economy is widely questioned. The main criticisms and concerns centre on the priority development status accorded to agriculture and solid minerals as the only two major ways of widening earnings and employment relative to other sectors of the economy. It is argued that this trend is clear evidence of the continued marginalization of the tourism sector, the development of which is described as 'a mockery' and serious concerns are expressed about its poor funding over the years. As tourism has some considerable comparative advantages over other sectors, these realizations are not lost on the lawmaker, who ends by drawing the government's attention to the potential of tourism and how to harness it. Essentially, tourism 'is a faster way to boost and jump-start the economy and yield faster results . . . would bring our people quick jobs which would bring about empowerment . . . it would showcase Nigeria in a good light' (Abuh, 2017).

This neglect is reflected also in the powerful article by Sessou (2019) which attempts to articulate four possible ways in which tourism is lagging behind in the Nigerian economy. The article clearly regrets the widening unemployment rate in the country, which, unfortunately, tourism has not been able to bridge. As the general public in Nigeria has very limited understanding of the link between tourism and employment, there are therefore calls for an awareness campaign to incorporate tourism into the school curriculum so as to increase public knowledge about tourism business. Further, the article indicts the Nigerian government, accusing it as neither ready for tourism nor making any serious attempt to grow it, despite the sector's huge development potential. The fundamental question posed is: what do you think about the fact that Nigeria does not have a Tourism Minister? The challenge is for the private sector to step in and steer the sector. But the key lies in collaborating with government, particularly in the area of policy formulation to guide the sector development objectives; a need for an enabling environment; as well as a mindset to grow the industry.

Several other writers (Octopus Admin, 2018; Tunde, 2019; Ojo, 2019; Akpan-Nsoh, 2019) have also joined the chorus of criticism of the sector, viewing tourism in Nigeria with something approaching contempt. For example, Ozigbo (Octopus Admin, 2018) has lamented the neglect of the Nigerian tourism sector by governments at all levels, both federal and state. This is demonstrated by the abysmal marketing of the sector by relevant ministries and agencies of government. He argues that, if well harnessed, the tourism sector, with its vast potential, could displace oil as the nation's highest revenue earner. He has, therefore, called for symbiotic relationships between the government and the private sector in the tourism business in which the government spearheads the marketing role while the latter drives the process.

Ojo (2019) has described the neglect of Nigeria's tourism sector as criminal, especially with respect to the country's monuments and tourist sites (see also Orekoya, 2018), which he says have become decrepit and taken over by weeds, reptiles and miscreants. He argues that many of these tourist sites lack a maintenance culture and the pilfering of the little income they generate has been a common phenomenon. He draws sharp contrasts between the immediate past administration of Goodluck Jonathan with that of current President Muhammadu Buhari at the federal government level. Under Goodluck there was a Ministry of Culture and Tourism, but under the current administration, the ministry evaporated into thin air. In effect it was scrapped, thus deepening the irrelevance of tourism in national development planning. For reasons of poor funding, the regulatory framework institutions, most notably the National Commission for Museums and Monuments, the Nigeria Tourism Development Corporation and the National Council of Arts and Culture institutions, have remained paper tigers as tourism is under the concurrent legislative list in which both the federal government and states can take on and develop the sector. Concern is further expressed at the way in which states which rely largely on monthly allocations from the federal government have never thought of diversifying their internally generated revenue by reviving their tourist sites to help preserve Nigeria's history and culture and enhance deeper understanding of the country's cultural heritage.

National parks are a useful resource input to tourism in some African countries where ecotourism is significant (Dieke, 2013). Akpan-Nsoh (2019) discusses this issue with reference to Nigeria, lamenting the neglect of the country's wildlife-based tourism. He refers to some African countries, such as the Democratic Republic of Congo (DRC), which are earning millions of dollars in foreign exchange from mountain gorilla-based ecotourism alone. Nigeria is, conversely, hunting its very rare species of gorilla for meat and destroying their habitats instead of conserving them to generate the much-needed revenue as is done in some parts of east and central Africa (Dieke, 1991). As a consequence, institutional investors, such as the Japanese Primate Research Organization, which has invested in gorillas and chimpanzees in Uganda, DRC, Rwanda and other African countries for up to 60 years, is now shunning West Africa with rarer species of those apes. Akpan-Nsoh (2019) then challenges the federal and state governments in Nigeria to take urgent steps in drawing up a Nigerian ecotourism map showing directions to all the attractive sites. Finally, he calls for the establishment of Marine Protected Areas and a Niger Delta Protected Area to serve not only as tourist sites but also for employment, climatic amelioration, balance of ecosystem services and safeguarding of biodiversity.

Writing in *The Glitterson Line*, Kazeem Tunde (2019) reports on tourism sites, but similar to Abuh (2017), as noted above, laments the poor performance of Nigeria's tourism sector. Specifically, he expresses disappointment at the lip service paid to the diversification of the economy in Nigeria when the potential for much-needed economic liberation is neglected. He views tourism as a potential catalyst to redress the economic downturn of the country, arguing that, based on the number of tourist attractions in Nigeria, the country has more potential than Mecca in Saudi Arabia and Jerusalem in Israel. He strikes a note to the effect that Africa in general, and Nigeria in particular, has been short-changed by western imperialists towards the quest for economic liberation through tourism. He pleads for an improved budget provision to strengthen the capacity to turn the sector around and, by implication, enhance the country's internally generated revenue.

There is an endless list of criticisms with every 'Tom, Dick and Harry' in Nigeria wanting to have a go at the tourism sector. However, is it fair to disregard the swipe by Bassey and Egon (2016) on the lack of tourism infrastructure, poor transportation and communication networks, poor participation of the private sector in tourism development, poor environmental conservation practices and poor marketing and promotion strategies for tourism growth in Nigeria? Or what about Ogbeidi (2012), who notes that rampant godfatherism in politics has also permeated the tourism agenda, with many non-professionals dominating the sector leading to the absence of experts in the industry? What about Agbebi (2014), who states that a lack of commitment and professionalism has affected the Nigerian tourism industry, drawing attention to elite domination of the tourism industry? This observation is consistent with an earlier conclusion by Tosun (2000) concerning elites dictating the nature and scope of tourism development as well as the level of community participation in many emerging economies.

The kind of thinking illustrated above has policy and planning implications – perhaps meaning that new tourism development directions are needed in Nigeria. In a wider international context, it is sufficient to say that the problems are generic, such as marketing, image, perception, institutional structures, financial, and expertise — all problems which can be found in some, if not many, African countries. One reflective question is appropriate in this regard: what accounts for this unbalanced approach?

An unbalanced approach

In seeking to account for this unbalanced approach, we must address two questions: first, who is to blame for the dysfunction of Nigeria's tourism sector? And second, what are the wider factors which are relevant to the development of tourism everywhere, the absence of which, in Nigeria, has evidently far-reaching implications for its tourism sector's policies and plans?

To start with, the Nigerian experience with tourism has generally pointed to one conclusion: that tourism development has been an unmitigated failure (Bankole, 2002; Ovat, 2003; Okon, 2014; Bassey and Egon, 2016). This is because many of the general problems afflicting Nigeria in nation-building are themselves barriers to development of the tourism sector. If Nigerians do some soul-searching, they will surely discover that both leaders and the led alike are all responsible for or guilty of the country's dysfunction with disastrous consequences for the tourism sector there. The list of such vices bedevilling the country is endless, most notably: encouraging some corrupt practices, some form of injustice and chaos, supporting predatory practices both in the public and private sector, all apparently jeopardizing long-term development (Falola *et al.*, 2019).

The fact remains that the capacity of Nigerian governments at various levels – federal, state and local – to implement development planning programmes and policies differs widely, lacks co-ordination and is unsustainable. As a consequence, many Nigerians are coming to the gloomy conclusion that many of these governments are simply not up to the task of development (Bukola and Olaitan, 2018). Unfortunately, some are pessimistic with regard to what the current policy for economic diversification can achieve (Ayeni, 2013). In the context of this chapter, we describe it as elusive and lopsided, with tourism playing second fiddle. Few others might see it as a bad thing, but they are not overly confident that it will make much of an impact on the operation of governments. Some point to the apparent contradiction of the twin processes of state retreat and democratization; in poor societies, states need to mobilize popular support for both democracy and state legitimacy, and are handicapped by the lack of resources from and from inaction by government.

Many other Nigerians seem to lose faith in the government's ability to provide even the most basic of welfare services – the provision of good road infrastructure, good drinking water, healthcare facilities – turning elsewhere for functions normally performed by government (Akinrinade, 2011). This further undermines the state's ability to play an effective role in development, regardless of whether that prescribed role is minimalist or maximalist. A few Nigerians, bewildered by state failure, throw up their hands in despair and say that the nation-state, imported from Europe by colonialism, is inappropriate technology in Nigeria and cannot hope to fulfil the requirements of development. One notable criticism has come from the Catholic Church hierarchy, especially the Catholic Bishops Conference of Nigeria in their recent statement: 'The level of insecurity in Nigeria today is such that whether at home or on the road, most Nigerians, in all the parts of the country, live in fear' (Akabeze and Umoh, 2020; see also Ajogbeje *et al.*, 2017).

Such growing criticism of development policies has not gone unnoticed by one public policy analyst who has recently added a further dimension to the depressed nature of Nigerian society. Writing in the *Vanguard*, Tabia Princewill (2019) characterizes Nigeria as an extreme case of class warfare whereby as long as the elite can continue to escape discomfort, they are happy to watch poverty and insecurity rise to the detriment of the majority poor in society. Putting it another way, while the profit motive is central to capitalism everywhere, consumer welfare is typically king. In contrast, the Nigerian brand of capitalism is perhaps even more predatory than its western counterpart, for two reasons: first, the consumer or customer experience means little to the capitalists; and second, there is the lack of regulation and enforcement of consumer protection laws. Paradoxically, we find that Nigerians applaud and encourage their oppressors, lavishing on them attention and accolades. Each one of them, lost in their personal, individual struggles, fails to truly connect the dots between the collapse of public administration in Nigeria and the hardship they encounter at various levels. This journey to self-destruction only intensifies with time; the stakes get higher and higher; poverty, violence and suffering are exploding all around them and they choose to pretend there is nothing they can do. They normalize injustice, blaming so-called culture and traditions, weak institutions and others, as if they themselves (the oppressed) are not equally responsible for upholding those practices or for corrupting those structures. By turning tradition into weapons, and arrogating themselves to the title of 'Giant of Africa', Nigerians lose the moral highground, refusing rational thought and creativity while, ironically, some of the 'smaller' African countries, such as Gambia and Ghana, that they look down on continue to leave them behind.

It is remarkable to find that a number of the most vibrant and well-educated Nigerian women remain politically passive. This is arguably nothing more than elite selfishness on their part; some, if not most, who are allowed into the 'boys club' seek political power, but without adopting a pro-people ideology to justify their quest. More generally, perhaps many people have ceased to believe there is another way of doing things, convinced as they are that this is it for Nigeria. For example, the majority of government functionaries and politicians are originally from poor, humble backgrounds; they have first-hand experience of poverty and deprivation. Yet, many do not work to revolutionize public services on behalf of those they left behind; rather, they simply work for their own share, believing 'it is their time' to partake in Nigeria's largesse. This scarcity mindset makes them focus greedily on wasteful frivolity and shy away from the impact of their behaviour.

Under such political and social circumstances, clearly tourism will not develop in Nigeria until these current development challenges, as described, are addressed. Otherwise, in using tourism for development, it would appear that the country is currently putting square pegs in round holes!

Some Prognostications

What, then, can we say about the future of the tourism sector in Nigeria? Here we identify and discuss the essential inputs to tourism development in general (see Jenkins, 2015), highlighting their strategic importance with respect to policy and planning principles, as a framework for exploring the future growth and direction of the tourism sector in Nigeria in particular.

Strategic resource allocation

The first issue is that of strategic resource allocation. The axiom that resources are limited and there are many competing needs for what is available is instructive (McKercher, 1993). Based on the economic concept of opportunity cost (or forgone alternative), it suggests the need to consider alternative use of resources, implying not only the competitive use of resources by others but, crucially, the efficiency with which such resources are used. Ideally, resource allocation, be it infrastructure, training or transport access, is based on the expectation of future gain, but for tourism, an additional dimension must be whether the sector is internationally competitive. Nigeria is a large country and as tourism must compete with other sectors in the economy for the scarce resources available for development, the country should, as carefully as possible, scrutinize resource allocation to the sector and carefully monitor its progress. It is not possible for the government to provide infrastructure in every location. Rather, infrastructure should be provided in those areas which have a high priority for development (or in the context of tourism, offer the most viable opportunities for development). Moreover, the development of tourism should proceed in full consultation with local communities, and the benefits of this development in terms of employment, income and entrepreneurial opportunities must accrue to these communities. The policy issue must be to devise an administrative structure to facilitate this objective. Part of planning consideration is to ensure enhanced operational standards and co-ordinate the various activities which impinge on tourism.

Human resource development

Another aspect in the future direction of tourism in Nigeria concerns human resource development (HRD). HRD encompasses both education and training, and is concerned with the manner in which an employee's characteristics are utilized in the performance of their roles. In the tourism development process, this involves the complex interaction of people and institutions; there exists a significant diversity of jobs which constitute the tourism 'product' amalgam and, hence a wide variety of social and other skills required to perform successfully in this globalized service sector. This is particularly important because tourism has the capacity to create more jobs than any other sector of the economy. In relating these considerations to Nigeria, the policy planning concern is to determine how current HRD practices limit tourism development and how these practices might be changed in the future. Certainly, there is usually no shortage of people available for work in Nigeria; rather. the problem is of the levels of skills available.

Two issues have already been signalled. First is the way in which working in tourism is perceived in Nigeria almost as servitude. As a consequence, the general public in Nigeria has very limited appreciation of the role of tourism in national development or of the benefits it can bring and the improvements it can make to daily lives. In order to change this perception, it might be useful to launch a sustained information and consultation programme with the aim of both creating greater awareness and understanding of tourism and its importance to the general economic development of Nigeria, and also, hopefully, to encourage wider participation in the industry. Such a campaign should focus primarily on sectors and in areas where tourism activity takes place. The target audience should be both politicians and senior government officials and the general public through a series of tourism awareness seminars, workshops and community awareness programmes.

The second, and related, issue is the widespread recognition that HRD problems, evident in all areas of the country's tourist industry, are a major barrier to upgrading service standards in the tourism sector. While some of the problems stem from the current lack of funding for improvements (discussed below), many others are related to the need for expertise in delivering training. There are two main training components: (i) the training to support private sector companies who will be expected to upgrade service standards for tourists; and (ii) training for the tourism commercial agencies in tour operating and retail travel management. Training programmes should be an ongoing requirement and supported by technical assistance. Technical assistance will support counterpart training to increase capacity for the indigenous management and delivery of future human resource development programmes.

Overall funding of the tourism sector

Another vital issue is the question of overall funding for the tourism sector. The importance of appropriate budget provision cannot be overemphasized. A budget allocation is needed for a variety of reasons: to upgrade service standards in the sector; to diversify products and markets; to provide training and recruitment of new workers to the sector; to secure physical facilities and infrastructure to support future expansion of tourism; to improve performance, particularly in the areas of market research; and to permit future sustained development of the sector.

At present there is a general concern throughout the Nigerian tourism industry about the level of government funding. Specifically, the existing government allocation to the NTDC is insufficient to support infrastructural investment, marketing and training. To some extent, this under-funding reflects the overall demands on the government's finances, the country's current stage of development and the significance of the tourism sector within the overall national economy. However, funding should also be assessed in relation to Nigeria's future tourism targets; without a realistic budget provision and the political support, NTDC will be unable to stimulate the development of tourism as its authority will be undermined and, consequently, tourism policies and plans as envisaged will not be achieved. In the future, the government should permit the NTDC to decentralize its activities. This development will permit revenue-generating possibilities (or internally generated revenue sources) for NTDC, make it more market-sensitive, and develop a range of services which could be sold. The policy planning direction is to identify funding sources and mechanisms which will provide adequate means to develop the potential of the tourism sector.

Public–private sector roles and conflicts

In tourism, the respective roles of the public and private sectors are well known; they are of equal importance but also complementary. In brief, the public sector is guided in investment decisions by the concept of 'national good'; it provides the basic infrastructure to facilitate the development of tourism, it provides financial incentives, infrastructure and legislative and institutional support, and it provides new opportunities and responsibilities for the private sector. Conversely, the private sector provides the main impetus for tourism growth through the provision of tourist amenities and facilities, such as accommodation, marinas, casinos, transport services, shopping, resorts and so on – in other words, the revenue-generating activities of tourism. Inevitably, there are at times conflicts between the two sectors in the performance of their tasks. This is a natural consequence of the diverse reasons for their involvement in the tourism sector, and may be reinforced by differences in political outlook. For example, the public sector champions a national tourism identity but, for the private sector, the main motive for investment is level of anticipated profit within a relatively short timeframe.

In Nigeria, the relationship between the public and the private sectors has not been cordial or close. Without a dynamic and supportive private sector, tourism will not develop in Nigeria. The policy and planning issues to consider will be how to devise an appropriate and representative structure which can bring together the many diverse stakeholders in the tourism industry in the country.

Conclusion

Tourism has been a neglected sector of the Nigerian economy. Nevertheless, it is recognized that tourism has substantial potential to contribute to future development and, therefore, cannot be ignored any longer. However, if the neglect continues, there are significant implications for tourism development in particular and, arguably, for the development of the country as a whole.

This chapter has discussed four broad implications for tourism development:

1. Resource allocation to tourism will become more important and, as a factor in strategic decision making, should often outweigh the allocations made to other sectors.
2. Human resource development is still an important concept in tourism analysis and the development of tourism in Nigeria cannot be regarded in isolation, in terms of viewing HR separately from wider labour force patterns. The emphasis should shift from what might be described as 'godfatherism' to getting 'the right people in the right jobs and at the right time'.
3. If tourism is to contribute its full potential to Nigeria's development, then it is evident that adequate funding provision must be an important input to the formulation of a policy for, and planning of, the tourism sector.
4. Further development of the tourism sector should be properly managed in order to narrow the public–private sector divide. In the final analysis, if there is an idea which guides our understanding of the possibilities for developing tourism in Nigeria, it must be that without determined and focused government support, not much development will occur. If the Nigerian government has an intention to invest and support tourism, then it has to focus on it in terms of policy, institutional support and funding. Nigeria is too large a country to expect much return from thinly spread resources.

References

Abuh, A. (2017) Ogbeide-Ihama decries neglect of tourism sector. *The Guardian*, 21 October. Available at: https://guardian.ng/saturday-magazine/travel-a-tourism/ogbeide-ihama-decries-neglect-of-tourism-sector/ (accessed 21 June 2019).

Achebe, C. (1983) *The Trouble with Nigeria*. Heinemann Educational Publishers, Oxford, UK.

Adejuwon, F.J. (1986) Trends of tourist demand in Africa. *The Tourist Review* 41(1), 20–24. DOI: 10.1108/eb057942.

Agbebi, P. (2014) Exploring tourism policy, planning and governance in the transformation of Nigeria's economy. *Journal of Tourism and Hospitality* 3(3), 1–4.

Ajogbeje, K., Adeniyi, O. and Folarin, O. (2017) The effect of terrorism on tourism development in Nigeria. *Tourism Economics* 23(8), 1673–1678. DOI: 10.1177/1354816617692477.

Akabeze, A. and Umoh, C. (2020) Prayer and procession for peace in Nigeria. *Letter from the Catholic Bishops Conference of Nigeria,* 26 February.

Akinrinade, S. (2011) Leadership, good governance and sustainable development. In: Popoola, L., Olaniya, O., Wahab, B., Ana, G. and Olayide, O. (eds) *Global Change and Sustainable Development: Challenges and Opportunities for South-South Cooperation in sub-Saharan Africa*. Centre for Sustainable Development, Ibadan, pp. 53–64.

Akpan-Nsoh, I. (2019) Expert decries neglect of ecotourism. *The Guardian*, 7 October. Available at: https://guardian.ng/news/expert-decries-neglect-of-ecotourism/ (accessed 10 November 2019).

Ayeni, D. (2013) Predicting the effects of economic diversification on sustainable tourism development in Nigeria. *American Journal of Tourism Management* 2(1), 15–21.

Bankole, A.S. (2002) The Nigerian tourism sector: contribution, constraints, and opportunities. *The Journal of Hospitality Financial Management* 10(1), 71–89. DOI: 10.1080/10913211.2002.10653762.

Bassey, F. and Egon, C. (2016) Nigerian tourism policy: a critical appraisal and re-appraisal. *Journal of Tourism, Hospitality and Sports* 20, 26–29.

Bukola, A.A. and Olaitan, L.L. (2018) Tourism development and sustainable economic development: evidence from South Western Nigeria. *South Asian Journal of Social Studies and Economics* 2(1), 1–9. DOI: 10.9734/sajsse/2018/v2i125816.

Dieke, P.U.C. (1991) Policies for tourism development in Kenya. *Annals of Tourism Research* 18(2), 269–294. DOI: 10.1016/0160-7383(91)90009-Z.

Dieke, P.U.C. (2013) Tourism in sub-Saharan Africa: production–consumption nexus. *Current Issues in Tourism* 16(7-8), 623–626. DOI: 10.1080/13683500.2013.785478.

Emery, E. (1995) No politics without inquiry! - A proposal for a class composition inquiry project 1996-7. *Common Sense: Journal of the Edinburgh Conference of Socialist Economists* 18, 1–11.

Falola, T., Hamilton, A., Kirk-Green, M., Udo, R. and Ajayi, J.F.A. (2019) Nigeria. Encyclopaedia Britannica. Available at: https://www.britannica.com/place/Nigeria (accessed 24 December 2019).

Government of Nigeria (1959) *Advisory Committee on the Promotion of the Tourism Industry, Sessional Paper No 5*. Federal Ministry of Commerce and Industry, Lagos.

Government of Nigeria (1962) *Nigerian Tourist Association:, Government White Paper on Tourism*. Nigerian Tourist Association, Lagos.

Government of Nigeria (1976) *Nigerian Tourist Board: Decree No 54 Establishing Nigerian Tourism Board*. Federal Ministry of Trade and Commerce, Lagos.

Government of Nigeria (1990) *National Tourism Policy for Nigeria*. Federal Ministry of Trade and Tourism, Abuja.

Government of Nigeria (1992) Nigerian Tourism Development Corporation (NTDC) Act, 1992. Available at: http://extwprlegs1.fao.org/docs/pdf/nig120668.pdf (accessed 24 March 2020).

Government of Nigeria, UNWTO and UNDP (2006) Nigeria Tourism Development Master Plan: Institutional Capacity Strengthening to the Tourism Sector in Nigeria. Available at: http://www.ng.undp.org/publications/environment/tourism_masterplan.pdf (accessed 26 February 2009).

Jenkins, C.L. (1991) Tourism in development – four case studies. In: Lickorish, L., Jefferson, A., Bodlender, J. and Jenkins, C.L. (eds) *Developing Tourism Destinations: Policies and Perspectives*. Longman, London, pp. 107–118.

Jenkins, C.L. (1994) Tourism in developing countries: the privatisation issue. In: Seaton, A., Jenkins, C.L., Wood, R., Dieke, P.U.C., Bennett, M., *et al.* (eds) *Tourism: The State of the Art*. John Wiley, Chichester, pp. 3–9.

Jenkins, C.L. (2015) Tourism policy and planning for developing countries: some critical issues. *Tourism Recreation Research* 40(2), 144–156. DOI: 10.1080/02508281.2015.1045363.

Kendell, P.J. (1990) Developing a tourism strategy for Nigeria. *Paper presented at the National Conference on Development of Tourism in Nigeria*, Lagos, 10-12 July.

McKercher, B. (1993) Some fundamental truths about tourism: understanding tourism's social and environmental impacts. *Journal of Sustainable Tourism* 1(1), 6–16. DOI: 10.1080/09669589309450697.

National Bureau of Statistics (2019) *Nigerian Gross Domestic Report*, 3rd Quarter. NBS - National Bureau of Statistics, Abuja.

Nigerian Tourism Development Corporation (1992) *Handbook on the Nigerian Tourism Development Corporation*. Farl Enterprises General Nigeria, Abuja.

Nwodo, J.O. (2020) Restructuring: building a knowledge-based society. A public lecture delivered at the launching of the maiden edition of the Goke Omisore voice of reason (VOR), held in Lagos on 20 February. Available at: https://youtu.be/2KR8kp-rYBg/ (accessed 23 February 2020).

Octopus Admin (2018) Transcorp Hotels MD laments neglect of tourism sector. *The Octopus News*. 23 May. Available at: https://www.theoctopusnews.com/transcorp-hotels-md-laments-neglect-of-tourism-sector/ (accessed 20 November 2019).

Ogbeidi, M. (2012) Political leadership and corruption in Nigeria since 1960: a socioeconomic analysis. *Journal of Nigerian Studies* 1(2), 1–25.

Ojo, J. (2019) Criminal neglect of Nigeria's tourism sector. *The Punch*. 10 April. Available at: https://punchng.com/criminal-neglect-of-nigerias-tourism-sector/ (accessed 20 November 2019).

Okon, E. (2014) Inbound tourism and social factors in Nigeria: evidence from an Ardl-Ecm model. *Asian Journal of Economics and Empirical Research* 1(2), 40–47.

Okorafor, C.N. (1995) *Tourism in Nigeria: Progress, Problems, and Prospects*. African Educational Services.

Orekoya, I. (2018) Nigerian tourism industry: a key value driver in employment generation. *Journal of Tourism, Hospitality and Sports* 36, 2312–5187.

Ovat, O.O. (2003) Tourism and economic development in Nigeria: an empirical investigation. *Global Journal of Social Sciences* 2(1), 33–44. DOI: 10.4314/gjss.v2i1.22765.

Princewill, T. (2019) Who is to blame for Nigeria's dysfunction? *The Vanguard*. 4 December. Available at: https://www.vanguardngr.com/2019/12/who-is-to-blame-for-nigerias-dysfunction-2/ (accessed 10 December 2019).

Sessou, E. (2019) Why tourism is lagging behind in Nigerian economy – Omolara Adagunodo. *The Vanguard*. 5th October. Available at: https://www.vanguardngr.com/2019/10/why-tourism-is-lagging-behind-in-nigerian-economy-omolara-adagunodo-md-jumia-travel/ (accessed 6 November 2019).

TPL Associates and Amhurst Tourism Services (1991) *Federal Republic of Nigeria – Tourism Development Strategy*. TPL Associates Ltd & Amhurst Tourism Services, London & Lagos.

Tosun, C. (2000) Limits to community participation in the tourism development process in developing countries. *Tourism Management* 21(6), 613–633. DOI: 10.1016/S0261-5177(00)00009-1.

Tunde, K. (2019) Expert laments neglect of tourism sites in Nigeria. *The Glitterson Line*. 28 September. Available at: https://theglittersonline.com.ng/laments/ (accessed 20 November 2019).

UNWTO (2018) *Yearbook of Tourism Statistics Dataset [Electronic]*. UN World Tourism Organization, Madrid.

UNWTO (2019) *Compendium of Tourism Statistics Dataset [Electronic]*. UN World Tourism Organization, Madrid.

Yusuff, M. and Akinde, M. (2015) Tourism development and economic growth nexus: Nigeria's experience. *European Journal of Hospitality and Tourism Research* 3(4), 1–10.

7 Social Sustainability: What Does It Say to Tourism Employment?

Tom Baum[1]* and Shelagh Mooney[2]
[1]*University of Strathclyde, Glasgow, Scotland;* [2]*Auckland University of Technology, Auckland, New Zealand*

Introduction

It is over 40 years since Emmanuel de Kadt (1979) highlighted tourism as a possible 'passport to development' for communities and countries in the Global South. While offering a balanced assessment of the benefits and limitations of tourism as a development tool, he brought together the experience of experts who had been involved globally in supporting the development of tourism in the Global South over the previous 20 years. The potential for tourism was presented in a predominantly positive way since when *Passport to Development* (without de Kadt's question mark) has been used as the intellectual springboard for international agencies and governments to drive aid and investment in tourism. The tone of debate at this time was substantively positive and constructive and tourism was promoted as bringing economic, social and, crucially, employment benefits that far outweighed any negative outcomes. More recent contributions have, however, been rather more critical in their approach (see, for example, Jenkins, 2015 for a historical perspective of this area).

The emergence of sustainability as a consideration in tourism development in the late 1980s initially did not much influence the tenor of this debate, although the subsequent emergence of sustainable tourism as a concept and aspiration has significantly changed approaches to tourism development globally. As Weaver (2011, p. 5) notes, 'since the mid-1990s, discourses about the tourism sector have become increasingly dominated, at least rhetorically, by the ideas and ideals of sustainability'. Likewise, there is a growing interest in what Rimanoczy and Pearson (2010, p. 11) describe as the 'new world of sustainability' in the context of human resource management at the level of the firm. Their argument is that for more sustainable practices to emerge in areas such as employee recruitment and selection, there is a need for fundamental change in organizational culture and leadership. This discussion at the level of the organization should also acknowledge a number of related themes, including the evolving nature of psychological contracts and the ways in which employees and employers are bound together (Coyle-Shapiro and Kessler, 2000), recognition of the importance of sustainability within organizational leadership (Hargreaves and Fink, 2006) and understanding of mutual responsibilities through seeking work-life balance (for example, Lee *et al.*, 2015). However, there is only limited evidence of such emergent interest in sustainability themes converging

*Corresponding author: t.g.baum@strath.ac.uk

 Tourism in Development: Reflective Essays (eds P.U.C. Dieke *et al.*)
DOI: 10.1079/9781789242812.0007

within the tourism workforce narrative. Indeed, although Miller *et al.* (2010) discuss the emergence of sustainable tourism as a concept, linking it to a combination of the growth of tourism, the intentions behind political discourse that has become more environmentally focused, and awareness of the meaning of the principles of sustainability, they make no mention of employment or workforce-related themes.

However, there is growing evidence that a sustainability approach to the management of people within the tourism sector is an increasingly important, emergent conceptual direction for research and practice, complementing or even replacing strategic HRM as a route to increased societal and individual well-being. Westermann *et al.* (2017) provide reasonable grounds for this positive outlook, highlighting the advantage of an approach to sustainable HRM proposed by Renwick *et al.* (2013) that links the enduring effects of internal systems with external symbiosis. In this way, they recognize that HRM systems are inextricably bound up with their external societal, economic and ecological environment. This approach, which Kramar (2014) terms the 'connections approach to sustainable HRM', moves beyond the earlier internally oriented focus of sustainable HRM.

In this chapter, we argue that while this optimistic vision may appear to be evident across some sectors, it appears to be a more challenging aspiration within the complex and varied environment that is international tourism. This is a significant issue. Not only is the sector a significant employer globally (WTTC, 2018), but such is its ecological and social footprint that a failure to implement sustainable approaches has major environmental and human well-being implications. For example, the unfettered increase in cruise ship tourism has had a pronounced negative effect on many destinations (Marsh, 2012), including costs to local infrastructures and the displacement of local businesses in favour of souvenir shops (MacNeill and Wozniak, 2018). On cruise ships and ferries, employment arrangements are largely unregulated and there is evidence to suggest that not only are working conditions poor and fundamental workers' rights not protected (Adams, 2017), but also that the national minimum wage levels of the countries where ships are based (but not necessarily registered) are not observed (RMT, 2018).

Two strands of research underpin our nuanced and critical assessment of employment in tourism: first, the presence of non-sustainable work practices in many different tourism locations and organizational contexts; and, second, the neglect of workforce dimensions in much of the sustainable tourism debate. Here, we explore research gaps relating to how the effectiveness of internal and external human resource processes and their tangible outcomes are measured in the tourism sector, as well as highlighting the ways that diversity management policies and practices, as key indicators of sustainability, are conceptualized and realized. A key tenet of this narrative is that espousing equality and equal opportunity for diverse groups is not just a 'tick box' legal compliance exercise; rather, it is a critical factor underpinning workers' well-being. In addition, we acknowledge the International Labour Organization's (ILO, 2016) aspirational goal of decent work for all as a key pillar in sustainable development while, overall, the chapter is framed in the context of emergent discussion about sustainability within HRM practice in tourism (see, for example, Baum *et al.*, 2016a; Baum, 2018) and, as such, seeks to contribute to this narrative.

The Context of Tourism Employment

The tourism sector comprises diverse businesses and enterprises of different types, many of whose defining characteristics could be considered representative of enterprises belonging to the service sector. It has a significant global presence, employing one in ten people worldwide (WTTC, 2018). It is also an industry which places notions of sustainability at the forefront of much of the narrative about its future (De Beer *et al.*, 2014). As noted above, however, until recently the discourse on sustainability in tourism has been framed by a narrative that substantially ignores its social pillar within which employment themes are located (Boström, 2012).

It is important to tackle this apparent neglect and, with it, the evident paradoxes in the light of an emergent literature (Jithendran and Baum, 2000; Gorenak, 2016; Baum *et al.*, 2016a; Baum, 2018) which advocates the incorporation of sustainable HRM principles within wider pillars of social sustainability (Boström, 2012; Vallance *et al.*, 2011). Poor working conditions, low rates of pay and precarious working conditions characterize employment in the sector across many different contexts in both developing and developed economies. In particular, women are over-represented in low-quality jobs in the sector, with a consequent wide gender pay gap across tourism (Santero-Sanchez *et al.*, 2015). In addition, concerns continue to be raised about the exploitation of migrants who compose a large part of the tourism workforce in western contexts (Janta *et al.*, 2011; Mooney *et al.*, 2016). For example, in Australia and New Zealand, systemic instances of 'wage theft' and very poor conditions for immigrant workers appear to be embedded in sectoral work practices (Stringer, 2016; Berg and Farbenblum, 2017).

Not surprisingly, labour turnover is very high in tourism for structural, primarily seasonal, and operational reasons and managers frequently offer the rationale that employees will not remain in post for long as justification for a lack of investment in training and development across the sector (Poulston, 2008; Baum, 2015). This may be something of an over-simplification given that, even in the fast-food sector, there is evidence that young employees' average tenure was two years (Mohsin and Lengler, 2015). More generally, poor HRM practices in tourism represent a paradox in light of the crucial importance of front-line employees to the success of customer-facing business in the sector (Madera *et al.*, 2017). Equally problematic is the way in which employee interests are excluded from the sustainable tourism debate, although it is clear that employees play a central role in the successful implementation of sustainability initiatives at organizational level (Guerci and Carollo, 2016).

Sustainable HRM

As noted earlier, sustainable HRM is garnering increasing attention as an approach to people management that is responsive to global societal concerns relating to responsible corporate governance and workplace practice. As such, it adopts positions that both complement and sit counter to strategic HRM, and remains an emergent amalgamation of ideas and principles. It is, therefore, important to appreciate the 'messiness' surrounding conceptualizations of sustainable HRM and how it translates into practice. Particularly, meanings of sustainable HRM are contested, and Mazur (2017, p. 221) observes that the field of sustainable HRM is 'fragmentary and incoherent, while the practical dimensions require developing'. Reflecting this, there is a lack of studies offering concrete guidelines as to how sustainable operating practices can be implemented in tourism organizations (Stansfield, 2018).

Sustainability, as a guiding principle, came to the fore in the late 1980s with the publication of the Brundtland Commission's report (WCED, 1987). It decried environmental degradation and the social costs associated with it that were, and continue to be, inflicted by the prioritization of economic goals (Dyllick and Hockerts, 2002). Kramar (2014), in a perceptive examination of the viability of sustainable HRM, divides the sustainability literature into three strands. The first is focused on capability reproduction, dependent on internal application of HRM processes, in Kramer's view rebranded as high-performance work practices, then a somewhat novel approach. Ehnert's (2006) influential model is assigned to this category and Kramar is critical of its neglect of the negative consequences of HRM. Ehnert's *et al.* (2016) updated definition of HRM specifically addresses negative effects outside the firm:

> the adoption of HRM strategies and practices that enable the achievement of financial, social and ecological goals, with an impact inside and outside of the organization and over a long-term time horizon while controlling for unintended side effects and negative feedback.

This definition is a distillation of Ehnert's (2009) earlier depiction of sustainable HRM in practice that includes

- attracting and retaining talent and being recognized as an 'employer of choice';
- maintaining employee health and safety;
- investing into the skills of the workforce on a long-term basis by developing critical competencies and lifelong learning;
- supporting employees' work-life balance and work-family balance;
- managing ageing workforces;

- creating employee trust, employee trustworthiness and sustained employment relationships;
- exhibiting and fostering (corporate) social responsibility towards employees and their communities; and
- maintaining a high quality of life for employees and communities.

In a second strand of sustainability literature, Kramar (2014) considers the defining orientation to be the 'triple bottom line'. This corporate social responsibility (CSR) orientation is frequently presented as a responsible attitude to sustainability, characterized by a broader interpretation of 'social and environmental health' and including external outcomes (Kramar, 2014, p. 1076). Kramer judges the philosophical orientation of the organizations engaging with CSR to be similar to the first strand in that the main driver is efficiency. In his view, external contextual factors are not taken adequately into consideration, and financial advantage remains the lens through which beneficial effects on the internal and external environment are viewed. Baum *et al.* (2016a) suggest that sustainability principles in tourism and elsewhere can also be viewed in terms of a local (destination), regional or national workforce. This supports Zaugg *et al.* (2001) in their framing of sustainable HRM as 'long term socially and economically efficient recruitment, development, retainment and disemployment of employees', which places employees at the centre of the sustainable HRM discourse. Interestingly, in a review of corporate social responsibility models used in European countries, only in one location, Switzerland, was economic performance not prioritized (Zaugg *et al.*, 2001).

Kramar's third sustainability strand is titled 'connections'. Here the common element lies in the ways that organizational practices impact their social and environmental context, both internally and externally, while factoring in the financial outcomes at firm and societal levels. What distinguishes this strand from both previous orientations is that the underlying sustainability principles are driven by moral or ethical imperatives rather than 'rational' economic interests. Kramar notes approval of Avery's (2005) work on sustainable leadership practice, which riles against the short-termism of much Anglo-American practice. Ethics and business have a somewhat tortious relationship; for example, Bratton and Gold (2015), among others, argue (with some justification) that ethical management behaviour is neglected in business school education. Mumford and Fried (2014) counsel against the conflation of ethics and leadership, arguing that there are considerable methodological and substantive concerns with many proposed management models. For example, the perceptions of the ethical behaviour of the leader viewed by one group of stakeholders only – the followers – may ignore the interests of other stakeholders, resulting in potentially negative outcomes for those outside the organization.

Green HRM (GHRM) is included in the third strand of sustainability grounded on ethical principles. Renwick *et al.*'s (2013) overview of the GHRM literature suggests that 'green HRM, defined as the HRM aspects of environmental management (EM) is relatively diverse and piecemeal'. However, they identify areas where GHRM practices have tentatively been established. These include attracting and selecting 'green' employees; motivating them through performance management and appraisals; targeted pay and reward systems; as well as privileging employee voice and participation, in a supportive culture. Notably, they include unions in such initiatives. The GHRM field has evolved considerably to include contributions that have investigated more realistically how sustainability practices and principles can be incorporated in organizational processes (for example, Dron *et al.*, 2018).

Significant insights into the complexities of embedding sustainability initiatives in organizations have been provided by Guerci and Carollo's (2016) study of firms implementing green HRM initiatives in Italy. They found paradox theory useful in interpreting their conflicting findings from interviews carried out with managers tasked with implementing sustainable practices in six Italian companies. There was a difference between the 'symbolic' espousal of sustainability principles and the 'everyday' practices that accomplished more discrete concrete goals. The authors identified eight pivotal paradoxes ranging from the objectives of the GHRM system to how its principles framed organizational strategies and actions and, ultimately, they highlighted myriad contradictions and ambiguities concerning the role of the HR manager in implementing green practices. They conclude that it may be unrealistic to expect organizations to have a perfect fit between the principles of sustainable HRM and their organizational practices. A failure to recognize the inherent difficulties of embedding sustainable principles in organizations could mean that valuable learning opportunities are overlooked, thus lessening employee motivation to further such initiatives. This point is of note when we review sustainability practices in tourism organizations.

The Application of Sustainable HRM in Tourism

Low levels of job satisfaction, motivation and engagement are revealed in overviews of HRM practices in the tourism industry (Kusluvan *et al.*, 2010). More generally, poor HRM practices have historically existed in many western contexts, such as the UK (Knox and Walsh, 2005) and the USA (Hinkin and Tracey, 2000). In Spain, many hotels practise a 'hard' HRM approach that focuses on minimizing labour costs, as opposed to a more humanistic human resources approach emphasizing team work and employee commitment (Marco-Lajara and Úbeda-García, 2013). The short-term contingent labour model is prevalent globally (McDowell *et al.*, 2009; Baum *et al.*, 2016b), in part the consequence of the pressures of stochastic demand, notably seasonality. Davidson and Wang (2011), when discussing sustainable HRM in Australian hotels, argue that hotels need to move beyond a cost-reduction focus and a reliance on precarious, casualized labour by adopting a more strategic approach. In sum, although it appears that the adoption of sustainable principles by tourism organizations may be emergent in debate, the evidence suggests that the practice of HRM in the sector frequently contravenes the principles of sustainable HRM (Baum *et al.*, 2016a; Baum, 2018).

These concerns are mirrored by highly critical reviews of further specialized areas of tourism workforce research. For example, at the operations level of the firm, Legrand *et al.* (2016) suggest that, when considering a sustainable approach that incorporates environmental dimensions, economic dimensions and social dimensions, there must be practical outcomes. These do not just relate to environmental impacts and go beyond maximizing internal and external profit but, as a third pillar, emphasize that the social well-being of all stakeholders should be maximized. However, very few studies suggest how these principles should be operationalized. Additionally, given the multicultural workforce upon which global tourism companies and hotel chains depend for their operations, it is reasonable to expect that a wide variety of HRM studies tested across diversity management performance metrics would demonstrate the value of inclusion and advancement for historically disadvantaged groups. Diversity and inclusion in this context includes the intersection of young people (Lashley, 2016), women (Santero-Sanchez *et al.*, 2015; Mooney *et al.*, 2017) and ethnic minorities (Ayres, 2006), demonstrating the potential to contribute benefits to organizational performance. Regrettably, albeit unsurprisingly, two recent reviews of diversity management research in tourism echo the same damning critique of HRM research. Both Manoharan and Singal (2017) and Kalargyrou and Costen (2017) criticize the lack of studies that focus on specific groups, such as the multi-generational workforce, the LGBT+community and people with disabilities. The absence of theoretical models or sophisticated research methods, usually found in HRM studies, indicates that previous research may not have adequately responded to the requirements and problems of the sector.

Ultimately, for the purposes of this review, it is clear that research into sustainable HRM in the context of tourism lags significantly behind that found in the wider management and HRM literature. This may explain the limited impact that employment researchers seem to have on the field of tourism and its workforce (Baum, 2018). Unbridled tourism development does not have a normatively inclusive effect. Tourism in developing locations brings associated problems such as the exploitation of sex workers tourism and human trafficking (Hawke and Raphael, 2016), which impacts the most vulnerable members of society, notably women and children belonging to marginalized social and ethnic groups. In the global south, tourism brings economic benefits to developing economies; however, it puts pressure on scarce resources such as water, including squeezing out access for locals. For example, Cole (2017), in a study of how tourism development in an Indonesian town impacts the local community's access to water, found that hotel owners and wealthy business owners were easily able to afford the expense of buying trucked water. However, the poorest women and girls belonging to ethnically marginalized groups experienced further deprivation and hardship from queuing for hours to obtain limited water available from municipal supplies for a restricted period of time each day. The effects of local cultural factors can often mean that the illusionary trickle-down, positive effects of tourism on employment opportunities for marginalized members of society are not easily realized.

An Agenda for Change

Overwhelmingly, the preceding review suggests that the successful implementation of sustainable HRM begins at the organizational level, while also recognizing the interaction of the economic environment on the organization's processes and its effects on its local environment, economically, socially and environmentally. HR managers bear the responsibility of ensuring green practices are instigated and of recruiting, selecting and rewarding according to these principles (Alcaraz *et al.*, 2017). They can both promote the 'symbolic' aspects of sustainability as well as managing the sophisticated challenges in establishing systems and dealing within the complexity of the Asia Pacific region as it appears that there is a move in the service sector, especially tourism, to outsource the HRM function (Solnet *et al.*, 2015).

In considering a 'best fit', contemporary approach to sustainable HRM, our aim is to consider the tourism context and its broader reflection of service sector characteristics. Our discourse thus seeks to address the research agenda at the organizational level, while emphasizing that sustainable HRM is not a closed HRM system. The wider macro context is of the utmost importance when evaluating which model may be most appropriate to examine tourism employment. In a positive step to influence the impact of tourism development, researchers have identified that more cohesive efforts between town planners, local government and tourism organizations are required to overcome the negative effects caused by 'overtourism' on destinations such as Barcelona (Blanco-Romero *et al.*, 2018). The issue of tourism development thus impacts all societies; few destinations are immune to its effects. Therefore, we will first address the macro level before turning our attention to the organizational level.

Macro-level considerations in implementing a sustainable HRM approach

In response to issues of tourism development, there is a lack of will at governmental level to tackle some of the widespread abuses connected with work in the tourism sector (Baum, 2018). Considerable gaps prevail at both the policy and organizational level in most contexts. At policy level, government actions, such as government warnings to migrant associations in New Zealand, are inadequate to tackle abuse that is designed to prevent abuses of migrant workers (Searle *et al.*, 2015; Stringer, 2016). At organizational level, the negative effect on individuals and, consequently, firm performance is visible in the universally high turnover statistics. In developing economies such as Kenya, there is evidence of sexual discrimination in hotels (Orido, 2017) and even in the most enlightened economic systems, such as Norway, the tourism sector observes very poor HRM practice in their treatment of young kitchen apprentices (Dagsland *et al.*, 2015) and older employees (Furunes and Mykletun, 2007). Therefore, the more far-reaching principles of sustainability are more observed in the breach than the observance, even in community terms (Higgins-Desbiolles and Powys Whyte, 2014).

From the evidence presented in this chapter, there is a strong case that an ethical sustainability-focused approach can assist in addressing the identified problems in tourism workforce planning. In order for the sustainable approach to work, researchers must focus on research at all three levels of society as recommended by Mazur (2017) and Renwick *et al.* (2013). HRM strategies greatly affect their immediate environment, as well as their local and sometimes wider national setting (Boxall, 2018). Supporting this perspective, Ehnert *et al.* (2016), reporting on sustainability practices among the world's largest companies, found attitudes towards sustainability did not really vary according to national context, regardless of the differing national economic systems, level of state control and regulations on employment, suggesting that notions of sustainability are reasonably consistent across diverse environments. Certainly, the 'messiness' of approaches to sustainability means that no one model is likely to work across all organizations, environments and locations. In any case, as observed previously, the operationalization of 'pure' sustainable green principles may be illusionary considering the paradoxes at HRM strategy level.

Is it reasonable to assume that the economic rational approach to businesses' performance will moderate, given the prevailing neo-liberal discourse (Siddiqui, 2012; Howell, 2016)? Probably not. Therefore, accepting that ambiguity is inherent in sustainable HRM practices (Guerci and Carollo, 2016), as an overarching framework we will draw on Jithendran and Baum's (2000) representation of a three-dimensional HRM development model that incorporates tourism industry personnel, the host community and the tourists. This can be set alongside the common approach to HRM in tourism, which just focuses on the organizational (meso/micro) level and the effects of practices on individuals (micro). In this application, the 'host community' label represents both local and state interests, so our contemporary iteration divides this focus into two separate dimensions – state interests and host community. The tourism industry personnel category is renamed the organization's employees.

We propose a paradigm shift in research design to refresh traditional models commonly used in motivational studies to show how actors respond within an organizational context to societal forces. A good example of the limitations inherent in much current work is Murray's (2018) longitudinal study, which examined the ways in which local context influenced Canadian lodging employees' preferred job rewards over a nine-year period. The analysis showed that, although the top six rewards remained the same over time, significant changes were reported for 'good working conditions', 'job security' and 'appreciating of work'. Unless researchers factor in the effects of the emergent significance of non-standard work arrangements at a local or sectoral level, research design will not reflect the prevailing employment conditions on employee's motivational factors.

Researchers can also adopt a mutuality focus by designing studies that link macro, meso and micro levels. Our model supports the perspective that community interests are a primary driver of environmental sustainability. Taking a 'social innovation' approach to social and environmental challenges by harnessing the synergy of community groups, tourism companies, tourists and academic interests is proving successful in dealing with tourism-related pressures on local and regional infrastructure in Majorca (Batle *et al.*, 2018). While some tourism researchers have adopted multi-disciplinary approaches in addressing problems caused by tourism development, many community groups, especially indigenous communities, lack voice and continue to be marginalized in the pursuit of the tourism 'dollars', with grassroots activists rather than academic research addressing inequities at local community level (Higgins-Desbiolles and Powys Whyte, 2014). In research terms, there are implications for how community and local tourism and hospitality businesses co-operate. New generations of academics, especially indigenous researchers, are sought to fill these essential roles between local communities and research institutes. It is important that researchers adopt multi-disciplinary and critical theorizing to overcome recognized research 'blind spots'. Therefore, research that can drive policy to redress societal inequities and ensure organizations are accountable for adopting sustainable HRM principles is urgently required.

Organizational Level

In order to facilitate the adoption of a sustainable approach to HRM, we now centre on one model to address elements that have been highlighted as significant in the tourism literature. We propose the adoption of dimensions of Mazur (2017) application of the principles of the De Prins's *et al.* (2014) holistic model of sustainable

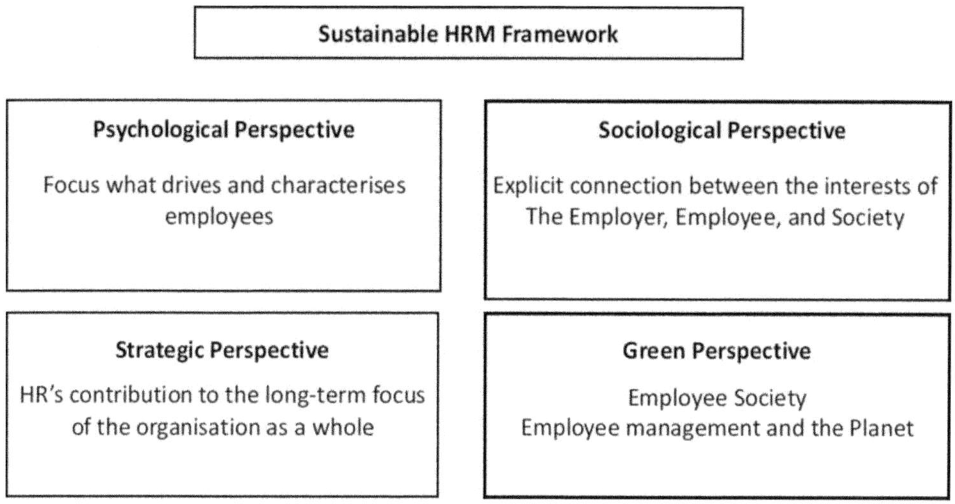

Fig. 7.1. Mazur's elaboration on the basis of De Prins *et al.*'s holistic model.

(Adapted from Mazur, 2017, p. 219)

HRM (Fig. 7.1) to provide a logical and coherent framing for the implementation of a sustainable HRM approach at firm level. Mazur's refinement responds to the significant stakeholder perspectives involved in approaches to sustainable HRM. This model can form a prism to reflect both the current state of tourism workforce research and what is required for a realistic approach to the implementation of its principles. However, we make one vital philosophical change; we remove 'firm success' as the overarching goal in this holistic approach. In its place, we suggest that green principles, based on moral rather that CRS arguments, should drive HRM policies and practices as recommended by Renwick *et al.* (2013). Significant additional orientations required by customer-facing service organizations, such as 'expressive hospitality' characteristics, where an employee derives intrinsic satisfaction from offering service (Poulston, 2015), may be concurrently embedded in HRM practices and policies.

Mazur (2017) cites Rompa (2011) when enlarging on the four core elements of the holistic model for implementing sustainable HRM, which will be addressed in turn:

i. the sociological perspective;
ii. the psychological perspective;
iii. the strategic perspective; and
iv. the green perspective.

The sociological perspective

First, the sociological perspective encapsulates the objective of embedding sustainable HRM as an enduring and social practice to reflect the interests of diverse groups at societal level. In Mazur (2017) adaptation, three agencies are involved: the employee, the employer and society. We further adapt this model to address the interests of the four core groups we identified earlier: the state interest; the host community interests; the employees of the organization's interests; and the tourists. Batle *et al.*'s (2018) social innovation approach indicates the benefits of this orientation. De Prins *et al.* (2014) advocate that to make sustainable HRM a more social practice it is necessary to incorporate insights from all disciplines.

We recommend the application of innovative methodologies and theoretical frameworks to ensure that the mutual interests of the four sections, including individuals belonging to non-dominant groups in society, are considered (Calás *et al.*, 2014). Therefore, linking back to our state and community vectors, organization-based interdisciplinary research in areas such as the unrealized employment opportunities for older workers (Jenkins and Poulston, 2014) could usefully inform policy at a local and national level. Likewise, when investigating questions on how society can best respond to the unprecedented numbers of refugees and economic migrants in many different societal contexts, we echo the call of economic geographers (Crawley and Skleparis, 2018) for post-positivist theoretical framings that allow the voices of asylum seekers and refugees to inform government policy, especially regarding visa application processes that facilitate employment opportunities and assimilation. Rydzik *et al.*'s (2012) research shows that tourism work leads to speedier social integration for new immigrants. To this end, collaboration between tourism researchers, refugee organizations and city authorities, as exemplified by partnerships between the Hilton hotel chain and Munich city authorities, to examine workplace barriers for refugees, can lead to positive outcomes for individuals and organizations.

The psychological perspective

Secondly, in Mazur (2017, p. 220) model, the psychological perspective addresses what employees consider relevant to drive a sustainable competitive advantage, because employees 'act in time, seek intention and have a soul'. Work-life balance, autonomy, employability, dialogue and worker participation are all important, and are all elements currently lacking in the tourism sector (Deery and Jago, 2015). There is increasing recognition that we have much to learn about how the motivations of individual employees contribute to organizational success (Madera *et al.*, 2017). In tourism businesses, this is even more crucial as responsive front-line employees are integral to the success of a business.

Critical paradigms, for example, intersectionality, that were developed to further social justice aims (Crenshaw, 2011) thus form a good philosophical fit with the 'connected' sustainable HRM approach. Intersectional theorizing is a valuable tool to examine the ways employees interact with and respond to HRM processes, according to different – sometimes competing – aspects of their intersecting identities (McBride *et al.*, 2015). The discrete effects and outcomes of HRM practices on diverse individuals and groups are not always considered; age (youth vs ageing) or gender (heteronormative vs LGBT+ orientation) effects are not experienced separately, or in an additive way, but simultaneously (Holvino, 2010). Yet youth, ageing, gender and/or sexual orientation will moderate what factors employees consider are important for their work-life balance. Here, diversity management studies are required with a critical orientation that goes beyond 'celebrating diversity' to present equal opportunities and equitable rewards to all employees.

The strategic perspective

Thirdly, the strategic perspective addresses the connectedness between strategic and sustainable HRM. Ehnert *et al.* (2016, p. 103) suggest that sustainable HRM research could explore how HRM and sustainability departments respond to the challenges of having to act 'in concert' on managing people in organizations and how this impacts the performance of HRM roles. This concern is reiterated by many of the researchers cited previously (for example, Boxall, 2018). In a sense, however, this is the perennial problem of evaluating the effectiveness of the HRM function. Again, more interdisciplinary research is required to uncover some of the underlying causes of employee behaviours and the effect of specific HRM policies on firm performance. Kalargyrou and Costen (2017) argue that certain areas are neglected in tourism research; for example, research on the combining of multi-generations in the workforce is not generally available. Therefore, researchers can usefully respond to existing calls in the HRM and management literature for more robust studies that combine quantitative and qualitative frameworks capable of clarifying the effects of individual HRM strategies on individuals and organizations (Legrand *et al.*, 2016; Boxall, 2018).

Although evidence supports our recommendation of Jithendran and Baum's (2000) updated HRM development approach, further studies are required to test its efficacy in diverse environments. As Baum *et al.* (2016b, p. 18) trenchantly observed when suggesting a new typology for tourism workforce research: there is a need to move beyond 'a "problem solving" managerial perspective on workforce research and seek to engage with explanation'. Madera *et al.*'s (2017) research highlights that in the HRM literature, although high impact work systems (HIWS) have attracted a great deal of academic interest (in excess of 300 articles), the processes by which HRM practices influence organizational performance are not clear. In response to growing research interest in examining factors that mediate the effect of HIWSs on performance, such as human and social capital and motivation, they suggest the integration of functional and strategic domains, moving from a focus on the individual to the effects of HRM strategies at organizational level.

Madera *et al.* (2017) emphasize that previous studies prove conclusively that enlightened HRM processes and practices, such as talent attraction and management strategies, effective employee socialization practices and equitable reward systems, benefit customer-facing tourism organizations. Regrettably, as we have indicated, these studies have, however, failed to modify the high-turnover culture of the sector. New research directions may produce more convincing research results to enable the development of an organizational culture that encourages innovation and agility when responding to seismic changes in competitive environments, for example, the rise of Uber and Airbnb (Dredge and Gyimóthy, 2015). We propose that Madera *et al.*'s (2017) recommendatioans credibly inform the research agenda for sustainable HRM. Studies which depart from the vague surveys of front-line managers' attitudes to HRM could instead focus on the identification of line managers' core accountabilities for employee recruitment, retention and reward and test for their efficacy. The incorporation of green HRM strategies can address these organizational-level concerns.

The green HRM perspective

The fourth focus of Mazur (2017) is the green HRM perspective based on Renwick *et al.*'s (2013) model of green HRM activities. The literature on sustainable HRM recognizes the paradoxes involved in implementing the approach (Guerci and Carollo, 2016). However, both De Prins *et al.* (2014) and Mazur (2017) use Renwick *et al.*'s outline when discussing how green principles can be imbedded in the core HR managerial functions, attracting and selecting 'green' employees. They are motivating employees through performance management and appraisals, targeted pay and reward systems, and by introducing a system that encourages employee voice and participation.

Green HRM processes address sectoral HRM challenges identified by Burke (2018), namely: being selective in staffing; offering competitive fair pay; providing empathetic supervision, job enrichment, consistent training and development, employee empowerment and participation; and improving current socialization and onboarding processes. Of significance when ensuring that all employees access the opportunity to develop their latent potential, De Prins *et al.*'s (2014) approach identifies weaknesses with approaches to talent development theory and thus recommends making career development a core constituent in their 'green' approach. Some of de Prins *et al.*'s concerns echo enduring concerns of critical management researchers. For example, Thunnissen *et al.* (2013) critique the focus on high achievers found in much of the talent management literature, arguing as Boxall (2018) does for a pluralist approach that takes the interests of all employees, not solely organizational goals, into account.

A significant problem with implementing a green HRM approach is that, as Solnet *et al.* (2015) reveal, in some locales, such as the Asia Pacific region, the HRM function is now outsourced to outside agencies. Therefore, rather than the HRM manager implementing sustainability initiatives, front-line managers will be dealing with the 'grey areas' and the ambiguities of sustainability. Solnet *et al.*'s (2015) conclusion expresses grave reservations about the ability of line managers to apply the required 'soft skills' in the dynamic service sector. It is unrealistic to imagine that line managers can effectively be charged with any further HRM responsibilities without specialized training and development. In sustainable approaches, the HRM function plays an essential, central role. Without a committed HR manager who will assume responsibility for embedding and monitoring sustainable practices, tourism organizations will be unable to adopt a sustainable HRM approach.

The Focus of Future Sustainable HRM Research in Tourism

We advocate for research that focuses on how HRM can usefully deal with the sustainable HRM paradoxes in sectors where the HRM function has been relegated back to its personnel administration roots (Boxall and Purcell, 2003). Sustainable HRM processes are particularly appropriate for the tourism sector's scope of business. For example, managing external supply chains in an ethical way is enormously important to global businesses such as hotels. As franchise arrangements appear to be implicated in abuses of migrant workers' rights (Stringer, 2016), research into the design and implementation of robust hotel supply chain arrangements is urgently required. Likewise, as is increasingly demanded by ethical consumers, animal and environmental abuses must be verified as absent from any part of the supply chain.

There are basic imperatives to be addressed before sustainable HRM is introduced at the level of the firm. Madera *et al.*'s (2017) core recommendations for the sector concur with Kusluvan *et al.*'s (2010) earlier calls for improvements in how workforce interests were managed; the advanced conceptual models routinely applied in other sectors must inform tourism organizational research. In particular, multi-level studies, as recommended by Baum *et al.* (2016a) and Boxall (2018), are required that specifically examine the effect of targeted actions on organizational success, and measure the impact on individuals within the organization, taking the wider social and economic context into account. Finally, tourism studies will need to test the effect of strategy outcomes beyond customer satisfaction by examining how targeted initiatives to improve employee well-being affect financial outcomes, in addition to demonstrable employee satisfaction. Added to these aspirations is the urgent necessity for researchers to measure the costs and benefits of discreet sustainability initiatives in real business settings, for example restaurants and airlines. While hotels have long focused on reducing energy consumption and trumpeted about savings in this regard, they have provided little proof of the strategic impact of employees' motivation on sustainability, yet employee involvement is crucial if any sustainability targets are to be researched (Renwick *et al.*, 2013; De Prins *et al.*, 2014; Guerci and Carollo, 2016).

Conclusions

In conclusion, we build from the premise that Renwick *et al.*'s (2013) approach to sustainable HRM, incorporated in green HRM practices, is founded on an ethical or moral perspective. To reprise the Brundtland Commission report (WCED, 1987), the goal of economic success – at all costs – has had punitive effects on our ecological environment and societal well-being, in its most extreme forms creating a privileged managerial elite and a deprived precarious underclass of (underemployed) workers. These class divisions, which disadvantage women and migrants as well as younger workers in diverse tourism organizations, are marked and visible. It would be an obvious, yet somewhat superficial, assessment of contemporary tourism research into sustainable HRM to conclude that the sector is beyond rescue.

The current situation is, as we observed in the introduction section, bleak, with many examples of poor, non-existent or unenlightened HRM policies or practices visible across the fragmented sector. Worker voice has hitherto been ignored and absent from the debate on sustainable tourism. Notwithstanding, there is cause for a cautious optimism. As the emerging critical tourism sustainability literature clearly identifies, no sector exists independently of its economic and social context. Increasingly, community-based activists and indigenous communities are joining forces to demand greater control over their natural environment. Tourism researchers are joining forces with town planners to examine and offer solutions to the impact of tourism on their environments (Blanco-Romero *et al.*, 2018). Increasingly, the influence of young people concerned about the damage that may be wrought on their physical environment and on the social fabric of society is being voiced through social media campaigns promoted by activist groups such as Avaaz (Neslen, 2018); although, as we have explained, the corporate social responsibility route appears to be the one most likely to be adopted by tourism organizations and such strategies are designed to enhance firm marketing and financial performance. An organization's CSR actions are highly visible and, once again, increasingly organizations are accountable for their actions as their sophisticated social media-savvy and fickle consumers demand greater transparency and accountability. The damage suffered by organizations seen to act in an unethical fashion are real, significant and enduring.

In research terms, even in the notoriously conservative tourism academy, we believe we may be witnessing the beginnings of the critical research and paradigm shift required to achieve the objectives of sustainable HRM, which will benefit society and the natural environment. We have illustrated the clamour of calls for more robust and critical research methods that cross disciplinary silos to measure targeted sustainable HRM initiatives in a variety of forums. Crucial to this call is the recognition that researchers must find new and better ways to engage with the communities and sectors on whose behalf they labour. The success of this two-way communication is the key to sustainable HRM, even if at times the critical academic voice is not welcomed by organizations who remain purely focused on delivering financial returns to one set of stakeholders, their owners. It is our duty and our honour as researchers to ensure meaningful engagement becomes a priority.

References

Adams, S.-A. (2017) Sustainable cruise ship employment for the Caribbean region – a conceptual perspective. *Worldwide Hospitality and Tourism Themes* 9(1), 60–69. DOI: 10.1108/WHATT-11-2016-0060.

Alcaraz, J., Susaeta, L., Suarez, E., Colón, C., Gutiérrez-Martínez, I. *et al.* (2017) The human resources management contribution to social responsibility and environmental sustainability: explorations from Ibero-America. *The International Journal of Human Resource Management* (Advance online publication).

Avery, G. (2005) *Leadership for Sustainable Futures: Achieving Success in a Competitive World*. Edward Elgar Publishing, Cheltenham, UK.

Ayres, H. (2006) Career development in tourism and leisure: an exploratory study of the influence of mobility and mentoring. *Journal of Hospitality and Tourism Management* 13(2), 113–123. DOI: 10.1375/jhtm.13.2.113.

Batle, J., Orfila-Sintes, F. and Moon, C.J. (2018) Environmental management best practices: towards social innovation. *International Journal of Hospitality Management* 69, 14–20. DOI: 10.1016/j.ijhm.2017.10.013.

Baum, T. (2015) Human resources in tourism: still waiting for change? – a 2015 reprise. *Tourism Management* 50, 204–212. DOI: 10.1016/j.tourman.2015.02.001.

Baum, T. (2018) Sustainable human resource management as a driver in tourism policy and planning: a serious sin of omission? *Journal of Sustainable Tourism* 26(6), 873–889. DOI: 10.1080/09669582.2017.1423318.

Baum, T., Cheung, C., Kong, H., Kralj, A., Mooney, S. *et al.* (2016a) Sustainability and the tourism and hospitality workforce: a thematic analysis. *Sustainability* 8(8), 809. DOI: 10.3390/su8080809.

Baum, T., Kralj, A., Robinson, R.N.S. and Solnet, D.J. (2016b) Tourism workforce research: a review, taxonomy and agenda. *Annals of Tourism Research* 60, 1–22. DOI: 10.1016/j.annals.2016.04.003.

Berg, L. and Farbenblum, B. (2017) *Wage theft in Australia: Findings of the National Temporary Migrant Work Survey*. Migrant Worker Justice Initiative, Sydney, Australia, p. 1056.

Blanco-Romero, A., Blázquez-Salom, M. and Cànoves, G. (2018) Barcelona: housing rent bubble in a tourist city. Social responses and local policies. *Sustainability* 10(6), 2043. DOI: 10.3390/su10062043.

Boström, M. (2012) A missing pillar? Challenges in theorizing and practicing social sustainability: introduction to the special issue. *Sustainability: Science, Practice and Policy* 8(1), 3–14.

Boxall, P. (2018) The development of strategic HRM: reflections on a 30-year journey. *Labour & Industry: A Journal of the Social and Economic Relations of Work* 28(1), 21–30. DOI: 10.1080/10301763.2018.1427423.

Boxall, P. and Purcell, J. (2003) Strategic HRM 'best fit' or 'best practice'. In: Burrell, G., Marchington, M. and Thompson, P. (eds) *Strategy and Human Resource Management*.Palgrave Macmillan, London, pp. 47–70.

Bratton, J. and Gold, J. (2015) Towards critical human resource management education (CHRME): a sociological imagination approach. *Work, Employment and Society* 29(3), 496–507. DOI: 10.1177/0950017014545266.

Burke, R. (2018) Human resource management in the hospitality and tourism sector. In: Burke, R. and Christensen Hughes, J. (eds) *Handbook of Hospitality Human Resource Management*. Edward Elgar Publishing, Cheltenham, pp. 3–39.

Calás, M., Smircich, L. and Holvino, E. (2014) Theorizing gender-and-organization: changing times … changing theories. In: Kumra, S., Simpson, R. and Burke, R.J. (eds) *The Oxford Handbook of Gender in Organizations*. Oxford University Press, Oxford, UK, pp. 17–52.

Cole, S. (2017) Water worries: an intersectional feminist political ecology of tourism and water in Labuan Bajo, Indonesia. *Annals of Tourism Research* 67, 14–24. DOI: 10.1016/j.annals.2017.07.018.

Coyle-Shapiro, J. and Kessler, I. (2000) Consequences of the psychological contract for the employment relationship: a large scale survey. *Journal of Management Studies* 37(7), 903–930. DOI: 10.1111/1467-6486.00210.

Crawley, H. and Skleparis, D. (2018) Refugees, migrants, neither, both: categorical fetishism and the politics of bounding in Europe's 'migration crisis'. *Journal of Ethnic and Migration Studies* 44(1), 48–64. DOI: 10.1080/1369183X.2017.1348224.

Crenshaw, K.W. (2011) From private violence to mass incarceration: thinking intersectionally about women, race, and social control. *UCLA Law Review* 59, 1418.

Dagsland, Å.-H., Mykletun, R. and Einarsen, S. (2015) 'We're not slaves – we are actually the future!: A follow-up study of apprentices' experiences in the Norwegian hospitality industry. *Journal of Vocational Education & Training* 67(4), 1–22.

Davidson, M.C.G. and Wang, Y. (2011) Sustainable labor practices? Hotel human resource managers' views on turnover and skill shortages. *Journal of Human Resources in Hospitality & Tourism* 10(3), 235–253. DOI: 10.1080/15332845.2011.555731.

De Beer, A., Rogerson, C.M. and Rogerson, J.M. (2014) Decent work in the South African tourism industry: evidence from tourist guides. *Urban Forum* 25(1), 89–103. DOI: 10.1007/s12132-013-9199-8.

de Kadt, E. (1979) *Tourism: Passport to Development*. Oxford University Press, Oxford, UK.

De Prins, P.D., Beirendonck, L.V., Vos, A.D. and Segers, J. (2014) Sustainable HRM: Bridging theory and practice through the 'Respect Openness Continuity (ROC)' model. *Management Revue* 25(4), 263–284. DOI: 10.5771/0935-9915-2014-4-263.

Deery, M. and Jago, L. (2015) Revisiting talent management, work-life balance and retention strategies. *International Journal of Contemporary Hospitality Management* 27(3), 453–472. DOI: 10.1108/IJCHM-12-2013-0538.

Dredge, D. and Gyimóthy, S. (2015) The collaborative economy and tourism: critical perspectives, questionable claims and silenced voices. *Tourism Recreation Research* 40(3), 286–302. DOI: 10.1080/02508281.2015.1086076.

Dron, M.-D., Müller-Camen, M. and Obereder, L. (2018) Green HRM. In: Covarrubias Venegas, B., Thill, K. and Domnanovich, J. (eds) *Personal management Wiesbaden*. Springer Fachmedien Wiesbaden, Germany, pp. 41–55.

Dyllick, T. and Hockerts, K. (2002) Beyond the business case for corporate sustainability. *Business Strategy and the Environment* 11(2), 130–141. DOI: 10.1002/bse.323.

Ehnert, I. (2006) Sustainability issues in human resource management: Linkages, theoretical approaches, and outlines for an emerging field. *Paper presented at the 21st EIASM SHRM Workshop*, Aston, Birmingham, UK, 28-29 March.

Ehnert, I. (2009) *Sustainable Human Resource Management*. Springer, Berlin.

Ehnert, I., Parsa, S., Roper, I., Wagner, M. and Muller-Camen, M. (2016) Reporting on sustainability and HRM: a comparative study of sustainability reporting practices by the world's largest companies. *The International Journal of Human Resource Management* 27(1), 88–108. DOI: 10.1080/09585192.2015.1024157.

Furunes, T. and Mykletun, R.J. (2007) Why diversity management fails: metaphor analyses unveil manager attitudes. *International Journal of Hospitality Management* 26(4), 974–990. DOI: 10.1016/j.ijhm.2006.12.003.

Gorenak, M. (2016) Sustainable development of careers in tourism. *Teorija in Praksa* 53(4), 959–975.

Guerci, M. and Carollo, L. (2016) A paradox view on green human resource management: insights from the Italian context. *The International Journal of Human Resource Management* 27(2), 212–238. DOI: 10.1080/09585192.2015.1033641.

Hargreaves, A. and Fink, D. (2006) *Sustainability Leadership*. Jossey-Bass, San Francisco, California.

Hawke, A. and Raphael, A. (2016) *ECPAT Report: The Global Study on Sexual Exploitation of Children in Travel and Tourism*. ECPAT International and Defence for Children, Netherlands, Bangkok, Thailand.

Higgins-Desbiolles, F. and Powys Whyte, K. (2014) Critical perspectives on tourism. In: Lew, A., Hall, M. and Williams, A. (eds) *The Wiley Blackwell Companion to Tourism*. John Wiley & Sons, Chichester, UK, pp. 88–98.

Hinkin, T.R. and Tracey, J.B. (2000) The cost of turnover. *Cornell Hotel and Restaurant Administration Quarterly* 41(3), 14–21. DOI: 10.1177/001088040004100313.

Holvino, E. (2010) Intersections: the simultaneity of race, gender and class in organization studies. *Gender, Work & Organization* 17(3), 248–277. DOI: 10.1111/j.1468-0432.2008.00400.x.

Howell, C. (2016) Regulating class in the neoliberal era: the role of the state in the restructuring of work and employment relations. *Work, Employment and Society* 30(4), 573–589. DOI: 10.1177/0950017015595954.

ILO (2016) Decent work. International Labour Organization. Available at: http://www.ilo.org/global/topics/decent-work/lang--en/index.htm (accessed 13 April 2016).

Janta, H., Brown, L., Lugosi, P. and Ladkin, A. (2011) Migrant relationships and tourism employment. *Annals of Tourism Research* 38(4), 1322–1343. DOI: 10.1016/j.annals.2011.03.004.

Jenkins, C.L. (2015) Tourism policy and planning for developing countries: some critical issues. *Tourism Recreation Research* 40(2), 144–156. DOI: 10.1080/02508281.2015.1045363.

Jenkins, A. and Poulston, J. (2014) Managers' perceptions of older workers in British hotels. *Equality, Diversity and Inclusion: An International Journal* 33(1), 54–72. DOI: 10.1108/EDI-10-2012-0096.

Jithendran, K.J. and Baum, T. (2000) Human resources development and sustainability? The case of Indian tourism. *International Journal of Tourism Research* 2(6), 403–421. DOI: 10.1002/1522-1970(200011/12)2:6<403::AID-JTR239>3.0.CO;2-3.

Kalargyrou, V. and Costen, W. (2017) Diversity management research in hospitality and tourism: past, present and future. *International Journal of Contemporary Hospitality Management* 29(1), 68–114. DOI: 10.1108/IJCHM-09-2015-0470.

Knox, A. and Walsh, J. (2005) Organisational flexibility and HRM in the hotel industry: evidence from Australia. *Human Resource Management Journal* 15(1), 57–75. DOI: 10.1111/j.1748-8583.2005.tb00140.x.

Kramar, R. (2014) Beyond strategic human resource management: Is sustainable human resource management the next approach? *The International Journal of Human Resource Management* 25(8), 1069–1089. DOI: 10.1080/09585192.2013.816863.

Kusluvan, S., Kusluvan, Z., Ilhan, I. and Buyruk, L. (2010) The human dimension: a review of human resources management issues in the tourism and hospitality industry. *Cornell Hospitality Quarterly* 51(2), 171–214.

Lashley, C. (2016) Corporate social responsibility – emancipating wage slaves. *Hospitality & Society* 6(1), 3–7. DOI: 10.1386/hosp.6.1.3_2.

Lee, J.-S., Back, K.-J. and Chan, E. (2015) Quality of work life and job satisfaction among frontline hotel employees: a self-determination and need satisfaction theory approach. *International Journal of Contemporary Hospitality Management* 27(5), 768–789.

Legrand, W., Sloan, P. and Chen, J.S. (2016) *Sustainability in the Hospitality Industry: Principles of Sustainable Operations*. Routledge, Abingdon, UK.

MacNeill, T. and Wozniak, D. (2018) The economic, social, and environmental impacts of cruise tourism. *Tourism Management* 66, 387–404. DOI: 10.1016/j.tourman.2017.11.002.

Madera, J., Dawson, M., Guchait, P. and Belarmino, A. (2017) Strategic human resources management research in hospitality and tourism: a review of current literature and suggestions for the future. *International Journal of Contemporary Hospitality Management* 29(1), 48–67.

Manoharan, A. and Singal, M. (2017) A systematic literature review of research on diversity and diversity management in the hospitality literature. *International Journal of Hospitality Management* 66, 77–91. DOI: 10.1016/j.ijhm.2017.07.002.

Marco-Lajara, B. and Úbeda-García, M. (2013) Human resource management approaches in Spanish hotels: an introductory analysis. *International Journal of Hospitality Management* 35, 339–347. DOI: 10.1016/j.ijhm.2013.07.006.

Marsh, E. (2012) The effects of cruise ship tourism in coastal heritage cities: a case study of Charleston, South Carolina. *Journal of Cultural Heritage Management and Sustainable Development* 2(2), 190–199.

Mazur, B. (2017) Sustainable human resource management and its models. *International Journal of Contemporary Management* 16(3), 211–223.

McBride, A., Hebson, G. and Holgate, J. (2015) Intersectionality: are we taking enough notice in the field of work and employment relations? *Work, Employment and Society* 29(2), 331–341. DOI: 10.1177/0950017014538337.

McDowell, L., Batnitzky, A. and Dyer, S. (2009) Precarious work and economic migration: emerging immigrant divisions of labour in greater London's service sector. *International Journal of Urban and Regional Research* 33(1), 3–25. DOI: 10.1111/j.1468-2427.2009.00831.x.

Miller, G., Rathouse, K., Scarles, C., Holmes, K. and Tribe, J. (2010) Public understanding of sustainable tourism. *Annals of Tourism Research* 37(3), 627–645. DOI: 10.1016/j.annals.2009.12.002.

Mohsin, A. and Lengler, J. (2015) Exploring the antecedents of staff turnover within the fast-food industry: the case of Hamilton, New Zealand. *Journal of Human Resources in Hospitality & Tourism* 14(1), 1–24. DOI: 10.1080/15332845.2014.904169.

Mooney, S.K., Harris, C. and Ryan, I. (2016) Long hospitality careers – a contradiction in terms? *International Journal of Contemporary Hospitality Management* 28(11), 2589–2608. DOI: 10.1108/IJCHM-04-2015-0206.

Mooney, S., Ryan, I. and Harris, C. (2017) The intersections of gender with age and ethnicity in hotel careers: still the same old privileges? *Gender, Work & Organization* 24(4), 360–375. DOI: 10.1111/gwao.12169.

Mumford, M.D. and Fried, Y. (2014) Give them what they want or give them what they need? ideology in the study of leadership. *Journal of Organizational Behavior* 35(5), 622–634. DOI: 10.1002/job.1921.

Murray, W. (2018) Motivational fluidity: exploring the preferred job rewards of of Canadian hotel workers. In: *Proceedings of the Council for Hospitality Management Education (CHME).* Bournemouth University, Bournemouth, UK.

Neslen, A. (2018) Monsanto demands Avaaz hands over all of its campaign data. *The Guardian*, 23 February. Available at: http://www.theguardian.com/environment/2018/feb/23/monsanto-demands-avaaz-hands-over-all-of-its-campaign-data (accessed 3 April 2020).

Orido, C. (2017) *Challenges faced by female chefs in the Kenyan hospitality industry: A study through an African oral tradition of storytelling*. Master's thesis. Auckland University of Technology, New Zealand.

Poulston, J. (2008) Hospitality workplace problems and poor training: a close relationship. *International Journal of Contemporary Hospitality Management* 20(4), 412–427. DOI: 10.1108/09596110810873525.

Poulston, J. (2015) Expressive labour and the gift of hospitality. *Hospitality & Society* 5(2), 145–165. DOI: 10.1386/hosp.5.2-3.145_1.

Renwick, D.W.S., Redman, T. and Maguire, S. (2013) Green human resource management: a review and research agenda. *International Journal of Management Reviews* 15(1), 1–14. DOI: 10.1111/j.1468-2370.2011.00328.x.

Rimanoczy, I. and Pearson, T. (2010) Role of HR in the new world of sustainability. *Industrial and Commercial Training* 42(1), 11–17. DOI: 10.1108/00197851011013661.

RMT (2018) Ireland-UK unions join forces on Irish Sea Ferries. National Union of Rail, Maritime and Transport Workers. Available at: https://www.rmt.org.uk/news/ireland-uk-unions-join-forces-on-irish-sea-ferries/ (accessed 2 April 2020).

Rompa, I. (2011) Explorative research on sustainable human resource management. Master's thesis, VU University, Amsterdam. Available at: http://www.innovatiefinwerk.nl/sites/./sustainable_hrm.pdf (accessed 30 June 2018).

Rydzik, A., Pritchard, A., Morgan, N. and Sedgley, D. (2012) Mobility, migration and hospitality employment: voices of central and eastern European women. *Hospitality & Society* 2(2), 137–157. DOI: 10.1386/hosp.2.2.137_1.

Santero-Sanchez, R., Segovia-Pérez, M., Castro-Nuñez, B., Figueroa-Domecq, C. and Talón-Ballestero, P. (2015) Gender differences in the hospitality industry: a job quality index. *Tourism Management* 51, 234–246. DOI: 10.1016/j.tourman.2015.05.025.

Searle, W., McLeod, K. and Stichbury, C. (2015) *Vulnerable Temporary Migrant Workers: Hospitality Industry*. Ministry of Business, Innovation and Employment, Wellington, New Zealand.

Siddiqui, K. (2012) Developing countries' experience with neoliberalism and globalisation. *Research in Applied Economics* 4(4), 12. DOI: 10.5296/rae.v4i4.2878.

Solnet, D., Kralj, A. and Baum, T. (2015) 360 degrees of pressure: the changing role of the HR professional in the hospitality industry. *Journal of Hospitality & Tourism Research* 39(2), 271–292.

Stansfield, M.L. (2018) Hospitality artisans and sustainability. *Hospitality Insights* 2(1), 7–9. DOI: 10.24135/hi.v2i1.29.

Stringer, C. (2016) *Worker Exploitation in New Zealand: A Troubling Landscape*. University of Auckland, Auckland, New Zealand.

Thunnissen, M., Boselie, P. and Fruytier, B. (2013) Talent management and the relevance of context: towards a pluralistic approach. *Human Resource Management Review* 23(4), 326–336. DOI: 10.1016/j.hrmr.2013.05.004.

Vallance, S., Perkins, H.C. and Dixon, J.E. (2011) What is social sustainability? A clarification of concepts. *Geoforum* 42(3), 342–348. DOI: 10.1016/j.geoforum.2011.01.002.

WCED (1987) *Our Common Future*. Oxford University Press, Oxford, UK.

Weaver, D. (2011) Can sustainable tourism survive climate change? *Journal of Sustainable Tourism* 19(1), 5–15. DOI: 10.1080/09669582.2010.536242.

Westermann, J., Vanka, S., Bhaskara Rao, M.B. and Gupta, M. (2017) *Sustainable HRM: Perspectives, practices, and prognoses*. Human Resource Management Review, Call for papers. Available at: https://www.journals.elsevier.com/human-resource-management-review/call-for-papers

WTTC (2018) *Evaluation of Job Creation in G20 countries*. World Travel and Tourism Council.

Zaugg, R.J., Blum, A. and Thom, N. (2001) *Sustainability in Human Resource Management: Evaluation Report*. Institute for Organisation and Human Resource Management (IOP) and Swiss Federal Office of Personnel (EPA), Bern, Switzerland.

8 Understanding Project Success or Failure

Stephen Wanhill*
Global Tourism Solutions (UK)

Tourism Project Environment

In frequent instances, it is not easy with tourism projects to define success or failure. Many investments, particularly for creating public space, attractions, entertainments and public transport, are often part of a wider development plan and do not have a conventional 'bottom line', since they commonly include both tangible and intangible capital and are frequently either not-for-profit or subsidized in some way. For example, where old industrial buildings, disused market halls, railway stations and docks are located close to urban centres, it has been quite popular to convert them into tourist zones as part of regeneration schemes which serve both visitors and residents alike. The base is often heritage and cultural activities through establishing museums and art galleries and arranging events (Brown and Geddes, 2007; Bracalente *et al.*, 2011; Viu *et al.*, 2008). However, since leisure shopping is an increasingly important visitor and resident activity, there has been a focus on specialty retailing, as in the case of the Faneuil Hall Marketplace, Boston (opened 1976) which, with over 15 million visitors per annum, is now one of the most popular tourist attractions in the USA. The same can be said for Covent Garden, Central London, which opened in 1980. Such developments may also be intermingled with hotels, further leisure attractions and business facilities, such as a convention centre, an exhibition hall or trade centre and offices, in order to attract commercial developers.

In this way, tourism has replaced manufacturing and distribution industries which have left their city locations for more spacious and cheaper sites on the outskirts. In so doing, it has become the 'glue' that holds the area together for urban regeneration, particularly where there is little else the local authority can do with such assets, as in the development of Baltimore's Inner Harbor. By the 1950s, Baltimore was suffering from the economic decline caused by changing business practices, something that was common to many industrial cities in the USA. Between 1958 and 1965, the waterfront was gradually transformed with award-winning parks and plazas surrounded by office buildings, hotels and leisure attractions, which reversed the city's decline and became a model for urban renaissance in harbour cities around the world. The object in most cases is to initiate public sector investment in order to stabilize economic 'shrinkage' by raising place image in order to attract private investment. For example, in 1981 the Merseyside Development Corporation (MDC) was created for the task of regenerating Liverpool's dock area in the UK. In 2003, the last remaining plot of the Albert Dock in Liverpool was brought into use with the opening of a Premier Inn hotel,

*stephen@wanhill.force9.co.uk

DOI: 10.1079/9781789242812.0008

demonstrating as in other instances the longevity of such grand schemes.

In the situation of Darling Harbour, Sydney, much of the land had been the site of the New South Wales Railways central marshalling yards and freight consolidation centre. By the mid-1980s, it had become largely derelict and the decision was made in 1984 to designate half the area for tourism businesses and residential developments, with the rest for open public space and urban renewal. Although opened officially in 1988, the development was not expected to be completed until 2020, though inevitably there is a dynamic in this as tastes change and businesses come and go.

Likewise, despite the enduring appeals of the seaside (Jarratt and Sharpley, 2017), the cold-water resorts of America and Europe have had to provide for additional leisure and business activities to counter the 'flight to the sun' by their traditional markets. However, many coastal towns do not have the resources to embark on long-term regeneration and continue to be marked as areas of economic deprivation (House of Lords, 2019).

Partnership

From the above, it may be deduced that the ownership of regeneration development normally results from a private–public sector partnership, if only because in the modern world the availability of sites for such land-extensive visitor complexes is limited and they are either owned by public authorities or are subject to strict planning controls. In this partnership, it is now thought best that, with a focus on developing revenue-earning activities, the governing body should act as facilitator rather than owner-operator by bringing the fragmented elements of the tourist industry together. The reason for this is that the rules of public accountability do not make government-run hotels, restaurants, attractions and entertainment facilities good performers. Thus, the commercial aspects of projects are generally in the hands of the private sector, with the rationale for public participation vested in the wider economic, social and environmental benefits that are bestowed. In this, the state has an important role in the provision of infrastructure, as tourism tends to follow infrastructure rather than lead it, hence the need for tourism planning.

Table 8.1. Reconciling private–public sector project selection.

Project classification	Commercial failure	Commercial success
Community success	QI Subsidize	QII Accept
Community failure	QIII Reject	QIV Reformulate

Table 8.1 indicates the general strategy for reconciling the public role in development with the commercial aspects of project selection. The significant quadrants in Table 8.1 are QI and QIV. Where the project bestows considerable public benefits, the rational choice for implementation is one of subsidy that is best achieved through capital grants (Wanhill, 1986). The cost structure of tourism projects makes them inherently risky; the dominant position is one of a high operating leverage, that is, a high ratio of fixed to variable costs, which results in businesses that are very sensitive to variations in demand. In most instances, this feature is the outcome of the relatively high initial capital investment and where much of this investment is borrowed, the principal concern of the project's owners usually becomes one of servicing the debt. Hence, capital grants act as substitute equity to mitigate risk.

QIV in Table 8.1 illustrates the flexible nature of tourism design, in that projects that do not meet public approval can be reformulated to meet local success criteria; for example, high-rise hotels in cities and low-rise in rural areas. Unlike, say, in manufacturing, there is no single best-practice technology for a given category of tourism projects, and the scope for innovative changes is considerable. However, for attractions this opportunity is often limited by the fact that the majority of today's attractions have not been brought into existence for visitor purposes and alterations in the cultural capital stock are, in the main, irreversible. The innovation process is, therefore, a graduation from a situation of no adaptation (but rather controlled management) to visitor attractions that are fashioned for purpose. The government sector is more likely to take on a stewardship role of what are considered to be national assets, where innovation is constrained, while private sector firms may be much more proactive because of the need to maintain contact with consumers and their status in the marketplace.

The new product or process that disrupts exiting competencies and creates new markets, collapsing old structures (Schumpeter, 1934), happens but rarely. Before the occurrence of such upheavals, it is more common to find cumulative changes in the product, the opening up of new markets and the gradual implementation of new ideas, so today it is customary to classify innovations into a variety of categories: marketing, organization, product, design, service, and process (Hjalager, 2010, 2015; Backman *et al.*, 2017; Divisekera and Nguyen, 2018a, b). Thus, the standard amusement park model lasted 62 years before Disney broke the mould at Anaheim in 1955, in so doing setting the agenda for the theme park developments that are so familiar today. Confounding its critics, the park brought in 3.8 million visitors in its first year, a figure that reached 13.9 million by the millennium (18.7 million in 2019) as the area became a fully-fledged resort.

Development Process

From a business standpoint, to reiterate the famous dictum about hotels that has been attributed to Conrad Hilton: 'There are only three things you need to know about hotels: location, location and location!' Location has general applicability and reflects the importance of real estate in developing tourist facilities, although Hilton (1994) makes it clear that he had a thorough understanding of the processes that were appropriate for the chosen location. The three aspects to consider in developing tourism are indicated on Fig. 8.1.

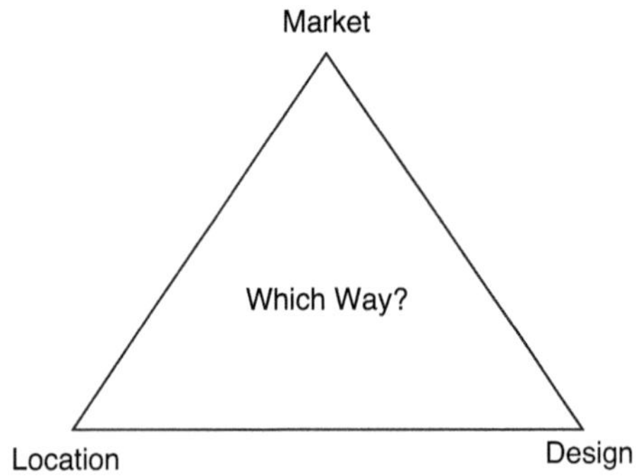

Fig. 8.1. Development process.

Commercial projects

Commercial logic dictates that the optimum path is Market > Design concept > Location, but this can only be applied to what are termed 'footloose' businesses within the different tourism industries – as defined by the United Nations World Tourism Organization (UNWTO, 2008) – that make up the total picture, which have flexibility across all three aspects in order to ensure economic success. 'Accommodation', 'Restaurants and similar' and 'Travel, tour and guide services' largely possess this ability, subject to consent of site use by the planning authorities. Conversely, 'Cultural, sporting and recreational activities' are more restricted given the fact that many attraction developments are already proscribed by their nature and location. Thus, a mining museum that has a mine visit as part of the attraction is tied by its location (Wanhill, 2000). This brings the debate back to ownership patterns and the organization's objectives because once the commercial pathway to development is abandoned, financial viability may be difficult to attain.

In contrast, businesses that can present subjects that are the focus of attention in a broad range of ways, so that it is only a matter of choosing the appropriate design concept and scale, have frequent opportunities. For example, Sea-Life Centres, which are modern aquariums, are not too sophisticated and require relatively small numbers to succeed, and so may be found in many coastal locations. A key aspect of such footloose attractions is that they are built for purpose, as in the case of urban leisure centres and amusement parks. The former are principally targeted at local residents, while the latter are seeking to maximize their visitor potential, which is functionally related to the population (including tourists) catchment area within a specified drive time of up to two hours for cars and three to four hours for coach, bus or train (Wanhill, 2013). For instance, the Japanese company, The Oriental Land Co. Ltd, which has 100% ownership of Disneyland Tokyo (opened 1983), was the first to acknowledge that their success was in large measure due to their location so close to Tokyo, giving a catchment area of 30 million people within one hour's drive time (Jones, 1994).

Site availability may limit commercial attraction developers to a second-best pathway that runs Design concept > Location > Market. This incurs the risk of 'talking up the market', as in the Disneyland Paris case where the site was offered at 1971 agricultural prices (d'Hauteserre, 1997). What is important here is to recognize that location, market assessment and design are bound to each other and the recursive process can become iterative over the development period of any project. Once the site is selected the development can come into place, but the further the location is away from the optimal market position, the more appealing and exciting has to be the design to 'pull' customers in. Alternatively, the market assessment may be changed, which in turn can affect the concept. Either way, the calculations for the feasibility study will need to be revised at this stage and most likely continuously to final completion stage.

With regard to passenger transport, modern cruise ships source factor inputs globally, are very flexible in terms of ports of call, and so may be considered a footloose product (Bull, 2013). However, land and air transport are tied to infrastructure that is commonly in the public domain and contentious to develop. Airports have a sizeable environmental footprint and obtaining planning permission for adding to capacity with a new runway, new build or expanding an existing airport can take many years and several government investigations. Even then, the final decision may not be the best one, as was the enquiry into selecting London's third airport (Hall, 1982a). It is felt to be no longer acceptable by many that the flight paths for take-off and landing in respect of today's passenger airliners should be over the centre of cities. Hence, South Korea downgraded its international airport at Seoul and built a new one on the coast at Inchon. Added to this is the fact that despite improvements, air transport is still not sustainable and green movements are proposing that it should not be used for short-haul journeys under 1000 km (625 miles); passengers should use bus or train instead, though even here there is environmental opposition.

Regeneration projects

The urban renewal model exemplified earlier typically follows a reverse pathway, Location > Design concept > Market. Where sites have been available within or on the edges of city boundaries, modern retail malls have been built to provide an entertainment experience for the whole family through adding to traditional shopping facilities themed areas and amusements, health clubs, cinemas, performance venues, restaurants and bars in an enclosed space, as in the West Edmonton Mall in Alberta, Canada (opened 1981, but last phase finished in 1999) and the Mall of America (opened 1992 and completed in 2015) close to Minneapolis–Saint Paul International Airport, USA. This type of development has spread from North America into Europe, the Middle East and Asia, where the malls have become noted visitor attractions in their own right, with high levels of repeat visits giving attendance levels for some in tens of millions.

By having a mixed development, regeneration schemes are spreading the risk, but where the weight has been focused on visitor attractions, the reverse can be the case in that they can result in an 'outside-in' project and/or project inflation due to 'optimism bias'. The former is used to describe an investment that goes from the physical structure to the design concept. Thus the structure of London's Millennium Dome (now successful as the O2 Arena) was finalized long before the content was known, so it had to be designed to give maximum exhibition space. The creativity of the different zones, and how they linked together, were not effectively communicated to the general public and allowed the media to satirize the project as ersatz and of no substance (NAO, 2000). Most visitor attractions that appear to be flourishing are 'inside-out' projects that they create the structure around the known design concept.

Where public authorities are involved, there is always the danger of project inflation in response to civic pride and the vainglory of local politicians. This results in an exaggeration of employment creation to obtain development grants, increased complexity, which boosts consultants' fees, and substantial capital structures to the benefit of the architects. Of the 12 millennium investments sponsored by the Heritage Lottery Fund in the UK, seven went this way and ended up in financial difficulties, whereas adjusting for optimism bias as the appraisal develops provides a more realistic assessment of the initial estimates of costs, benefits and the time taken to implement the project (Treasury, 2018).

In recent years, the growth of online shopping has encouraged the flight of general retail from urban centres and, in part, from large malls, which has forced local authorities to look further into providing leisure facilities as a means of urban renewal. In contrast, the under-regulated growth of the 'sharing' economy in the form of Airbnb in some important tourist cities has led to complaints by residents about so-called 'overtourism' owing to the intrusion of visitors into local neighbourhoods, as highlighted in the case of Barcelona (Milano, 2017). This not only defeats the object of planning tourist zones, but it also creates negative stereotypes about tourist behaviour that can be spread rapidly around the world via today's social media. It is not difficult today to get the impression that a country's cultural and environmental heritage is now just a backdrop for taking a selfie.

Industrial projects

Industrial visitor attractions tend to pursue the path of Location > Market > Design concept. In the first instance, natural association with the place of production dictates the location, as, for example, Cadbury World, Bournville, Birmingham, UK (opened 1990), Universal Studios, Hollywood (opened 1964), Ford at Dearborn, USA (opened 1999) and Glenturret Whisky, Crieff, Perthshire, Scotland (opened 1980). The attraction is a case of reverse product development, as it is a created experience built on consumer interest in the final output of the company and its history. This may be termed 'brand stretching' which is 'reproductive' of the product so that visitors know what to expect, unlike the more 'avant-garde' ideas presented in London's Millennium Dome. There is similarity here with chain hotels and restaurants that offer a defined standard that the public can recognize and judge in terms of value for money, which is amplified by online reviews from websites such as Tripadvisor.

Table 8.2. Evaluating the design concept.

	Design concept Current/New/Future	
Market		
Current	QI 'Me Too' project	QII 'Great Idea' project
New/Future	QIII 'New Version' project	QIV 'Mega' project

Design Concept – Market Mix

Experience shows that flexibility in technology and design within tourism allows similar investments to exist at different positioning levels within the market place, ranked according to the experience they offer to their customers. This is shown in Table 8.2, where it is proposed that by positioning projects in terms of the above matrix, much can be understood about the success criteria, because many of the elements that would be included in a feasibility study are subsumed within it.

'Me Too' is the most common way in which new investments arise and is a consequence of the spatial distribution of demand, competition and the fact that the tourism product is consumed at the place of production. Hence, it is common for chain hotels to establish capital-city presence to support their brand and give managers large city experience. Projects in QI involve least risk, because it is possible to look at parallel projects to obtain reliable operating data to combine with overall market trends to see what the market absorption capacity is likely to be. Innovation is minimal here as it is a matter of replicating an accepted concept that has already demonstrated its potential worth. Such developments may not get public-sector support because of questions over displacement of demand from other enterprises and the inherent danger of competitive replication saturating the market.

One of the difficulties of evaluating investments in the QII is weighing up whether the 'Great Idea' will work in relation to the market or whether it is simply the 'single genius' approach to project development, which could be an indulgence that is unrealistically costly in the context of what the market can afford. Such schemes are difficult to evaluate as there is no past track record, and so organizations have to rely on test marketing. In some cases, this is relatively easy, such as the launching of a new holiday package by a major tour operator; in other cases, however, it may involve considerable prior investment without really knowing whether the 'opening' will be a success. A common evaluation strategy here is to reverse the project evaluation sequence by estimating the volume of customers needed to make the project both feasible and viable at a price the market is prepared to pay. Alternatively, the great idea could be a prestige investment, for instance the redevelopment of Adare Manor, Republic of Ireland (opened 2018) as a luxury country house hotel, where the owner wrote off some €250–300 million of capital.

The development process in QIII may be the opening up of new market opportunities while preserving the existing concept in content and format, rejuvenating the existing product because the current public has become too familiar with it and the market has moved on, or a combination of the two. The spatial division of markets can be important and, thus, old concepts can work in new destinations while new concepts are needed to move forward in established destinations. An example of this, other than the well-known expansion of international hotel chains through ownership, franchise or management contracts (Dev *et al.*, 2007), is the globalization of Legoland from its home in Billund, Denmark (opened 1968) to New York in 2020. Risk is low to medium provided a thorough market assessment is made to ensure cultural adaptation.

'Mega' projects are those very large investments that have major economic impacts on their location and are eagerly sought after as 'flagship' enterprises, since they create a raft of new jobs and skills. Maximum uncertainty holds in QIV because of the number of unknowns and the scale of the project. Such projects can only be funded by very large multi-national operators, the government sector or a partnership of the two; proceeding with public sector support so as to spread the risks and help draw in external finance. For example, 80% of the investment funds for the Hong Kong Disneyland (opened 2005) came from the Hong Kong governing authority.

Market assessment for mega projects can be notoriously difficult because developers are moving into 'uncharted waters'. There is little data to hand and surveys are often inadequate as the market has no experience of the product. Developing alternative project scenarios so as to give a thorough understanding of what is being proposed and the risks involved are more important than the actual projections, though the latter are required to give dimensions to the project and to assess its impact on the economy. Thus, estimates of visitor numbers for the Millennium Dome in London ranged from 8–17 million; 12 million was the figure that the government was prepared to accept and budget for, on the basis that it was meant to be a public festival, so that everyone who might want to come should be able to do so. The 12 million was a design standard and not a market forecast, but was determined in the press as a target number which was then set against the achieved result of 6.5 million as the basis for judging the success or failure of the project.

Project Realization

Table 8.3 identifies in broad terms the possible range of outcomes when the design concept is examined against project management to viability stage, which includes timings, finance and the capacity of the organization to deliver their

Table 8.3. Project realization.

Management plan	Design concept Clear/Unclear	
Clear	QI Successful projects	QII Completed project failures
Unclear	QIII Seemingly never-ending projects	QIV Abandoned projects

mission. The objective is to move away from the uncertainty that surrounds failure and the abandonment of a project in QIV to a clear concept and planning process against which to judge success, QI.

The remaining two quadrants in the table are associated with organizational behaviour, which shields the project from being judged a failure by changing the situation of the design concept. Governments, local authorities and like-minded not-for-profit organizations are often the main culprits here, though major corporations (who do have a bottom line) can be equally guilty. From the public-sector perspective. the confusing influences that give rise to these difficulties come from masking political decisions as economic ones for acceptability and pandering to the press, giving rise to cost overruns and/or optimism bias.

Successful projects

Given the varied nature of tourism projects, both in type and ownership, it is difficult to be exact concerning what constitutes success but, given a clear concept and planning process as a necessary condition of success, such projects:

- demonstrate creative but realistic vision;
- show awareness of trends;
- know the scale of development;
- meet stated and honestly intended goals and targets;
- add more value than they consume;
- do not stand in the way of more valuable projects;
- show that operational performance justifies starting the project;
- enhance the survival of the organization.

Some examples that have been recognized as successful are the Guggenheim in Bilbao, Spain (opened 1997), the Tate Modern in London (opened 2000) and the Baltic Centre, Gateshead, UK (opened 2002). Creativity is important in the development of attractions but, as the artist Grayson Perry stated in his 2013 Reith Lecture, the watchword is 'MAYA' – Most Advanced, Yet Acceptable! (Perry, 2013). Thus, a major criticism of London's Millennium Dome was that it was inviting visitors to spend large sums for a product that they did not really understand or know how to consume. This was then compounded by the opening date of 31 December 1999, which meant that all faults in the following days had to be corrected under the full gaze of the paying public and the media. First impressions count and so, for all projects, it is customary to have a 'soft' opening some weeks or months before the official launch to iron out teething problems.

Abandoned projects

The converse of any of the above criteria may be associated with project collapse. If the uncertainties are such that it is not possible to move the project concept from just ideas and vague notions of viability, then it should be abandoned. Once a project is seen to be failing it should be corrected. This requires the source of failure to be identified and corrective action to be taken. If it is not possible to take remedial action, then depending upon the severity of the failing activity, the project has either to be reformulated or rejected. Unfortunately, unwillingness to accept defeat often means that they are abandoned after completion, as in the situations of the Earth Centre in Doncaster, UK (opened 2001: closed 2004), and the National Centre for Popular Music in Sheffield, UK (opened 1999: closed 2000), achieving only 30% and 25% of their design level of 500,000 and 400,000 visitors, respectively (Richards and Wilkes, 2008). On the other hand, London's intended theme park for the 1980s, 'The Battersea', to be based at the redundant power station situated on the south bank of the river Thames, was discarded before any new works began. This was a mega project for which the capital cost was unaffordable from the standpoint of raising equity to match debt, and the site is currently residential development.

Completed project failures

These are projects that are completed to plan, but often in the face of negative evidence about the value of their outcomes. Typically, they add less value than they consume owing to cost overruns and/or underperformance, and need refinancing to continue operating. Good examples here are the National Botanic Garden of Wales (opened 2000) and the Royal Armouries in Leeds, Yorkshire (opened 1996), which had vastly inflated attendance forecasts (NAO, 2001). For the latter, the design level was 1.3 million paying visitors with a break-even of 550,000, but it was quickly realized that this would not be achieved. Its best attendance level was some 344,000 visitors in 1997, falling to 158,000 in 2000 and now settling in the range 200,000–250,000 upon the abolition of charges for national museums by the UK government in 2001.

Continuation of such projects in both QII and QIII frequently lies in maintaining image credibility by reversing the 'sunk' cost argument, so that the investment becomes too important to fail. Within the public realm, multiple and often conflicting goals keep the concept vague, so goal diffusion or redefining outcomes allows almost any result to be reconciled. This disrupts any feedback system to recognize failure, thus avoiding change. Civil servants may also be given a political directive, which leaves them powerless to remedy mistakes.

The Channel Tunnel (opened 1994), a project mooted for generations, did not have major delays in construction (1½ years), but had an 80% cost overrun and, with financing costs 140% higher than forecast, private equity lost most of its share value. There were limited impacts on the wider economy because of displacement of ferry services and few developments associated with the project.

The creativity of arts and cultural projects means that they can be valued for their own sake, so the consequences of economic evaluation may be ignored. In the second half of 2012, Valencia, Spain's third largest city, had to ask the government in Madrid for a €2.5 billion (2020 prices) 'bailout'. Access to cheap money in the boom years, local vanity and competition for international tourists saw politicians trying to outspend each other in cultural and arts projects, with Valencia leading the way. The centrepiece was the opera house, completed in 2005 at a cost of €1.7 billion (2020 prices), which was supposed to rival the Guggenheim in Bilbao and the cityscape of Barcelona. The recent recession cruelly exposed this folly with hospital wards closing, local taxes rising and many of the city's young people out of work.

Seemingly never-ending projects

These involve time delays with the result that they may never be completed or are completed, having been a continual drain on funds, with debt write-offs. Delays in building and changes in project specification are familiar causes of cost escalation. Project survival depends upon similar arguments for QII, but with an added defence of their uniqueness, which immunizes them from criticism since the great leap forward proposed is beyond the vision and understanding of outsiders. Results are just around the corner, so more funding is essential for going on. In many instances, the project is a statement of organizational policy, say, a political manifesto, so what is expressed must be maintained irrespective of costs and what is required is an extension of the time horizon.

Some of the classic examples that could fit into this category are the development of Concorde (in service 1976), and the Sydney Opera House (opened 1973), which seemed at the time to go on and on until the investment was written off. Over a ten-year period, the cost of Concorde rose by a factor of 12 (Hall, 1982b) and, with only 20 aircraft built, the Concorde programme was a substantial economic loss. Air France and British Airways received considerable government subsidies to purchase them and the aircraft was finally retired in 2003 owing to a general downturn in the aviation industry after the plane's only crash in 2000. The Sydney Opera House came in ten years late and 14 times over budget (Hall, 1982c), but is now a World Heritage Site. Economic folly can be forgotten with the passing of time and the public learns to treasure such developments. Today, innovative architecture is evolving as an integral part of new tourism development.

On a smaller scale was the Wonderworld theme park proposed in 1981 for the old steel plant in Corby, UK, and scheduled to open in 1985. The design standard was for 5.2 million visitors at a time when Alton Towers, the largest theme park in the UK, had 1.6 million attendances, rising to 1.9 million in 1985. The defence against optimism bias was the uniqueness of the various park zones and lack of vision by sceptics. To put it mildly, after prolonged selling 'hype' and expenditure of regional assistance money, the proposal was abandoned in 1991 without on-site works being carried out. By this time, the local community had lost faith in the whole scheme. Conversely, many developers have experienced and recognized the ability of local pressure groups to 'kill off' sound project proposals because political leaders 'keep kicking the can down the road' by calling for further evaluation studies.

Conclusions

While it might be easy to define the criteria for project success, it is more complex to achieve them. Economic success dictates the commercial pathway, subject to meeting community objectives, but the further one looks into the future as with mega schemes, the greater is the number of unknowns and hence the risk. Consequently, it is not uncommon for such investments to be managed as private–public partnerships, as seen in area-regeneration models. However, the latter are set by their location, which may not be ideal, with the result that political pressure may dominate economic realism, giving rise to cost overruns and/or optimism bias in forecasting. The lesson from several UK millennium investments was that major capital projects should not be undertaken unless their market function is clear, visitor displacement from other businesses has been considered and a pragmatic feasibility study has been undertaken, so that the risks involved are thoroughly understood and accepted.

It is important that the planned design concept should be 'reproductive'; that is, about creating experiences which have strong associations with potential buyers, enabling value for money to be easily recognized. The concept can be distinctive, as in the architecture, but not too different so that the public does not know how to consume it, one example being chain hotels and restaurants that have known brands. It is the same for attractions. For example, the London Eye, a Ferris wheel on the bank of the river Thames, is a tried-and-tested product drawing in well over 4 million visitors, or Universal Studios only creating themed zones in its parks from films that have experienced major box-office success.

Avant-garde ideas are difficult to evaluate in the marketplace because there is no recognition at large of their value and so they run the risk, as in the case of London's Dome, of being lampooned in the media as the emperor's new clothes, or causing public outrage when they run counter to what is considered to be good taste. Accordingly, they have a high probability of economic failure, both commercially and also in the wider sense of attracting visitors and their expenditure to an area. This lends weight to the argument often put forward by cultural institutions that the arts have an intrinsic value, and the consequences when they vary in their financial outcomes are unimportant, so that evaluation is irrelevant. From a welfare perspective, it is accepted that cultural activities are public goods that generate consumption externalities for society as a whole, such as adding to the creativity of the population, quality of life, identity, social criticism, aesthetics, pluralism and so forth. These are legitimate arguments for public provision or subsidy, but evaluation is still possible using surrogate market techniques to assess willingness to pay (Epstein, 2003; Treasury, 2018).

Yet within the creative industries, past success is no guarantee because of the 'nobody knows' factor for new products. This can be seen in the 2008 production of the musical of *Gone with the Wind*. Musicals are a very popular art form and this was the story from a best-selling book, a 'blockbuster' movie, and produced under an internationally acclaimed director, Trevor Nunn. It closed in London after 79 performances (despite two rewrites), losing about £8 million (2020 prices). But the musical *Les Misérables*, based on the story by Victor Hugo that was not well known outside France, has gone on to international acclaim.

At the end of the day, economic success requires the avoidance of community pride, prestige and barely concealed electoral calculations. However, it must be recognized that economic calculus cannot cover all aspects of the project evaluation process. If it were to, then there would be no need for a 'decision maker', so, in the final analysis, all publically supported investments are subject to socio-political assessment to some degree.

References

Backman, M., Klaesson, J. and Öner, Ö. (2017) Innovation in the hospitality industry: firm or location? *Tourism Economics* 23(8), 1591–1614.

Bracalente, B., Chirieleison, C., Cossignani, M., Ferrucci, L., Gigliotti, M. *et al.* (2011) The economic impact of cultural events: the Umbria jazz music festival. *Tourism Economics* 17(6), 1235–1255. DOI: 10.5367/te.2011.0096.

Brown, K.G. and Geddes, R. (2007) Resorts, culture, and music: the Cape Breton tourism cluster. *Tourism Economics* 13(1), 129–141. DOI: 10.5367/000000007779784524.

Bull, A. (2013) Cruise tourism. In: Tisdell, C. (ed.) *Handbook of Tourism Economics*. World Scientific Publishing, Singapore, pp. 339–359.

d'Hauteserre, A.-M. (1997) Disneyland Paris: a permanent economic growth pole in the Francilian landscape. *Progress in Tourism and Hospitality Research* 3(1), 17–33. DOI: 10.1002/(SICI)1099-1603(199703)3:1<17::AID-PTH59>3.0.CO;2-J.

Dev, C., Brown, J. and Zhou, K. (2007) Global brand expansion: how to select a market entry strategy. *Cornell Hospitality Quarterly* 48(1), 13–27.

Divisekera, S. and Nguyen, V.K. (2018a) Determinants of innovation in tourism evidence from Australia. *Tourism Management* 67, 157–167. DOI: 10.1016/j.tourman.2018.01.010.

Divisekera, S. and Nguyen, V.K. (2018b) Drivers of innovation in tourism: an econometric study. *Tourism Economics* 24(8), 998–1014. DOI: 10.1177/1354816618794708.

Epstein, R.A. (2003) The regrettable necessity of contingent valuation. *Journal of Cultural Economics* 27(3/4), 259–274. DOI: 10.1023/A:1026375220210.

Hall, P. (1982a) *London's Third Airport. Great Planning Disasters*. University of California Press, Berkeley, California, pp. 15–55.

Hall, P. (1982b) *The Anglo-French Concorde. Great Planning Disasters*. University of California Press, Berkeley, California, pp. 87–108.

Hall, P. (1982c) *Sydney's Opera House. Great Planning Disasters*. University of California Press, Berkeley, California, pp. 138–151.

Hilton, C. (1994) *Be My Guest*. Simon & Schuster, NewYork.

Hjalager, A.-M. (2010) A review of innovation research in tourism. *Tourism Management* 31(1), 1–12. DOI: 10.1016/j.tourman.2009.08.012.

Hjalager, A.-M. (2015) 100 innovations that transformed tourism. *Journal of Travel Research* 54(1), 3–21. DOI: 10.1177/0047287513516390.

House of Lords (2019) *The Future of Seaside Towns*. House of Lords, London.

Jarratt, D. and Sharpley, R. (2017) Tourists at the seaside: exploring the spiritual dimension. *Tourist Studies* 63(3), 349–368. DOI: 10.1177/1468797616687560.

Jones, T. (1994) Theme parks in Japan. In: Cooper, C. and Lockwood, A. (eds) *Progress in Tourism, Recreation and Hospitality Management*, Vol. 6. John Wiley & Sons, Chichester, UK, pp. 111–125.

Milano, C. (2017) *Overtourism and Tourismphobia: Global Trends and Local Contexts*. Ostelea School of Tourism & Hospitality, Barcelona, Spain.

NAO (2000) *The Millennium Dome*. The Stationery Office, London.

NAO (2001) *The Re-negotiation of the PFI-type deal for the Royal Armouries Museum in Leeds*. The Stationery Office, London.

Perry, G. (2013) Playing to the gallery. The Reith Lectures. BBC. Available at: www.bbc.co.uk/programmes/p01kr984 (accessed 1 June 2019).

Richards, S. and Wilkes, K. (2008) Attraction failure and success. In: Fyall, A., Garrod, B., Leask, A. and Wanhill, S. (eds) *Managing Visitor Attractions: New Directions*, 2nd edn. Elsevier, Oxford, UK, pp. 39–58.

Schumpeter, J. (1934) *The Theory of Economic Development: An Inquiry into Profits, Capital, Credit, Interest and the Business Cycle*. Harvard University Press, Cambridge, Massachusetts.

Treasury, H.M. (2018) *The Green Book: Central Government Guidance on Appraisal and Evaluation*. H.M.Treasury, London, pp. 89–99.

UNWTO (2008) *2008 Tourism Satellite Account: Recommended Methodological Framework (TSA: RMF 2008)*. Madrid, Spain.

Viu, J.M., Fernández, J.R. and Caralt, J.S. (2008) The impact of heritage tourism on an urban economy: the case of Granada and the Alhambra. *Tourism Economics* 14(2), 361–376. DOI: 10.5367/000000008784460481.

Wanhill, S.R.C. (1986) Which investment incentives for tourism? *Tourism Management* 7(1), 2–7. DOI: 10.1016/0261-5177(86)90052-X.

Wanhill, S. (2000) Mines—A tourist attraction: coal mining in industrial South Wales. *Journal of Travel Research* 39(1), 60–69. DOI: 10.1177/004728750003900108.

Wanhill, S. (2013) The business of amusement parks: their development and operation. In: Tisdell, C. (ed.) *Handbook of Tourism Economics*. World Scientific Publishing, Singapore, pp. 317–337.

9 Human Capital Issues and Challenges in the Hotel Industry: The Case of Indonesia

Sienny Thio[1]* and Brian King[2]
[1]*Hotel Management Program, Petra Christian University, Surabaya, East Java, Indonesia;* [2]*School of Hotel & Tourism Management, The Hong Kong Polytechnic University, Kowloon, Hong Kong*

Introduction

Human resources are a leading challenge for hotel executives in their pursuit of operational effectiveness (Enz, 2001, 2009) and are a long-established topic of investigation among tourism and hospitality scholars (Guerrier and Deery, 1998; Baum, 2007, 2015; Baum and Szivas, 2008). In particular, it has been observed that labour shortages are the most intractable challenge for hotels in almost all of the world's regions (Wang, 2009), with about 90% of the hospitality industry commonly experiencing under-staffing (Poulston, 2008). Such ongoing human capital problems may arise from the widespread deployment of young, female and unskilled/semi-skilled staff that characterizes the tourism and hospitality sector and which presents particular challenges for managing the workforce (Kusluvan *et al.*, 2010).

With the World Travel and Tourism Council asserting that tourism accounts for some 10% of employment globally (WTTC, 2019), it is unsurprising that human resource concerns loom large in developing countries as they seek to make the leap towards developed status. The challenges are diverse, because there is a global hunt for talent to lead the sector at the most senior levels, including among prominent transnational hotel companies. At the same time, the many developing countries with growing youthful populations need to find work for both the low-skilled and also for those seeking rapid upward social mobility. Such complexity places considerable pressure on political and industry leaders to strike an appropriate balance between attracting top talent, while ensuring sufficient opportunities for an existing and potential local labour force.

Indonesia is a notable example of a country that is confronting such challenges. Located in South-east Asia, with a population of some 264 million (2017 figures) and host to the world's largest number of Muslim residents, Indonesia is a test case for the potential of tourism to support national development. The rapid construction and opening of hotels across the country has been driving demand for qualified human resources to fill vacancies, thereby enhancing the importance of acquiring qualifications and experience for the workforce. Negara (2014) has noted that Indonesia's aspiration to be innovation-driven underscores the urgency

*Corresponding author: sienny@petra.ac.id

DOI: 10.1079/9781789242812.0009

of producing highly educated and well-trained human resources. Indonesia will undoubtedly need additional investment in human resources if it is to maintain future economic growth prospects and to extend social welfare. In the face of rapid industry expansion, particularly across Asia, hoteliers in particular will need to deploy all available resources in preparing for an increasingly global market. However, although studies on human capital issues and concerns in the hotel industry have been undertaken in various countries (Enz, 2001, 2009; Watson *et al.*, 2002; Qiu Zhang and Wu, 2004; Duncan, 2005; Poulston, 2008; Yang and Cherry, 2008; Jauhari, 2012a; Bharwani and Butt, 2012), few researchers have investigated the challenges in the context of Indonesia and how these may be addressed. Context is important because, as noted by Davidson *et al.* (2011), any assessment of hospitality human resource trends should consider the local circumstances which influence various people management issues, notably concerning cultural aspects.

If the Indonesian hotel industry is to respond effectively, this will involve acknowledging current and potential challenges. Debrah and Budhwar (2004) observed that the development of any human resources management (HRM) system should consider the regional and/or country context, while similarly, Enz (2009, p. 197) commented that different regions in the world pose 'different management challenges'. The present chapter contends that explorations of human capital challenges in hospitality should consider such contexts. In the case of Indonesia, the hotel industry will need to strengthen workforce qualifications if it is to address the growing competition for labour.

It is acknowledged here that hospitality labour-related issues are nothing new and HRM hospitality roles have evolved over recent decades (Ubeda-Garcia *et al.*, 2013). Nevertheless, the limited research on current and future human capital issues and challenges in Indonesia in particular suggests the need for more thorough investigation. On this basis, this chapter will investigate the current and future human capital issues and challenges that confront hotel managers in that country. By identifying areas of concern and providing a future platform, the empirical human capital-focused research discussed below offers the prospect of raising the awareness and enhancing the future-oriented knowledge of hotel operators, academics and government bodies. Before proceeding, however, it is appropriate to provide a brief review of human capital issues in the hotel sector both generally and in Indonesia.

Human Capital

Lado and Wilson (1994, p. 705) have defined human capital as the 'set of knowledge, skills, and abilities that are embedded in the firm's human resources'. Meanwhile, Hyun (2010) describes the same concept as the 'productive capacity of the people'. Hyun (2010) has also further described human capital as the skills and abilities of the workforce which can be mastered through education and can transform materials and capital into goods and services. On this basis, human capital is central to the image and reputation of an organization, particularly in the case of service industries such as hospitality.

Tourism remains a labour-intensive sector (Baum, 2010) and concerns about human capital are high on the list of problems identified by hospitality operators (Enz, 2001). Respondents in Enz's (2001) survey of hospitality managers revealed a reluctance to invest in human resources because they would be unable to benefit from their investment as a result of rapid labour turnover and the prospect of imminent departures. This pattern sometimes prompts hotel managers to underestimate worker capabilities and to withhold potential training and development opportunities. Solnet and Hood (2008) endorsed such views by highlighting that even the most enlightened hospitality training provision may be rendered useless when employees opt for job offers elsewhere.

The prevalence of labour shortages in hospitality, compounded by the difficulty of attracting new recruits, has been noted by many tourism and hospitality researchers (Enz, 2001, 2009; Duncan, 2005 Yang and Cherry, 2008; Poulston, 2008; Bharwani and Butt, 2012; Jauhari, 2012a). In many settings, an increase in the number of hotel properties has not been accompanied by an equivalent number of additional human resources, leading ultimately to labour shortages. Notably, although they are a prime target for the hotel sector, young people often opt to work in other industries which are perceived as providing better remuneration and/or allowances (Qiu Zhang and Wu, 2004). Thus, human resource professionals face the challenge of considering the perspectives of both current and prospective employees and of the company. The former are deemed a crucial source of competitive business advantage (Ulrich and Brockbank, 2009).

Human capital issues and concerns in the hotel industry

As has already been noted, human resources are a significant challenge for hotel managers and hospitality professionals (Enz, 2009), a challenge that is exacerbated by the dynamism of the business environment, including the workforce. Hospitality establishments are confronting a diverse proliferation of competitors, all pursuing the same group of potential talent (Chen and Choi, 2008). Such conditions, along with prevailing economic, political, socio-cultural and technological changes, have impacted on employment patterns in the sector (Baum, 2015), while Jauhari, 2012b has noted that changing socio-economic contexts are compounding the complexities for industry leaders. In the face of such business turbulence, Jauhari (2012b) has urged hospitality organizations to be more proactive when addressing emerging challenges. There is global development of the hospitality industry, with certain regions growing at an accelerating rate (Jayawardena *et al.*, 2013). In particular, the increasing supply of hotel rooms is placing pressure on the demand for labour while, simultaneously, the needs, wants and expectations of customers, employees and companies are also changing (Jayawardena *et al.*, 2013). There is a pressing need to engage in regular updating of salient industry issues and concerns to ensure that companies are adequately prepared for change.

Numerous studies have examined people management-related issues and concerns in various settings such as the USA, Canada, New Zealand, Taiwan and China (Enz, 2001, 2009; Watson *et al.*, 2002; Qiu Zhang and Wu, 2004; Duncan, 2005; Poulston, 2008; Yang and Cherry, 2008; Jauhari, 2012a; Bharwani and Butt, 2012). These studies have highlighted the urgent need for executives to take full account of human capital concerns. In reviewing the relevant extant literature, the authors of this chapter have identified 11 papers that focus explicitly on human resource issues and concerns in different world regions. The various sources are summarized in Table 9.1.

As is evident in Table 9.1, hotel industry organizations across the world confront similar human resource issues and challenges. However, the handling of such issues will be influenced by the specific socio-economic backgrounds of companies and regions/countries (Baum, 1993). The research discussed in this chapter provides a broad perspective that draws upon the views of human resource practitioners and industry leaders in anticipating upcoming challenges for Indonesia's hotel industry.

Human capital in Indonesia's hotel industry

International tourism flows to Indonesia have been expanding rapidly, with most visitors emanating from short- and medium-haul sources such as Singapore, Malaysia, Australia, China and Japan (OECD, 2014b). Domestic travellers are also contributing to tourism growth. With its thousands of islands and relatively unconstrained by space-related concerns, Indonesia offers undoubted potential for substantial tourism development. Regrettably, however, the destination is unlikely to be effective when competing globally unless human capital deficiencies are remedied promptly (Widodo, 2016a); in 2015, a ranking of human resources in tourism placed Indonesia in fifth place among the ASEAN countries after Singapore, Thailand, Malaysia and Philippines (Widodo, 2016b). According to Ahman Sya, Indonesia's Deputy Head of Institutional Development of Tourism and as cited by Widodo (2016a), tourism human resources in Indonesia are still lagging in three aspects, namely, (i) mastery of foreign languages, especially English; (ii) information technology (IT); and (iii) managerial skills. These are viewed as the main problems producing low competitiveness within the ASEAN region.

Table 9.1. Human resource issues and challenges in the hospitality industry.

Author (s)	Region	Sample	Issues and challenges
Enz (2001)	USA and Europe	Senior managers	Attracting talented people, lack of strong employees' skills, the need for more training, labour shortages due to low unemployment rates, raising compensation and employee morale, professionalism and career opportunities for managers
Watson *et al.* (2002)	Across 32 countries	HR directors and managers	Service quality, training and development, staff recruitment and selection
Qiu Zhang and Wu (2004)	China	Hotel managers, travel managers, and academics	Employee retention, human resource shortages, shortages of qualified managers and the expectations gap between education and industry
Duncan (2005)	Australia, UK, USA, Canada, NZ, Europe, Ireland (worldwide)	Hospitality and hotel association websites around the world	Increase in minimum wages, increasing flexibility of working hours, and increasing reliance on overseas workers
Poulston (2008)	Auckland	Staff, supervisors and managers	Under-staffing and high staff turnover, poor training, employee theft and sexual harassment
Yang and Cherry (2008)	Taiwan	Hotel managers	Shortage of employees, training & development internship employment, employee turnover, outsourcing, flexible/ casual employment, downsizing, the new labour retirement pension system, employment alternatives, workforce diversity
Enz (2009)	60 countries within six different world regions	Hotel executive and managers	Attraction, retention, training and morale of employees
Bharwani and Butt (2012)	India	Hospitality practitioners, roundtable of experts	Attraction and retention, training and development issues, employee engagement, work-life balance and growth opportunities and career progression
Jauhari (2012b)	India	Hospitality professionals	Skilled manpower shortage, rising labour costs, high attrition (higher training costs), managing millennials
Adler and Rigg (2012)	Jamaica	Expatriate hotel GM	Punctuality, educational levels, productivity, and low skill levels
Jayawardena *et al.* (2013)	Canada	Industry leaders and hospitality educators	Recruitment, training and retention of employees, seasonality

Following the establishment of the ASEAN Economic Community (AEC) which brings the prospect of a flow of ASEAN workers migrating freely to Indonesia, it is anticipated that tourism and hospitality employees in Indonesia will struggle to compete. On the positive side, there is optimism that tourism has the capacity to compete with other sectors, such as gas and oil, in Indonesia's economy because the tourism sector is 'pro-job', 'pro-growth' and 'pro-poor' (Widodo, 2016a). Accordingly, collaboration among all stakeholders (organizations, government, and education providers) in the tourism and hospitality industry will be necessary to improve the quality and performance of the Indonesian workforce. Meanwhile, there is an urgent need to standardize the education and training system with a view to developing skilled labour for the hotel industry. Key issues include English language proficiency, managerial skills, technology literacy, and general business skills (ILO, 2009). Following acknowledgment by the government of Indonesia of the need to enhance such skills, there has been a recent emphasis on vocational education and training with a view to strengthening the job opportunities for youth who dominate Indonesia's labour market.

With about 50% of Indonesia's population being aged under 30, the country is not short of human resources (Indonesia-Investments, 2016). Unfortunately, however, the volume is also responsible for various labour-related issues. These include: low education levels, unskilled workers and poorly remunerated occupations. According to a survey on Indonesia by the UNWTO (ILO, 2009), the main factor in reducing the competitiveness of Indonesian tourism is the shortage of a skilled workforce as a result of poor wages, unfavourable working environments, and low barriers to joining the industry. The OECD (2014a) also highlighted improving job quality and productivity as major challenges for emerging countries, including Indonesia. It is evident that education and skills are deemed as essential for enhancing employment outcomes and improving the qualifications among the workforce.

The prevalence of low wages relative to other sectors is a major cause of the shortage of qualified labour in the tourism and hotel industry in Indonesia (ILO, 2009). Hotel establishments typically offer their employees only the basic monthly salary which equates to the minimum wage regionally. Regional minimum wages are commonly determined by provincial governments and range from IDR 1.100.000 to IDR 3.100.000 or USD 83–232 as of February 2016 (National Wages and Productivity Commission, 2016). Hotel employee allowances in Indonesia consist of three main elements – basic salary, service charge and tips (Swan, 2015). The applicable basic salary and service charge vary by company, depending on the nature of the role and prevailing policies enacted by management. The service charge may itself exceed the basic salary for employees and varies according to hotel occupancy rates. This means that employees who are working in hotels with higher occupancy rates will receive a higher service charge. Many Indonesian hotel services are inclusive of 21% tax, of which 10% is deemed as government tax and the other 11% is a service charge. In light of the distinct features of the industry in Indonesia that have been noted in the previous section, human resource executives are expected to recruit and retain their workforce by designing and formulating effective human resource planning and development.

The Research: Challenges in Indonesia's Hotel Industry

Methodology

In order to investigate the issues and challenges confronting Indonesia's hotel industry, a qualitative research method was used based on semi-structured in-depth interviews with hotel practitioners. As was mentioned previously, context is important in the Indonesian context. Chadwick *et al.* (1984) have noted that qualitative research has the benefit of involving the researcher in observing behaviour in a 'natural setting', thereby acquiring greater in-depth understanding of the research subjects with greater flexibility. Using such an approach suits the purposes of the present investigation by ensuring that richer explanations and descriptions enhance the meanings attributable to information and statements. In this case, the observations relate to human capital issues in hospitality. Semi-structured interviewing has been selected because it is 'fluid in nature and follows the thinking processes of the interviewee' (Jennings, 2001). Interviews were deemed to be an appropriate method for in-depth explorations because each interview is a dynamic process between the interviewer and interviewee which allows particular issues or problems to unfold (Brotherton, 2015).

Data collection

Indonesia's reputation as a tourism destination depends predominantly on Bali. Java has a secondary influence, with its scenic diversity and historic monuments (Nuryanti, 2001). Indonesia has a population of approximately 253.6 million, of whom about 60% live on Java and Bali (Thadani *et al.*, 2015). The Javanese cities of Jakarta and Surabaya have been recognized by many hotel chains as prospective locations for expansion beyond Bali (ILO, 2009). The islands of Java and Bali were selected as interview settings for the present investigation owing to their established business and dynamic hotel growth.

For comparative purposes, most Surabaya hotels focus on business travellers, whereas their Bali counterparts are more leisure and resort-based tourism–oriented. The characteristics of hospitality employees in the two regions differ because of prevailing geographic and socio-cultural backgrounds. Located in eastern Java, Surabaya is frequented predominantly by Javanese people, with most being Muslim; in contrast the people of Bali are predominantly Hindu.

Although there is no particular rule about sample size in qualitative research, Kuzel (1992) has suggested that a sample of 12–20 participants may be appropriate when it is necessary to obtain more extensive information and insights into the research topic. For the purposes of this study, 20 participants were randomly selected by sending an invitation through emails and telephone calls to the general managers and/or human resource executives/managers of upscale hotels (four- and five-star) in Bali and Surabaya, inviting them to participate voluntarily in the study. The list of hotels in Bali and Surabaya was identified from Tripadvisor. Of 34 hotels contacted, 20 accepted to be interviewed over a three-month period in September–December 2010. All participants were Indonesian citizens, with six being from local (chain) hotels and 14 from international (chain) hotels. The respondent profile is presented in Table 9.2.

Data analysis

The researchers deployed a content analysis approach. This is a scientific tool that has been defined by Krippendorff (2013) as 'a research technique for making replicable and valid inferences from text (or other meaningful matter) to the contexts of their use', and is a method that has gained in popularity over recent years (Stringam and Gerdes, 2010). Research using content analysis examines texts by identifying and calculating a particular theme (Lu and Zhu, 2006) and focusing on language as communication with an emphasis on the content and contextual meaning of the text (Tesch, 1990).

Transcriptions were conducted in Indonesian in order to maintain the original meaning of each conversation with respondents, while the categorization and summarizing of interviews was in English. Field notes were used to transcribe other important information gathered during interviewing such as respondent genders, age, educational levels, employer name and work experience. Since the analysis of research content and the formulation of connections among themes relied on the researchers themselves and were not mediated via a computer software program, the researchers proceeded to organize and examine the qualitative data manually. Microsoft Word 2010 was used to generate codes, categories and themes for the purposes of further analysis. Applying manual analysis enabled the researchers to examine the data at close quarters and to conduct a rigorous identification of emerging themes across the transcripts. The terms 'issue' and 'challenge' are presented as a collective and are used interchangeably throughout the chapter. According to Miles and Huberman (1994), there are three elements of analysis, namely: data reduction, data display, and conclusion drawing/verification, and these were adopted for the purposes of analysing the qualitative data.

Results

The following human capital issues and challenge themes emerged from the interview data: (i) recruitment; (ii) strong competition; (iii) employee issues; (iv) retention; (v) government regulations; and (vi) local community.

Recruitment

Respondents in both Bali and Surabaya admitted that the challenge of recruiting potential talent

Table 9.2. Respondent profile.

Code	Position	Type of hotel	Gender	Age range	Current working experience (year/s)	Hotel working experience (year/s)
	SURABAYA					
Sub1	HR Manager	5*	Male	37–51	2	19
Sub2	Director of HR	5*	Male	37–51	4	23
Sub3	General Manager	4*	Male	>51	13	36
Sub4	HR Manager	4*	Female	37–51	3	21
Sub5	Director of HR	5*	Female	37–51	12	25
Sub6	General Manager	4*	Male	>51	13	19
Sub7	General Manager	4*	Male	37–51	7	21
Sub8	HR Manager	4*	Male	37–51	1	5
Sub9	HR Manager	4*	Female	37–51	1	15
Sub10	Training Manager	5*	Female	22–36	6	6
	BALI					
Bal1	Director of HR	5*	Female	22–36	1	7
Bal2	HR Manager	4*	Female	37–51	13	16
Bal3	General Manager	4*	Female	37–51	1	25
Bal4	Assistant HR Director	5*	Female	37–51	4	19
Bal5	General Manager	4*	Female	>51	3	22
Bal6	Area Director of HR	5*	Male	37–51	5	23
Bal7	General Manager	5*	Male	37–51	3	20
Bal8	HR Manager	4*	Female	22–36	1	4
Bal9	HR Manager	5*	Male	37–51	3	18
Bal10	Director of HR	5*	Male	37–51	10	25

has intensified because of competition between hotels. Consequently, hotel companies compete to offer better salaries and benefit packages that will attract talented employees who have relevant qualifications. On the other side, the increasing room supply is making a fall in occupancy inevitable. It is unsurprising that Indonesian hospitality leaders are confronting a recruitment problem.

A talent war among hotels is viewed as unavoidable and hijacking potential staff by offering them higher salaries and positions seems to offer a quick way to attract and recruit talented staff. This creates a particular dilemma for hotels that can only offer a standard- or low-salary package as they will be unable to recruit competent staff or managers due to budgetary constraints. This was observed by two respondents, as follows:

> Hotels are facing dilemma, if they want to offer lower package, they will have a problematic in quality. (Sub7)

> The challenge is talent acquisition, to find people who fit with company expectation. (Bal4)

Strong competition

Several Surabaya respondents mentioned that the significant growth of hotel construction has intensified competition among hotels by producing an imbalance between supply and demand. The supply of hotel rooms is viewed as excessive and has led to strong demand for labour. However, there are insufficient qualified employees in the Indonesian labour market to meet the requirements of the hotel industry. As a result, hotels have little option other than offering better pay and higher positions to attract talented employees from other hotels. Additionally, hotels are compelled to offer instant promotion to staff because they have no opportunity to fill potentially vacant positions, even though the staff have not reached the required level of quality provision. Two views about the impacts of severe competition were expressed as follows:

> Finding qualified personnel is rather difficult because too many hotels open opportunity. Staff is easier to get promotion even though they are not ready with that position. (Bal3)
>
> It is like talent war. Between one hotel and another hotel they find a way to hijack talents. (Bal8)

Employee issues

Hotels that have 20 years or more of history face particular problems with their more senior staff who are aged above 50 years and have been working in the establishment for over 15 years. This issue was most pronounced among respondents from Surabaya and was associated with the following problems: (i) lower staff productivity; (ii) decreasing stamina; (iii) less motivation; (iv) less creativity; (v) boredom due to doing the same work for many years; (vi) technologically illiterate. The ageing workforce is viewed by some HR leaders as being both a burden and a challenge. The issue of senior staff was primarily raised by respondents who are employed in older hotels because the ageing staff have been there ever since opening. Unsurprisingly, these older hotels usually have lower staff turnover. Older employees are those most likely to remain at the hotel since other properties will generally hire a younger workforce.

Moreover, both Bali and Surabaya respondents associated an underqualified workforce with a deficit of key competencies such as logical thinking. Inadequate skill levels and English proficiency have emerged as persistent problems at a time when job-related demands are increasing and hotels are expecting high standards. Two respondents expressed the following concern:

> The quality of human resources is lacking, their logical thinking is low, especially fresh graduates. (Bal2)
>
> There is a skill gap because job demands are getting high. Hotel has a lot of resources but those who can execute the job are uneven. For example, staff must be able to speak English but the requirement is not only English now but also other languages...in fact, not all staff can speak English. (Bal9)

Retention

Bali and Surabaya respondents identified the prevalence of high staff turnover, particularly in the case of millennials (often defined as those reaching young adulthood in the early 21st century) since they are prone to boredom and demand rapid advancement. The issue of high staff turnover is caused by the rapid growth of hotel development, and staff obviously have many options to apply for hotel-related jobs. As the competition between hotels intensifies, it is increasingly challenging to retain employees. Millennial staff are more tempted to move from one hotel to another within a short period because they are confident about moving higher. Unsurprisingly, they can readily move to other hotels which offer better positions and benefits. A Bali respondent believed that preparing and ensuring the attainment of agreed positions within a defined period may provide a mechanism to retain millennials for longer. The challenge relating to millennial staff was expressed as follows:

> Retaining new staff especially millennials are a challenge because they easily get bored and want to get quick promotion. (Sub10)
>
> ...ensure staff to keep staying by preparing next position, particularly for millennials. For millennials they do not want to remain in a position because within certain period they must get certain position. (Bal5)

Government regulations

Relative to their counterparts in Bali, Surabaya hotels were encountering more problems related to local government regulation, particularly concerning the UMK (Municipal Minimum Wage). Surabaya respondents stated that UMK has been increasing significantly each year and has been burdensome for companies. On the other hand, hotel revenues have been decreasing due to the increasing number of hotels being built and severe competition between hotels. The UMK increase has created a serious budget issue for hotels as they need to escalate their revenues in order to cover increased labour costs in the face of decreasing room occupancies.

A Bali respondent claimed that government was underprepared to implement its mandatory national health care initiative (BPJS *kesehatan/ Badan Penyelenggara Jaminan Sosial Kesehatan*) which was established at the beginning of 2014. The programme was intended to improve the provision of public health services. However, the implementation has not run smoothly. Inflexible hours of service and limited facilities have generated complaints from among hotel employees. The plan is mandated and has posed particular problems for hotel employers in Bali. On this basis, several Bali respondents urged their management to provide better health facilities for employees by offering double health insurance coverage to attract potential employees. Two respondents expressed their opinions concerning government policies as follows:

> Government regulation is inapplicable and difficult to be implemented, every year UMK increases significantly and as a result we cannot achieve revenue which only increase very little. (Sub2)

> We support government program, I am happy with BPJS healthcare, but the problem is the implementation...no flexibility and then puskesmas (government-mandated community health clinics) is not ready. (Bal5)

Local community

The local community issue was raised by respondents in Bali. Known for its unique local culture, Bali is characterized by the institution of the customary village (Desa adat) and village section (Banjar). These contribute to the definition of relationships and boundaries such as place, friendships and common interests (Geriya, 2003). Smooth hotel operations and management depend heavily on a close relationship with the Banjar. The dominant authority that is associated with the local community in areas where hotels are located expects hotel managers to adhere to a set of (unwritten) regulations that have been set by the surrounding community. For instance, new hotels in particular should employ a share of local people ranging between 20% and 40%. A problem that was noted by Bali respondents was that the qualifications of local people did not meet hotel standards due to low education levels and a lack of skills. A Bali respondent expressed her concerns as follows:

> Bali has its own challenge, in our area, Jimbaran is quite tough in its local community...we need to focus on recruitment as they set the quota and they have different way to push companies. (Bal1)

Noting the additional challenge associated with entry by foreign workers as a result of AEC and the increasing need for proper training, most respondents anticipate that these issues will continue. It is anticipated that the free flow of labour within ASEAN will threaten local workers since incomers may accept lower or equal pay relative to Indonesian workers. Noting a lack of output from hotel schools, respondents felt that hotels would need to provide more training themselves. Training programmes will be crucial for preparing qualified and talented staff that are capable of competing with foreign labour. Training is also deemed to be a key factor for the retention of competent employees (Choi and Dickson, 2009). Comprehensive training programmes will be needed to help hotel companies to achieve the best staff performance and the delivery of excellent service to hotel guests. In view of the various issues that have been noted in the preceding section, hotel leaders should consider the various issues and challenges affecting the labour market to take appropriate action.

Discussion

From the preceding section, it is an evident that the difficulty of recruiting potential talent has become the biggest challenge for hotel executives in Indonesia. International and domestic hotel companies have been competing with each other to attract talented employees by offering better salaries and benefit packages. On the other side of the ledger, the significant increase in room supply has made falling occupancies inevitable. Though each job vacancy attracts many applicants, finding qualified staff with high potential is still difficult. This has been highlighted by Allen's (2016) assertion that quality of employment remains a major issue in Indonesia. A lack of skills and mismatched skills are also considered to be continual challenges. It is likely that involvement from both government and academics will be required to play a role in addressing the human capital concerns encountered by the hotel industry.

Three fundamental causes of the identified human capital issues and challenges have been noted, namely: (i) significant hotel development; (ii) underqualified workforce; and (iii) the impact of AEC.

Rapid hotel development

The massive pace of hotel development, which continues unabated, is the root cause of concerns about strong competition, recruitment and retention. As reported by Gebbie (2016) in the Asia Pacific Market Report, the number of hotel rooms in Bali will increase from 70,000 to over 85,000 by 2020. Meanwhile new hotels under development in Surabaya over the next three to five years will generate an additional 3800 rooms. Intense competition between hotels has produced a demand-and-supply imbalance, exacerbated by increased demands on the hotel workforce due to aggressive hotel development. This massive demand for labour is not paralleled by an increased supply of people to work in the industry. Given the systemic nature of the challenge, it will be important for both local and central governments to be involved in addressing this concern.

There have been recurrent calls for a development moratorium, particularly in Bali which already has a surplus of some 10,000 rooms and had sufficient room supply up to 2015 (Atmodjo, 2010). Officials at Indonesia's Ministry of Culture and Tourism have urged the Bali administration to apply a temporary restriction on any new hotel construction to avoid unhealthy business competition. Although the moratorium was released by the local authority and though no new permits have been issued to build new hotels, some new construction is still proceeding. This is due to the issuance of some permits prior to the release of the moratorium. This condition partly explains the aspiration of the Indonesian government to replicate the success of tourism in Bali by launching '10 new Balis' by embarking on new destinations for further development (Chan, 2017). Of the ten proposed tourism destinations, two are located in the western part of Indonesia, Lake Toba (north Sumatra) and Belitung (Bangka Belitung); four are located in central Indonesia, Tanjung Lesung (Banten), Seribu Islands (Jakarta), Borobudur Temple (central Java) and Mount Bromo (east Java); and four are in eastern Indonesia, Mandalika Lombok (west Nusa Tenggara), Komodo Island (east Nusa Tenggara), Wakatobi National Park (south-east Sulawesi), and Morotai (north Maluku).

The involvement of central government is essential since it holds the highest authority to work with local governments in setting strict regulations to investigate issues and problems in each area depending on local conditions. Hence, a commitment from both local and central governments will be needed to overcome excessive room supply in the region.

An underqualified workforce

Lack of skills and low English proficiency, as well as lack of logical thinking, have become hotel workforce issues and challenges. Though hotels have abundant staff, the workforce lacks the capability to perform their jobs. Furthermore, older staff who have been working for over a decade are demotivated and less productive. This leads to underperformance. These various issues are causes of an underqualified workforce. Despite the abundant workforce within the labour market, it is challenging to find staff who meet necessary hotel qualifications.

The government has a vital role in preparing strategic programmes to improve the quality of its people and to compete with foreigners. Indonesia still lags behind other ASEAN countries such as Singapore, Malaysia, Thailand and

Vietnam. There is evidently a lack of government attention to enhancing the competitiveness of the local workforce (Fanggidae, 2016). Thus, the involvement of Indonesian government will be crucial to producing a qualified and competent workforce. To mobilize the full potential of its people, the Indonesian government should engage stakeholders such as professional/business leaders, academics and associations in a collective endeavour to prepare a talented workforce that is globally competitive. For a start, and as noted by Shodiq (2016), improving the quality of basic education should be the key challenge for the Indonesian government. Taty (2016) added that the government should develop a good education system, strengthen the role of religion for character-building, and provide capacity-building through training and competency.

The Indonesian government has enhanced its attempts to improve the quality of the workforce, notably in the hotel industry. Respondents expressed the view that the regulation requiring all hotel staff to possess specific competencies in their field offers a means of improving the quality of employees. However, the government must evaluate and control the implementation of this regulation. The evaluation of the material delivered and the implementation of the training need to be considered in order to provide a qualified workforce who possess knowledge and expertise in their domain.

The impact of AEC

The ASEAN Economic Community (AEC) was launched in late 2015; however, the implementation of the AEC agenda is still unclear. The implementation of AEC does not currently appear to be a threat to hotels within the wider context of Indonesia's hotel industry. It can generally be said that Indonesia is not yet ready to welcome foreign workers, particularly in the case of hotels. This is consistent with Taty's (2016) and Gunadi's (2016) study which concluded that there is a big question about whether Indonesia is ready for AEC or not. This conclusion refers to Indonesia's ranking which is relatively lower than other ASEAN countries, particularly Singapore, Brunei, Darussalam, Malaysia and Thailand.

Among the various ASEAN countries, Indonesia has both the largest population and a tremendous talent pool. The potential of abundant talent needs to be unleashed and government, business professionals, academic institutions and individuals must discern this issue by establishing policies (Shodiq, 2016). The implementation of AEC in Indonesia was deemed by respondents to be challenging. The country may suffer because Indonesian workers need to compete with their counterparts elsewhere in ASEAN. Most are inadequately equipped with skills and regionally recognized certifications (Gunadi, 2016). Hence, job training and other skills training efforts should be considered to improve the quality of human capital and to be prepared for future impacts of AEC.

Though AEC seems to have had minimal impact on the hotel industry to date, it may become a serious future problem. Ready or not, all national stakeholders should be prepared for the free flow of labour, including across the hospitality industry, since this is part of the commitment to being an ASEAN single market. The establishment of the Mutual Recognition Arrangement on Tourism Professionals (MRA-TP) is a critical joint initiative among ASEAN regions to improve tourism and hospitality professionals' standards within the countries.

Conclusions

This chapter has sought to provide an enhanced understanding of human capital issues and challenges and to broaden the scope of previous studies that were conducted in other jurisdictions (Enz, 2001, 2009; Watson *et al.*, 2002; Qiu Zhang and Wu, 2004; Duncan, 2005; Poulston, 2008; Yang and Cherry, 2008; Jauhari, 2012a; Bharwani and Butt, 2012). It is interesting to note that recruitment, retention and employee-related concerns have become commonplace globally across the hospitality industry. In addition, the worries that have been identified in this study relating to intense competition, government regulations and local community are evidence that local circumstances should be acknowledged. Local community concerns in

Bali have shown that cultural uniqueness brings its own human capital challenges thereby confirming Bali as exceptional within the Indonesia context.

The establishment of AEC is an upcoming issue for hotel practitioners and was found to be equally concerning for both Surabaya and Bali respondents. The free flow of labour within ASEAN countries generally and within the hotel industry in particular should awaken hotel operators to serious preparation of their workforce in the face of foreign competition. This concern will be most challenging for the various developing countries within the ASEAN region, which are challenged by poor-quality workforces, notably Cambodia, Indonesia, Laos, Vietnam, Philippines and Myanmar. Indonesia is a special case and worthy of dedicated study because it is both a developing country and large and populous enough to have a substantial domestic market. This makes it a significant emerging country example.

This chapter has also proposed that hotel stakeholders should develop strategies to bolster human capital development within the industry, particularly in the case of government and education institutions. Strong support from government officials and collaboration with academics offer the prospect of a strategic approach to the alleviation of HR problems. Although this chapter has documented human capital concerns across the hotel industry, the respondents were limited to hotel managers/professionals from upscale hotels in Surabaya and Bali. Future researchers are encouraged to involve a greater diversity of hotel stakeholders in their studies, notably government officers, employees and academics from leading Indonesian cities such as Jakarta, Bandung and Medan. The results might generate additional issues and challenges that are being encountered by the hotel industry across Indonesia in different urban settings. Further research could also be undertaken in other countries and contexts. The perspectives gained from different parties may potentially enrich and sharpen our understanding of human capital concerns. Examining the cultural and socio-demographic backgrounds of different regions with larger sample sizes should be a valuable strategic input to the design of appropriate HR planning and development in anticipating possible future challenges.

Acknowledgement

The authors wish to thank the Directorate General of Higher Education (DIKTI), Ministry of Education and Culture, the Republic of Indonesia, for their financial support without which the research could not have proceeded.

References

Adler, H. and Rigg, J. (2012) Expatriate hotel general managers in Jamaica: perceptions of human resource, organizational, and operational challenges. *Journal of Hotel & Business Management* 01(01), 1–9. DOI: 10.4172/2169-0286.1000104.

Allen, E. (2016) Analysis of trends and challenges in the Indonesian labour market. Manila: Asian Development Bank. Available at: www.adb.org/sites/default/files/publication/182935/ino-paper-16-2016.pdf (accessed 10 February 2017).

Atmodjo, W. (2010) Indonesian culture and tourism ministry urging moratorium on new hotel development in Bali through 2015. Available at: www.hotel-online.com/News/PR2010_4th/Dec10_BaliHalt.html (accessed 10 February 2017).

Baum, T. (1993) Human resource in tourism: an introduction. In: Baum, T. (ed.) *Human Resource Issues in International Tourism*. Butterworth-Heinemann, Oxford, UK, pp. 3–21.

Baum, T. (2007) Human resources in tourism: still waiting for change. *Tourism Management* 28(6), 1383–1399. DOI: 10.1016/j.tourman.2007.04.005.

Baum, T. (2010) Demographic changes and the labour market in the international tourism industry. In: Yeoman, I., Hsu, C., Smith, K. and Watson, S. (eds) *Tourism and Demography*. Goodfellow, Oxford, UK, pp. 1–19.

Baum, T. (2015) Human resources in tourism: still waiting for change? – a 2015 reprise. *Tourism Management* 50, 204–212. DOI: 10.1016/j.tourman.2015.02.001.

Baum, T. and Szivas, E. (2008) HRD in tourism: a role for government? *Tourism Management* 29(4), 783–794. DOI: 10.1016/j.tourman.2007.09.002.

Bharwani, S. and Butt, N. (2012) Challenges for the global hospitality industry: an HR perspective. *Worldwide Hospitality and Tourism Themes* 4(2), 150–162. DOI: 10.1108/17554211211217325.

Brotherton, B. (2015) *Researching Hospitality and Tourism*. Sage Publications, London.

Chadwick, B.A., Bahr, H.M. and Albrecht, S.L. (1984) *Social Science Research Methods*. Prentice-Hall, New Jersey.

Chan, F. (2017) Jokowi plans to replicate Bali's success in 10 other Indonesian spots. Available at: www.straitstimes.com/asia/jokowi-plans-to-replicate-balis-success-in-10-other-indonesian-spots (accessed 12 October 2017).

Chen, Po-Ju. and Choi, Y. (2008) Generational differences in work values: a study of hospitality management. *International Journal of Contemporary Hospitality Management* 20(6), 595–615. DOI: 10.1108/09596110810892182.

Choi, Y. and Dickson, D.R. (2009) A case study into the benefits of management training programs: impacts on hotel employee turnover and satisfaction level. *Journal of Human Resources in Hospitality & Tourism* 9(1), 103–116. DOI: 10.1080/15332840903336499.

Davidson, M., McPhail, R. and Barry, S. (2011) Hospitality HRM: past, present and the future. *International Journal of Contemporary Hospitality Management* 23(4), 498–516. DOI: 10.1108/09596111111130001.

Debrah, Y.A. and Budhwar, P.S. (2004) HRM challenges in the Asia-Pacific region: agenda for future research and policy. In: Budhwar, P.S. (ed.) *Managing Human Resources in Asia-Pacific*. Routledge, New York, pp. 269–279.

Duncan, T. (2005) Current issues in the global hospitality industry. *Tourism and Hospitality Research* 5(4), 359–366. DOI: 10.1057/palgrave.thr.6040034.

Enz, C. (2001) What keeps you up at night?: key issues of concern for lodging managers. *Cornell Hospitality Quarterly* 42(2), 38–45.

Enz, C. (2009) Key issues of concern in the lodging industry: what worries managers. *Cornell Hospitality Report* 9(4), 6–14.

Fanggidae, V. (2016) Indonesia's workforce in need of total overhaul. Available at: www.thejakartapost.com/academia/2016/09/03/indonesias-workforce-in-need-of-total-overhaul.html (accessed 10 January 2017).

Gebbie, M. (2016) Asia Pacific market reports: Indonesia hotel market performance. Horwath HTL Indonesia. Available at: www.horwathhtl.com/2016/10/25/indonesia-hotel-market-performance/ (accessed 6 January 2017).

Geriya, W.I. (2003) The impact of tourism in three tourist villages in Bali. In: Yamashita, S. and Eades, J. (eds) *Globalization in Southeast Asia: Local, National, and Transnational Perspectives*. Berghahn Books, New York, pp. 81–94.

Guerrier, Y. and Deery, M. (1998) Research in hospitality human resource management and organizational behaviour. *International Journal of Hospitality Management* 17(2), 145–160. DOI: 10.1016/S0278-4319(98)00017-6.

Gunadi, A. (2016) ASEAN economic community impact for Indonesia. *Jurnal Opinio Juris* 19, 8–30.

Hyun, H.S. (2010) Human Capital Development. ADB Economics: Working paper series No. 225. Available at: www.oecd.org/site/iops/researchandworkingpapers/48239408.pdf (accessed 30 December 2016).

ILO (2009) Implications of the Global Economic Crisis for Tourism Employment: Case study for Indonesia. Available at: www.ilo.org/wcmsp5/groups/public/@asia/@ro-bangkok/@ilo-jakarta/documents/publication/wcms_122154.pdf (accessed 4 January 2017).

Indonesia-Investments (2016) Unemployment in Indonesia. Available at: www.indonesia-investments.com/finance/macroeconomic-indicators/unemployment/item255 (accessed 4 January 2017).

Jauhari, V. (2012a) Summing up key challenges faced by the hospitality industry in India. *Worldwide Hospitality and Tourism Themes* 4(2), 110–117.

Jauhari, V. (2012b) Strategic growth challenges for the Indian hotel industry. *Worldwide Hospitality and Tourism Themes* 4(2), 118–130. DOI: 10.1108/17554211211217307.

Jayawardena, C., Lawlor, F., Grieco, J.C., Savard, M. and Tarnowski, M. (2013) Challenges and innovations in hotel operations in Canada. *Worldwide Hospitality and Tourism Themes* 5(2), 177–189. DOI: 10.1108/17554211311314128.

Jennings, G. (2001) *Tourism Research*. John Wiley & Sons, Queensland, Australia.

Krippendorff, K. (2013) *Content Analysis: An Introduction to its Methodology*, 3rd edn. Sage, Thousand Oaks, California.

Kusluvan, S., Kusluvan, Z., Ilhan, I. and Buyruk, L. (2010) The human dimension: a review of human resources management issues in the tourism and hospitality industry. *Cornell Hospitality Quarterly* 51(2), 171–214.

Kuzel, A.J. (1992) Sampling in qualitative inquiry. In: Crabtree, B. and Miller, W. (eds) *Doing Qualitative Research*. Sage, Thousand Oaks, California, pp. 31–44.

Lado, A.A. and Wilson, M.C. (1994) Human resource systems and sustained competitive advantage: a competency-based perspective. *Academy of Management Review* 19(4), 699–727. DOI: 10.5465/amr.1994.9412190216.

Lu, Z. and Zhu, F. (2006) The post-purchase evaluation of the China's star rated hotel service quality: A content analysis of guest comments on third party hotel booking websites. *The Journal of China Tourism Research, 2(1/* 2, 93–103.

Miles, M. and Huberman, A. (1994) *Qualitative Data Analysis: An Expanded Sourcebook*. Sage, Thousand Oak, California.

National Wages and Productivity Commission (2016) Comparative Wages in Selected Countries. Available at: www.nwpc.dole.gov.ph/pages/statistics/stat_comparative.html (accessed 15 January 2017).

Negara, S.D. (2014) Indonesia needs to invest more in human resources. *The Jakarta Post*. Available at: www.thejakartapost.com/news/2014/08/09/indonesia-needs-invest-more-human-resources.html (accessed 15 January 2017).

Nuryanti, W. (2001) Bali and beyond: Locational effects on the economic impact of tourism. In: Teo, P., Chang, T.C. and Ho, K.C. (eds) *Interconnected Worlds: Tourism in Southeast Asia*. Routledge, Amsterdam, pp. 228–237.

OECD (2014a) How does Indonesia compare? *OECD Employment Outlook,* September 2014. OECD publishing. Available at: www.oecd.org/indonesia/EMO-IDN-EN.pdf (accessed 10 January 2017).

OECD (2014b) Indonesia. *OEDC Tourism Trends and Policies*. OECD Publishing. Available at: www.ec.europa.eu/growth/tools-databases/vto/content/oecd-tourism-trends-and-policies-2014-indonesia (accessed 10 January 2017).

Poulston, J. (2008) Hospitality workplace problems and poor training: a close relationship. *International Journal of Contemporary Hospitality Management* 20(4), 412–427. DOI: 10.1108/09596110810873525.

Qiu Zhang, H. and Wu, E. (2004) Human resources issues facing the hotel and travel industry in China. *International Journal of Contemporary Hospitality Management* 16(7), 424–428. DOI: 10.1108/09596110410559122.

Shodiq (2016) Talent war in the ASEAN economic community. *The Jakarta Post*. Available at: www.thejakartapost.com/academia/2016/08/31/talent-war-in-asean-economic-community.html (accessed 9 February 2017).

Solnet, D. and Hood, A. (2008) Generation Y as hospitality employees: framing a research agenda. *Journal of Hospitality and Tourism Management* 15(1), 59–68. DOI: 10.1375/jhtm.15.59.

Stringam, B.B. and Gerdes, J. (2010) An analysis of word-of-mouse ratings and guest comments of online hotel distribution sites. *Journal of Hospitality Marketing & Management* 19(7), 773–796. DOI: 10.1080/19368623.2010.508009.

Swan, A. (2015) Gaji Kerja di hotel (hotel salary). Available at: www.kompasiana.com/annaswan/gaji-kerja-di-hotel_54f34fed7455137d2b6c702f (accessed 9 February 2017).

Taty, S. (2016) Human capital competitiveness in ASEAN economic community (AEC): the role of regulation and supervision, Indonesia experiences. *Journal of Business and Management* 18(10), 1–7.

Tesch, R. (1990) *Qualitative Research: Analysis Types and Software Tools*. Falmer, Bristol, Pennsylvania.

Thadani, M., Vadehra, B. and Sindarta, F. (2015) Indonesia hotel watch 2015. HVS Singapore. Available at: www.hvs.com/article/7462/indonesia-hotel-watch-2015 (accessed 9 February 2017).

Ubeda-Garcia, M., Marco-Lajara, B., Sabater-Sempere, V. and Garcia-Lillo, F. (2013) Does training influence organizational performance? Analysis of the Spanish hotel sector. *European Journal of Training and Development* 37, 380–413.

Ulrich, D. and Brockbank, W. (2009) The HR business-partner model: past learnings and future challenges. *People and Strategy* 32(2), 5–7.

WTTC (2019) *The Economic Impact of Travel & Tourism*. World Travel and Tourism Council, London.

Wang, J.J. (2009) Issues, challenges, and trends facing the hospitality industry. *Management Science and Engineering* 3(4), 53–58.

Watson, S., D'Annunzio-Green, N. and Maxwell, G.A. (2002) Human resource management issues in hospitality and tourism: identifying the priorities. In: D'Annunzio-Green, N., Maxwell, G.A. and Watson, S. (eds) *Human Resource Management: International Perspectives in Hospitality and Tourism*. Continuum, London, pp. 1–14.

Widodo, W.S. (2016a) SDM pariwisata RI ditargetkan ranking 1 di ASEAN tahun 2019 (Indonesia's tourism human resources is targeted ranked 1 in ASEAN in 2019). Available at: www.travel.detik.com/read/2016/03/31/100313/3176439/1382/sdm-pariwisata-ri-ditargetkan-ranking-1-di-asean-tahun-2019 (accessed 14 February 2017).

Widodo, W.S. (2016b) 3 Masalah SDM pariwisata Indonesia bersaing di tingkat global (3 problems of tourism human resources in Indonesia competing at the global level). Available at: www.travel.detik.com/read/2016/03/31/080407/3176358/1382/3-masalah-sdm-pariwisata-indonesia-bersaing-di-tingkat-global (accessed 14 February 2017).

Yang, H.-O. and Cherry, N. (2008) Human resource management challenges in the hotel industry in Taiwan. *Asia Pacific Journal of Tourism Research* 13(4), 399–410. DOI: 10.1080/10941660802420978.

10 Politics and Tourism within South-east Asia

Joan Henderson*

Introduction

The purpose of this chapter is to examine the interactions between politics and international tourism which are shown to be inextricably linked in assorted ways. The focus of the discussion is on members of the Association of South East Asian Nations (ASEAN), an intergovernmental organization dating from the late 1960s and which now comprises ten countries. After some general observations with regards to the tourism–politics relationship, the chapter looks back to the 1970s when political uncertainties and upheavals characterized parts of the region and impeded tourism. Such circumstances are then contrasted with those of the current era which tend to be more settled, reflecting the political shifts occurring in the intervening years, and which are more favourable for tourism. While seeking to offer interesting insights into how tourism reacts to political environments in a part of the world where the industry is of growing significance, it should be emphasized that the account presented here constitutes only a glimpse into a complex and multi-faceted subject and region.

Politics and Tourism

Political structures, processes, actors and events all affect tourism development, while tourism itself may shape domestic and international politics (Hall, 1994). Tourism can become a political issue, inspiring debate and arguments between those who welcome and resist its operation. It is also a potential domestic and foreign policy tool, employed to pursue wider objectives, such as those related to national identity and international relations, as well as hegemonic goals. Hence, government backing of tourism and supporting strategies are an important factor in destination development which are furthered by the presence of political order and stability. A stable political sphere has implications for economies and societies; in combination, these help to determine the attributes of a place both generally and as a tourist destination, the latter encompassing accessibility, the stock of attractions and amenities, industry practice and awareness and images. Moreover, such stability is not necessarily confined to liberal democratic systems but can also be found in authoritarian countries in which international tourism has sometimes flourished. Notable examples of this include Spain under General Franco and certain Gulf states. It is, however, unlikely to do so in violent totalitarian states which can be defined as inherently unstable and dysfunctional. Such situations tarnish perceptions of the place and impede the movement of people, goods and capital and, as a consequence, mainstream tourists and industry investment are deterred. Nevertheless, manifestations of unrest in more democratic and relatively stable nations can also be disruptive, prompting damaging publicity and falls in arrivals, as demonstrated by, for example, the

*jch3011@outlook.com

 Tourism in Development: Reflective Essays (eds P.U.C. Dieke *et al.*)
DOI: 10.1079/9781789242812.0010

terrorist attacks in Tunisia in 2015 (Jeffrey and Bleasdale, 2017). The impacts depend upon the scale and severity of incidents, but international tourism has shown resilience and a capacity for rapid recovery in the face of terrorist outrages and other politically inspired crises. The extent to which these general observations apply to South-east Asia is now discussed.

Politics and Tourism in South-east Asia: Then and Now

Contemporary politics in South-east Asian countries is the outcome of a modern history comprising instances of war, colonization and autocratic rule. The past has informed government types and their decisions and policy making, but there have been marked changes in general over the last 50 years which have had consequences for tourism. For example, in the 1970s and despite having gained self-rule in 1959, Brunei was still formally protected by the British while Indonesia and the Philippines were dictatorships. Indonesia witnessed internal strife in 1965–1966 when thousands were killed in the aftermath of an unsuccessful military coup blamed on the Indonesian Communist Party. Subsequently, Suharto became acting president in 1967 and, after being elected president the following year, held office for 30 years until his resignation in 1998. In the Philippines, President Marcos imposed martial law in 1972 and retained power until the People Power Revolution in 1986. Unlike most of the region, Thailand had not been colonized and avoided the dilemmas of dealing with a colonial legacy, but the government was in transition after the 1973 student uprising. This prompted the ousting of military leaders, a succession of whom had ruled for 40 years, but a violent coup in 1976 led to the reinstatement of the armed forces which were also in control of neighbouring Myanmar/Burma in the 1970s following the army takeover in 1962.

Political upheaval was similarly apparent in the Indo-Chinese nations of Cambodia, Laos and Vietnam. The Khmer Republic was declared in Cambodia in the early 1970s, but the Khmer Rouge overthrew the government in 1975 during the ensuing civil war while the Lao People's Democratic Republic was instituted by the Communists in the same year following the king's abdication. The invasion of South Vietnam by the north followed the withdrawal of American troops with subsequent reunification in 1976, resulting in the establishment of the Socialist Republic of Vietnam. In comparison, conditions were more settled in Malaysia, although the country was coming to terms with the 1969 race riots. Many ethnic Chinese died in the disturbances precipitated by election losses of the major Malay party, the United Malays National Organization (UNMO). UNMO was the dominant partner in the ruling coalition which had been in power since independence in 1957. Parliamentary government was suspended for two years and a broader-based coalition, Barisan Nasional (BN), which was still led by UNMO, then took charge. Fears about racial tensions were also present in Singapore where the population comprised large communities of Malays and Indians in addition to the majority ethnic Chinese. Managing race relations was a top priority of the People's Action Party (PAP) which had retained office since winning the 1968 election, with reference frequently being made to the need to avert a recurrence of the 1964 race riots.

This brief summary of the conditions prevailing across South-east Asia only 50 years ago suggests that formidable barriers faced the development of tourism owing to the political factors at work, both directly and indirectly. Considerations such as internal strife and the aftermath of warfare raised doubts about personal safety and contributed to negative destination images, which discouraged individual tourists and industry operators and investors. At the same time, many economies in the region were in a parlous state so that poor transport and lack of spending on infrastructure engendered practical problems of international connectivity and internal mobility. In addition, certain political ideologies were hostile to outsiders and, as a consequence, entry was made difficult. Discretionary leisure tourism is perhaps most vulnerable to these negative influences, but business and other forms of travel were adversely impacted as well. It must also be recalled that international movement was less extensive around the world during this period, especially long-haul travel from major western generators.

Table 10.1. Types of political regime in South-east Asia.

Country	Ranking (out of 165)	Score	Type
Cambodia	125	3.59	Authoritarian
Indonesia	65	6.39	Flawed democracy
Laos	=151	2.37	Authoritarian
Malaysia	52	6.88	Flawed democracy
Myanmar/Burma	118	3.83	Authoritarian
Philippines	53	6.71	Flawed democracy
Singapore	=66	6.38	Flawed democracy
Thailand	=106	4.63	Hybrid regime
Vietnam	139	3.08	Authoritarian

Adapted from EIU, 2019

Nevertheless, there are contrasts in performance which reveal the significance of relative political stability. Both Singapore and Malaysia, for example, recorded over a million arrivals in 1975/6 compared to just over and under half a million in Indonesia and the Philippines, respectively. Thailand too was establishing itself as a popular destination, attracting in excess of one million visitors that year despite the activities of the military. Figures for the remaining countries are not documented by UNWTO, but are likely to have been very low.

The political landscape of the twenty-first century in the region displays some similarities, but also marked differences, to that outlined above. Generalizations about the politics of South-east Asia are complicated by national differences, but the Economist Intelligence Unit (EIU, 2019) argues that regimes conform to certain categories as listed in Table 10.1. The categorization is based on measures of electoral process and pluralism, government functioning, political participation, political culture and civil liberties. Although it has its detractors, the index does afford insights into the region's politics and it is perhaps noteworthy that still no nation meets the criteria for full democracy status. Brunei is excluded from the exercise, but the powers of the Sultan equate to authoritarianism.

As indicated in Table 10.1, Indonesia and the Philippines have taken steps towards greater democracy since the 1970s. President Suharto was forced to resign by mounting popular opposition in the late 1990s, ushering in a period of constant leadership changes and outbreaks of civil unrest. A degree of democratization has since occurred and a new generation of politicians, without strong links to the old regime, has emerged. Politics in the post-Marcos Philippines has been defined by military and civilian coups, parties centred on personalities, inept governments and accusations and instances of corruption. While adhering to certain democratic practices, several of these complaints are echoed by critics of the incumbent President Duterte who was elected in 2016. Thailand's politics, too, has exhibited instability and susceptibility to military intervention, recently dominated by the split between supporters and opponents of Thaksin Shinawatra who was deposed by the military in 2006. The death of the much-loved king and ascent to the throne of his less-revered son in 2016 introduced another dynamic into the distribution and exercise of power. Laos remains a one-party state, as does Vietnam which is run by Communist regime, albeit one prepared to pursue economic liberalization. A constitutional monarchy and parliamentary democracy were enshrined in Cambodia's 1993 constitution but democratic politics are not deeply rooted. The long-serving prime minister, Hun Sen, in office in various coalitions since 1985, has been censured for his authoritarianism and crackdown on opposition.

Considerable reforms have taken place in Myanmar/Burma since 2010, signified by free elections in 2015. The new government installed in 2016 is headed by Aung San Suu Kyi, the Nobel Peace Prize winner and leader of the National League for Democracy, which had formerly been suppressed. While hailed as a step forward, sceptics claim that the military are still very influential in the country's politics while international condemnation, much of it directed at Aung San Suu Kyi, followed army moves against Muslim Rohingyas in 2017. The campaign resulted in half a million refugees fleeing Rakhine State into Bangladesh, a step described by the United Nations as ethnic cleansing. Political circumstances have been less volatile in Malaysia where Mahatir Mohamad led the coalition which had governed since independence until his retirement in 2003. Power struggles within and between various factions seem to have intensified and Mahatir partnered with erstwhile political opponents to oust the BN in 2018. The unexpected election outcome was partly the result of a scandal about money laundering and the misuse of public funds, and the previous prime minister and his wife are, at the time of writing, facing corruption charges. The case of neighbouring Singapore, epitomizing the rewards of political continuity and stability generally, and for tourism, is returned to later in the chapter. Finally, Brunei remains an absolute monarchy, with concerns expressed internationally in 2019 when the implementation of a Syariah Penal Code Order was announced.

Deficiencies in governance therefore persist and there is evidence of continued discord and human rights abuses; but generally, the region's politics appear to be less turbulent now than during the 1970s. This greater stability is linked to economic advances with the more stable nations benefitting from relatively vibrant economies, and advancement overall is reflected in patterns of tourist arrivals which have seen substantial increases. Indeed, Asia has emerged as a prime generator of and destination for tourism, both domestic and international, in recent decades. According to the United Nations World Tourism Organization (UNWTO, 2019), Asia and the Pacific as a whole was the fastest-growing region in 2018 when it recorded 348 international arrivals and US$435 billion receipts or 25% and 30%, respectively, of global totals. The emphasis is frequently on China, but South-east Asia has risen to prominence, and figures for individual ASEAN countries can be found in Table 10.2, which illustrates the range in volumes and value.

Table 10.2. ASEAN countries international tourist arrivals and receipts 2018.

Country	Arrivals (thousands)	Receipts (US$ millions)
Brunei	278	190
Cambodia	6201	4352
Indonesia	13,396	14,110
Laos	3770	734
Malaysia	25,948	19,143
Myanmar/Burma	3551	1969 (2017)
Philippines	7129	7461
Singapore	14,673	20,528
Thailand	38,277	63,042
Vietnam	15,498	10,080

Adapted from UNWTO, 2019

Another indication of comparative performance is the competitiveness ranking published by the World Economic Forum (2019) which scores 140 economies in terms of enabling environment (business environment, safety and security, health and hygiene, human resources and labour market, and ICT readiness); travel and tourism policy and enabling conditions (prioritization of travel and tourism, international openness, price competitiveness and environmental sustainability); infrastructure (air transport, ground and port and tourist services); and natural and cultural resources. The list of determinants confirms the key role of governments and wider political and economic factors in the effective operation of the tourism industry and successful functioning as a tourist destination. Singapore is ranked 17 globally with a score of 4.8 compared to 3.4 for both Laos and Cambodia at 97 and 98, respectively. The remaining countries are Malaysia and Thailand (4.5), Indonesia (4.3), Vietnam (3.9) and Brunei and Philippines (3.8). Myanmar is not included in the ranking.

Growth has been assisted by changes in the political arena and there is now widespread official appreciation of the economic benefits of international tourism and support for the industry. Policies have been formulated and administrative machinery put in place with often very active destination marketing organizations which engage in aggressive promotion. Even Brunei, where the wealth accrued from oil reserves renders the financial returns from tourism of less relevance, is appreciative of the industry's role in diversifying the economy. ASEAN itself acknowledges the importance of tourism to the region and it is an arena for collaboration and co-ordination, although countries are rival destinations which compete for visitors. The alliance has weaknesses, however, which are an outcome of political and economic conditions and divergences between national and regional interests. Shortcomings are mirrored in the field of tourism in which the number and effectiveness of joint projects have been disappointing. At both an ASEAN and country level, it should not be forgotten that capacities to manage growth and secure sustainable development are debatable. There are undoubted opportunities, but realization of potential and improvements in sustainability require political will and competences in addition to resources, which may not be forthcoming.

While tourism trends have been positive overall, it must be recognized that the generally upward movement has been interrupted by adverse political events and there have been some irregularities from year to year. Illustrations include the 2002 Bali bombings which led to travel advisories being imposed by foreign governments warning against visits to the island and also a fall in domestic Indonesian arrivals. Nevertheless, as noted earlier, damaging effects tend to be short-lived and tourism demand usually proves to be resilient and quick to recover. In the case of Bali, following a decline in 2003, total international arrivals the following year exceeded the number in 2001, the year before the bombing, and have since continued to increase (UNWTO, 2016). However, tourist reaction to the political ideologies and actions of destination regimes is harder to assess and represents a valid topic for further research, particularly embracing comparative studies of Asian and western responses. For example, variations are indicated by the calls for a boycott of tourism to Myanmar/Burma in the early 2000s as a protest against the military junta which were received more sympathetically outside of the region. For others in the West, the lure of unknown places seen as exotic, such as those in Indo-China, may be very powerful and overcome reservations about political matters. Many tourists, whether Asian or non-Asian, also seem unaware of political issues and are more interested in what the destinations have to offer and prices than government orientation and activities. Such attitudes and behaviour imply a weaker direct relationship between politics and tourism, although the indirect impacts of the former on the tourist experience cannot be denied.

Thailand is an interesting example where tourism has continued to thrive despite continuing eruptions of political unrest, such as the 2006 military coup, the occupation of the main entry airport by Thaksin supporters in 2008 and the 2014 street protests in Bangkok. There was a contraction in visitor numbers in the immediate aftermath of these happenings, but the industry soon rebounded to maintain growth. Such a trajectory can be partly explained, perhaps, by Thailand's popularity and familiarity as a destination founded on its wealth of very appealing tourist offerings, while, in addition, much of its tourism is coastal-based and, hence, urban-centred political upheaval may be less threatening and disruptive. Although a newer destination than Thailand for contemporary tourists, the current administration in Myanmar/Burma also appears to have resisted the potential damage to tourism caused by outrage expressed globally at the treatment of the Rohingyas. At the same time, political reforms had previously boosted tourism; signs of democratization and the opening up of the country to the outside world stimulated positive publicity, interest and inbound tourist flows. The longer-term consequences of any reversal in this process of reputation recovery and rehabilitation, for many observers represented by the transformation of Aung San Suu Kyi from hero to villain, remains to be seen.

Finally, the manner in which politics and political history has become something of a tourist attraction in itself is noteworthy. This use of politics is revealed in the packaging and presentation of events such as the Vietnam War and the Cambodian genocide for visitors. Imperial heritages can also have tourist appeal and are promoted by destination-marketing organizations and the industry. However, conflicts and contradictions are sometimes generated, as demonstrated by refurbished colonial-era grand hotels and their cultivation of feelings of nostalgia for a time defined by the privileging of colonial elites. This corresponds to a general tendency towards excessive commercialization and commodification of heritage through tourism, distorting ideas and depictions. Politicization too can affect how colonization is communicated to tourists. Governments may not wish to dwell on or glorify a colonial past, facets of which may be inconsistent with contemporary policies and national myth-making, or it may be felt necessary to sanitize narratives to avoid offending tourists of certain nationalities.

Current Political Issues and Tourism

Some of the political factors and forces at work in shaping the environment within which the tourism industry operates, and features of demand and supply, have been acknowledged in previous sections. These include the nature of the government, political institutions and processes alongside decision and policy making by individuals and groups in combination with wider economic and social conditions. Among issues of current significance is the role of China as a world power and generator of international tourism. The former is evidenced by the Chinese 'One Belt, One Road' project which has promised US$1 trillion spending on transport and energy infrastructure in over 60 countries. More specifically, a series of road and rail corridors, intended to stimulate trade and communications within Asia and between it and the Middle East and Europe, are planned. The concept echoes that of the ancient Silk Road which got its name from the silk trade and stretched over 4000 miles. Chinese companies and investment are already visible in South-east Asia and many entrepreneurs from China are active there. In addition to business travel, leisure tourism from Mainland China has increased dramatically. The phenomenon is well documented, with Chinese outbound tourists leading the global spending league in 2018. As a consequence, destinations around the world have gained, not least in South-east Asia, and continue to seek to attract the Chinese market. Demand from the Chinese is, however, sensitive to political developments, as evidenced by the sudden sharp drop in arrivals to the Philippines owing to heightened tensions between the two nations over the sovereignty of the Spratly Islands which are also claimed by Taiwan, Malaysia and Vietnam.

Separatist movements, too, are a cause of disturbance to tourism, as are the ethnic conflicts that are prone to erupt. The threat and occurrences of terrorism continue to be another critical consideration, especially that inspired by Islamic extremism, and the presence and violent actions of Muslim groups in both the southern Philippines and in Thailand have resulted in zones of danger which tourists are strongly advised not to visit. In addition, a rise in religious conservatism within societies and some political parties has been observed in the Muslim-majority countries of Indonesia and Malaysia. The trend could have ramifications for tourism by cultivating unfavourable destination images and exposing tensions between tourist and resident values and conduct. While a less pressing a concern than in the theocratic states of the Middle East, the adverse reaction among parts of the international community to Brunei's more rigorous application of Islamic law affords insights into the dilemmas for stricter Muslim nations of accommodating non-Muslim tourist demands.

Politics can thus be a source of barriers to tourism, yet South-east Asian destinations can profit from the political problems of locations elsewhere in Asia and beyond. For example, the wider region was reported to be benefitting from visitors shunning Hong Kong owing to the turmoil there in 2019. South Koreans also appeared to be avoiding Japan that year because of deteriorating relations and opting for destinations such as Malaysia and Thailand instead. Unsettled conditions and crises of a political character can therefore represent both opportunities and challenges.

The Case of Singapore

Singapore provides an appropriate case of how political stability, order and efficient government can facilitate tourism. As a busy cosmopolitan trading centre before and subsequent to appropriation by the East India Company in 1819, the island has a long history as a destination for travellers. Tourism on a significant scale began, however, in the modern era after the gaining of full independence from British rule. It joined the Federation of Malaysia, but was expelled in 1965, accused of breaking a pledge not to intervene in Malaysian domestic politics. There were doubts about the island's ability to survive because of its small size and limited resources, but the republic prospered under the PAP, which was led by Lee Kuan Yew. The island was physically transformed through urbanization and industrialization engineered by the comprehensive land-use Concept Plan, which sets the direction for the next 45–50 years and is reviewed every decade. The economy is now among the world's most open and competitive and Singaporeans have one of the highest standards of living in Asia. Society and culture have been modernized to an extent, but divisions persist underneath a surface of harmonious multi-culturalism.

Lee Kuan Yew's son is the current prime minister and head of the PAP, which has ruled uninterrupted since 1959. The administration's capabilities and competence are not in doubt and there has been some relaxation under the younger generation of leaders, yet it is still criticized for excessive controls. For example, critics speak of the impeding of opposition within and outside parliament and constraints on free speech. Indeed, it is misleading to imply that there are no complaints about the government domestically, but the populace as a whole has hitherto been largely accepting of the curbs on freedom in exchange for the material benefits delivered. Yet, although the death of Lee Kuan Yew in 2015 could herald a new political era in which there is less reluctance to query PAP supremacy, the party seems firmly entrenched and likely to remain so in the foreseeable future. It depicts itself as the constructor of the modern state, the rightful party of government and the only one capable of protecting Singapore against assorted threats and of maintaining the success for which it is responsible.

The PAP and its style of governing has incited some negative comments and censure internationally, but its existence and achievements have generally favoured tourism for the purposes of both business and leisure. Heavy spending on transport and communication networks has been possible with the award-winning Changi airport offering global connectivity, while business confidence has been boosted and the open economy has drawn companies and capital. At the same time, high standards of health and hygiene allied with low crime rates also minimize fears about personal safety. Yet Singapore is not free from the dangers of Islamic-inspired terrorism, particularly given its location between Indonesia and Malaysia and its own ethnic mix, but it would seem better placed than many destinations to manage the risks. These advantages help to offset weaknesses of a lack of natural and cultural attractions in contrast to much better-endowed neighbouring countries. However, perceptions have been fostered of a destination that is overly regulated by a controlling government and, consequently, that is rather dull. These notions do not always conform with realities and are challenged by new projects and marketing campaigns which endeavour to convince audiences of the city state's vibrancy and excitement. Such efforts have had some success, but stereotypes persist and perhaps discourage certain visitors.

Conclusion

The experiences in South-east Asia recounted in this chapter indicate how a degree of political order and stability are conducive to destination development and, in turn, contribute to healthy and sustainable economies – other supportive considerations. Actual and potential instability can disturb industry operation and depress demand, but the severity of damaging impacts depends partly on the form, scale and duration of such instability. Authoritarianism and the absence of democracy are not necessarily inimical to tourism and these political issues may be perceived to be of little relevance to decision making by external tourism businesses and their customers. However, the official corruption which is seemingly endemic in several South-east Asian countries undermines international standing and inhibits good governance, extending to the arena of tourism planning and administration.

Despite ongoing uncertainties and the possibilities of corresponding disruption to international tourism, further growth in arrivals into South-east Asia appears likely. Demand and supply will, nevertheless, change and aspects of these shifts and underlying dynamics can be best understood within the context of broader political and economic movements. Looking ahead to the next 50 years or so, it is to be hoped that advances will be made towards meeting ASEAN aims of greater peace and stability, regional economic growth, social progress and cultural development. There are grounds for some optimism about regional and national prospects generally and with specific reference to tourism, yet any such views must be tempered by awareness of the formidable obstacles to be overcome arising in the political and other domains.

South-east Asia thus affords numerous interesting instances of the manner in which a country's politics and international political circumstances impinge on and interact with tourism. Tourism can be both frustrated and encouraged by these influences while also displaying a capacity to resist and recover from harmful pressures of a political origin. Only a glimpse into these processes and their workings in the region has been provided here and the subject has already yielded a wealth of valuable and thought-provoking studies. It is a rich area for further reading and research, tracking the unfolding and fascinating stories of the region and individual countries as they continue to evolve as political entities and generators of, and destinations for, tourists.

References

EIU (2019) *Democracy Index 2018*. Economist Intelligence Unit, London.

Hall, C.M. (1994) *Tourism and Politics: Policy, Power, Place*. John Wiley & Son, Chichester, UK.

Jeffrey, H. and Bleasdale, S. (2017) Tunisia: mass tourism in crisis? In: Harrison, D. and Sharpley, R. (eds) *Mass Tourism in a Small World*. CAB International, Wallingford, UK, pp. 191–199.

UNWTO (2016) Indonesia Tourism Recovery: Lessons and Programme. United Nations World Tourism Organization. Available at: http://cf.cdn.unwto.org/sites/all/files/pdf/indonesia_0.pdf (accessed 6 November 2019).

UNWTO (2019) *Tourism Highlights 2018*. United Nations World Tourism Organization, Madrid.

World Economic Forum (2019) *The Travel and Tourism Competitiveness Report 2019*. World Economic Forum, Davos, Switzerland.

11 In Search of Alchemy: The Art and Science of Destination Management

Terry Stevens*
Stevens & Associates

The Philosopher's Stone

This essay was prompted by the encouragement and enthusiasm of 'Kit' Jenkins – a longstanding friend, mentor, wise counsel and all-round gentleman – the understated philosopher and touchstone for many working in the world of tourism. He has been an inspiration, a source of good advice and plenty of laughter and was the first academic to appreciate my radical views on destination management, born out of practical experiences, recognizing the value of opening up these ideas to scrutiny and debate among academic communities. In 2017, at The Visitor Economy: Strategies and Innovation conference organized by Bournemouth University, I presented a paper about my work on benchmarking DMOs. Kit enthused that these findings had to be shared with a wider audience. It is very appropriate that the vehicle for spreading the word is in this volume dedicated to him.

Over the past 35 years I have had the honour to work on over 400 projects in 56 countries, involving all aspects of destination development and management. In 2004, a leading strategist, James Gibbs, from Highlands and Islands Enterprise (HIE) in Scotland, asked me to answer what must be the most fundamental of tourism questions: Why do some destinations outperform others? From a HIE perspective, there was a clear rationale for investigating the issue of destination competitiveness relating to HIE's concerns about the future of Aviemore and the Cairngorm Mountains which, having received very significant levels of public and private sector investment to revitalize its tourism appeal, was struggling to compete with other mountain destinations in Europe (Butler, 1985; Getz, 1986; MacLellan and Somerville, 1998; Holden, 2000; HIE, 2002; Lennon, 2004).

The HIE work had a dramatic impact, transforming the organizational structure of tourism in the Aviemore area by implementing the lessons from international best practice, leading to a new form of private sector-led DMO and creating a fresh approach to destination management. This innovation was heralded by the World Tourism Organization (UNWTO) as representing 'a fundamental shift in thinking on rural development' while, recognizing the success of this approach, the Scottish government encouraged other destinations to follow the DMO model (MacLellan, 2011; Anastasiadou and Halkier, 2014; Visit Scotland, 2013). In the five years that followed the arrival of this new concept in the Highlands, Scottish Enterprise, HIE and Visit Scotland supported DMO development in Loch Lomond and the Trossachs, Loch Ness, Fort William/Lochaber, Dumfries and Galloway and Edinburgh.

*terry@stevensassoc.co.uk

 Tourism in Development: Reflective Essays (eds P.U.C. Dieke *et al.*)
DOI: 10.1079/9781789242812.0011

The HIE commission was the start of a personal journey of exploration involving the real-time exploration of destination management and the benchmarking of some of the world's leading DMOs. Without exception, everyone involved in these DMOs readily shared their knowledge, experiences and their stories of success and failure. They were keen to complete, to receive and to share the results of the benchmarking survey. They represented many thousands of professionals working at the sharp end of tourism management delivering great visitor experiences: an elite band who intuitively understand how to deliver competent, dynamic, efficient and effective destination management. They understand that success is hard-fought, characterized by dogged determination, a strong belief in getting it right, even if that means setting democracy to one side for the greater good, and requiring the development of highly customized, innovative, hybrid business models to suit the needs of their destination. It is their perseverance, philosophies and inherent wisdom which have created the body of knowledge summarized in this essay. Their influence has been game changing, significantly contributing to rethinking the topic of destinations and how they are managed.

Destinations and Destination Management: Taking Stock

A recent review of the 2019 release of Bob Dylan's 1975 *Rolling Thunder Revue* opens with the statement, 'Oceans of ink have been spilt, entire forests felled, in pursuit of a subject we like to think we know, but that is, the wiser of us know and eventually discover, unknowable' (Cains, 2019). This observation can similarly be levelled at the cloudburst of publications on tourism over the past 50 years since it became a fashionable area of academic study. Within this canon of literature there has been some work focusing on destination marketing, management and, since 2012, literature specifically exploring DMOs (Morrison, 2013; Varghese and Paul, 2014; Neqruşa and Coroş, 2016). Analysis of the literature on destination management shows that it has emerged as a focus of study since 2000, with almost 40,000 published papers appearing between 2000 and 2016 (68% of them published between 2010 and 2016).The more recent publications on the concept of the destination, including those by Beritelli *et al.* (2014), Hristov and Zehrer (2015) and Pechlaner *et al.* (2014), and those about destination management and marketing (e.g. Adeyinka-Ojo *et al.*, 2014; McKercher and Prideaux, 2014; Pike and Page, 2014), reveal a lively discussion taking place trying to define a destination and the role of the DMO. In particular, however, the work on destination management by Hölzl (Pechlaner *et al.*, 2005) supported by others (Luštický and Kincl, 2012; Zlatković, 2016; Seraphin *et al.*, 2016; Klimek, 2017) helped generate the innovative thinking behind the St Gallen Destination Management Model. These findings, together with the real-time benchmarking work by Stevens & Associates' (S&A) (Stevens, 2018), raise fundamental questions about the usefulness of present definitions of destination management and the way we consider the future work of DMOs.

As a result, the UNWTO's definition below of the tourism destination is regarded as being symptomatic of the shortcomings in dealing with the challenges related to destination formation, operation and impact measurement:

> A local tourism destination is a physical space in which a visitor spends at least one overnight. It includes tourism products such as support services and attractions, and tourism resources within 1 day's return travel time. It has physical and administrative boundaries defining its management, images and perceptions defining its market competitiveness. Local tourism destinations incorporate various stakeholders often including a host community and, can nest and network to form larger destinations. (UNWTO, 2002)

The 2014 *St. Gallen Consensus on Destination Management* (Reinhold *et al.*, 2015) concluded that the UNWTO definition falls short in providing a meaningful conceptualization of the tourist destination for the future, demanding more research on all aspects of destination planning, development and management (Beritelli *et al.*, 2015). This resulted in the group proposing an alternative definition based on an understanding of the demand-driven mechanisms that shape a more variable and multi-faceted view of the destination, with Reinhold *et al.* (2015, p. 403-417) stating that:

> A tourism destination is a market-oriented productive system. By means of their behaviour in space and time and household production, tourists, aggregated as flows, activate this productive system. Actors on the supply side support tourists in producing their holiday experience by providing marketable resources and together, tourists and supply side actors co-produce an experience – from origin to destination(s) and back.

'The Times They Are a-Changin': From Destination Marketing to Destination Management

In 1994, Gunn recognized the need for destination management, describing the need for a more integrated destination-wide approach involving all stakeholders in order to deliver a better matching of supply and demand in order to maximize the use of destination resources (Gunn, 1994). Subsequently, Laws (1995) produced one of the earliest references to include destination management in the title of a book while acknowledging that others had previously alluded to the need for a destination management approach and going beyond destination marketing activity (Gunn, 1972; Leiper, 1979; Inskeep, 1991).

It took a further ten years to legitimize the idea of destination management when Ritchie and Crouch (2003) produced a management framework for the 'complex and multi-faceted nature of all the factors that affect destination competitiveness'. Four years later, the UNWTO (2007) established a platform to promote destination management. It was clear that established destination organizations were coming under increasing pressure to evolve (Presenza, 2006) and, specifically, to move away from marketing to become management focused (UNWTO, 2011) with inclusive stakeholder engagement, more robust governance and stronger private-sector leadership (Gretzel *et al.*, 2006; Hristov and Zehrer, 2015) and with greater consideration for the overall quality of life of local people (Timur and Getz, 2008; Morrison, 2013; Pike and Page, 2014). These factors have immense prescience given today's issues associated with so-called 'overtourism', concerns about the quality of visitor experience (NECSTUoR, 2018), the need for more innovation (Hjaglar, 2015; UNWTO, 2019) and creating shared community value and wealth (Porter and Kramer, 2011; Serra *et al.*, 2017).

In 2009, the British Chambers of Commerce argued for DMOs, their concerns directed at Visit Britain (VB) and the UK government in a hard-hitting report about the inadequacies of the structure of tourism, especially below the sub-regional level (BCC, 2009). Two years later, Robinson (2011, p. 4-5) argued that:

> The spotlight is on destinations, as the key entities to implement the Government's tourism strategy.....Tourism growth needs much more than marketing, but few DMOs have embraced tourism management. The Government's Strategy prescribes that DMOs cover tourism management, and be private sector led. This means that DMOs must be formalised and strengthened.

Despite this unambiguous call for change, there was little evidence of a shift away from marketing towards a management focus and, with destination management appearing as an agenda item in international tourism-related conferences throughout 2011, British Chambers of Commerce continued to bemoan the absence of destination management, directing criticism in particular towards Visit Britain (Visit Britain, 2010).

Clearly, destination management is not an easy job. Success is based on the complex combination of the stewardship of tangible and intangible assets within a defined geographic area (Neqruşa and Coroş, 2016). In order to assist destination organizations to deal with their 'new' role as managers, a library of advisory materials has emerged to help make the shift in a systematic way. The advice takes the form of: simple checklists (Brooks and Forman, 2003); detailed guides on destination management (UNWTO, 2007; Scottish Enterprise, 2008; Australian Government, 2016; Visit Wales, 2019); reviews of international best practice (The Communications Group, 2018); and large compendiums covering all aspects of the subject such as Morrison's magnum opus (2019) built on his earlier work (2013).

Neqruşa and Coroş (2016, p. 267) concluded that:

> Destination success is a combination of tangible and less intangible attributes... All these factors

> determine the visitor experience. While some of these are beyond the control of the DMO, for a large-part, organisations can assure their managerial process. The DMO managers have a strong influence on destination success (and) through DMO performance evaluation, there can be identified areas for economic improvement, efficiency, capacity and effectiveness in achieving strategic and operational needs, which can further increase the destinations competitiveness.

Yet, despite a decade of advocacy for change, the emergence of DMOs has been slow to take place and destination management is still in its infancy with only few competent DMOs. In the UK, the public sector exhibits an entrenched, stubborn reluctance to take on board the evidence of how successful private sector-led DMOs can be organized, governed and resourced. The national tourism bodies, often in collusion with local authorities, continue to maintain a highly centralist 'command and control' approach over destinations, demonstrating a reluctance to shift from government to governance (Cooper and Hall, 2008). There is fear of allowing the private sector to take the lead in delivering destination management reinforced by a misguided desire to retain some illusion of 'power' (Shikira and Qirici, 2013; Serra *et al.*, 2017; Qirici and Theodhori, 2013), flagrantly ignoring the lessons from international best practice. Indeed, in the Czech Republic, Poland, Romania and the Slovak Republic, recent research (Derco, 2013; Neqruşa and Coroş, 2016) has clearly illustrated the failure of destinations to be competitive in attracting tourists and investors in tourism, resulting from 'inconsequential organisations of tourism at the destination level coupled with decisional centralisation' (Derco, 2013, p. 354-364).

Nevertheless, the UNWTO (2019) recognizes there has been increasing awareness of the need for management of tourism destinations, with DMOs having a broader mandate that includes strategic planning, co-ordination and management of a full range of activities within an adequate governance structure that integrates different stakeholders, including the community, and operating with common goals: 'These developments stem from the urge to achieve an optimal management of the destination and a collective vision in pursuit of a common goal, the competitiveness and sustainability of the destination' (UNWTO, 2019, p. 1). The main goal is a strategic approach to link up the complex and often very isolated, fragmented and occasionally diverging elements for better management of the destination.

Yet much of the literature implies that there has been no real breakthrough in terms of the application of a consensual approach to destination management, with some concluding that there is a distinct lack of DMOs (Adeyinka-Ojo *et al.*, 2014), and pointing to a shortage of empirical evidence about the impact of destination management in creating competitiveness. S&A's research addresses these concerns, revealing a significant number of DMOs with a competent managerial approach bravely forging ahead with new dynamic models of governance, getting to grips with contemporary issues by breaking the rules associated with collaborative working in order to drive new forms of partnerships and innovation. The experiences of these exemplar DMOs are now discussed illustrating that their approach is based on science and art: the alchemy of destination management.

Benchmarking Successful DMOs: Revealing the Art and Science of Destination Management

Benchmarking is an important strategic activity to be undertaken in pursuit of the continuous search for best practice. It is critically important where destination management itself is a key factor of competitiveness. Benchmarking aims to help those involved to objectively consider where they are at, help shape ambition, direction and to stimulate, provoke and inspire innovative ideas. Kozak (2004) pulls together the ideas developed to create a framework for destination benchmarking, recognizing that 'destination competitiveness is not an individual concept. Rather it is totally dependent on social, economic and political developments in the tourist-generating countries as well as the tourist-receiving countries' (Kozak, 2004, p. 185).

S&A's destination benchmarking embraced this holistic, symbiotic systems approach in the analysis of destinations of all types and locations from around the world. In addition to examining performance and process measures of the destination, S&A deployed a specially designed Simple Multi-Attribute Rating Technique to assess the opinion of key tourism and community stakeholders about the reputation, competency and efficacy of the way their destination was managed using a Likert technique to derive an overall score for the destination with sub-scores for different dimensions of the destination and its management.

This model was based on the pioneering research developed for the evaluation of winter tourism destinations in Canada by Crouch and Ritchie (1999), subsequently refined by Hudson in collaboration with Ritchie and Timur (Hudson, 2000, 2003; Hudson *et al.*, 2004). Using this approach, in discussion with Hudson, S&A now has data for over one hundred destinations collected between 2003 and 2018, making it possible to benchmark these destinations both at a particular point in time and/or over an extended period of time, and permitting destinations to objectively benchmark with their competitor set.

In 2018, the results from the benchmarked destinations were anonymized and collated into a set of graphs revealing some obvious, consistent patterns, showing that the destinations are doing the same things well and the same things badly. This was surprising considering that the information reflects a large number of destinations, each with unique settings, different assets and characteristics and surveyed over a 15-year period. The results illustrate how the management of the destination is perceived by their key stakeholders:

The activities they all do particularly well are:

- visioning with strategy and action plan;
- clear positioning;
- strong leadership with a well-resourced, well-governed, private sector-led DMO;
- innovative approach to market-focused product development;
- dynamic funding models;
- regular reviews of competitors;
- investment in market research and trend analysis;
- good community engagement; and
- commitment to sustainable and ethical development.

The activities they all intend to improve upon in the future are:

- human resource development;
- accessing new sources of capital investment;
- new methods of harnessing social media for marketing; and
- enhancing local access and transport for residents and tourists.

In many ways the conclusions from studying these DMOs can be summarized as follows. Interestingly, these findings are comparable to the conclusions of the 2018 *St. Gallen Consensus on Destination Management* and their five 'domains of destination management'. They are:

a. the need to reconsider and to review the prevailing definition and geography a destination;
b. the importance of destination management and a well-run DMO;
c. the need to regularly review the role, function, organization and funding of DMOs;
d. destination success is highly dependent upon the clarity of its positioning, the nature of its tourism offer, and sustainable and ethical foundations;
e. commitment to innovation, creating sustainable experiences for the visitor and quality of life for the residents in line with the United Nations Development Programme's Sustainable Development Goals (UNWTO/UNDP, 2017; also NECSTUoR, 2018).

It is now possible to draw some conclusions about the success factors for destinations. A discussion of some of these factors follows together with reflections of the likely future agenda for destination management and DMOs.

Successful, Sustainable Destination Management: Some Reflections

What constitutes a destination? The simple truths

Destinations are places where people want to be for whatever reasons and motivations. The destination must be accessible; there must be a critical mass of things to see and do as well as places to stay and, for the destination to work successfully, it must be professionally and competently managed. This means that a 'managerial' approach involving a DMO should be adopted, taking into account the size and scale of the destination's geography, its tangible and intangible assets and resources, its stakeholder base and its visitor experiences. These simple truths resonate with Vanhove's (2005, pp. 132–138) observations that the fundamental product of tourism is the destination experience. Competition, therefore, is centred on the destination and, for most tourists, this experience takes place in a rather small geographical area – an entity which, from a tourism management point of view, is managerial.

The analysis reveals ten shared, dominant, managerial characteristics:

1. *Compact:* the destinations are relatively small in scale averaging 1800–2500 km^2 in size and all less than 3000 km^2 but having the flexibility to expand and grow geographically according to the strategic needs of a particular visitor experience, event or product promotion – an approach that is defined as the dynamic geometry of destinations.
2. *Coherent:* the destination's geography, geomorphology, topography, natural and built heritage and infrastructure makes them coherent from visual, social and cultural perspectives, creating a strong sense of place and form that is easily understandable to the visitors.
3. *Customer focused:* destination managers are well-informed, through systematic research, annual benchmarking and frequent customer engagement, about tourism trends, visitor behaviour, perceptions and expectations, allowing them to develop dynamic strategies responding to this market intelligence on a frequent basis.
4. *Cohesive:* the destination makes sense as a logical construct for all stakeholders as perceived from different dimensions (topographical, territorial, cultural, historic, political, economic and strategic).
5. *Critical mass:* the destination must have a clear vision, strategy and costed action plan ensuring the development of an appropriate, competitive, sustainable and balanced range of products, experiences, facilities and infrastructure to reflect its brand position and meet the needs and expectations of its residents and visitors.
6. *Competent:* the destination must be competently managed by a dedicated, well-resourced, well-organized and committed team of local people. These teams tend to be structured and functioning as an efficient, not-for-profit company with community intent, led by a board of private sector and community representatives with a strong leadership ethic, operating collaboratively with strong networks and communication channels, driving investment using innovative methods of raising capital and revenues.
7. *Consistent:* the destination must ensure that the 'promise' made, implicitly or explicitly, to visitors through its marketing are delivered on a consistent basis. The complicated, multi-faceted nature of the tourism system makes the co-ordination and management of the components the most challenging aspects of destination management.
8. *Creative and innovative:* creating an environment that stimulates, fosters and supports innovation and creativity is, as will be discussed in greater detail later, of increasing importance to achieving successful, sustainable destinations.
9. *Competitive:* driving success is the need for a destination to be competitive in a national and international context over time. Creativity and innovation are key factors of competitiveness and DMOs need to be acutely tuned to recognizing and applying innovative thinking to all aspects of their work.
10. *Collaborative and co-created:* 'together stronger' is often used to encourage a collegiate, open, inclusive and transparent approach to destination management characterized by common trust and a shared set of objectives.

Competent: the importance of destination management delivered by well-run DMOs

There exists a considerable amount of research on the importance of destination management and its critical success factors, usefully summarized by Pike and Page, 2014. Their study references over 450 articles, concluding that destination management delivered by appropriately organized and governed organizations is critical to successful tourism development. A more focused review (MBIE, 2017) on effective destination management concluded that there is a high degree of agreement about the factors that positively affect successful tourism destination development. It demonstrated that the most important factor for successful destination development was the presence of an effective, efficient and well-resourced destination management body structured as a private–public partnership. In 2019, a study on behalf of Fáilte Ireland examined 200 transformational tourism projects from around the world that had positively impacted on their host destinations. The most important findings of this work showed that the majority of projects exist in destinations where there is an existing DMO or, as a result of the transformational project, a newly activated DMO has been created.

How Successful Destination Management Organizations Are Organized, Governed and Resourced

For the editor of 'Wired' magazine (Rowan, 2013), experimentation will be the key *modus operandi* for all involved in DMOs: 'The next 10–15 years will be a time of unprecedented experimentation; where old formulas no longer hold true; where traditional forms of partnership and collaboration will have to change and where hybrid solutions created by hybrid thinkers will prove to be winners' (Temperton, 2013). The stand-out feature will be the ability of DMOs to react quickly to the changing tourism environment.

At the 2010 UNWTO seminar on the subject of public–private partnerships in tourism, it was noted that there needs to be innovation in the governance models of DMOs. It was particularly advocated that the private sector needs to take the lead in order to eliminate weaknesses that prevail in most destinations where the public sector dominates, such as: the public sector being risk averse; an unacceptable tolerance of bad results; and, finding it difficult to raise quality in tourism and to lead innovation while struggling to positively shape brand and positioning. The DMO model that appears to be most efficient in this environment is one that has the following governance set-up:

- The DMOs are always to adopt a partnership approach involving the community, the public, third and private sectors with the public sector playing a support role enabling the private sector to drive destination management.
- Being a private sector-led, not-for-profit, community interest company with a self-appointed, relatively small-scale board with the skillset, expertise and talent to run a complex, growing, business.
- The chair of the board is generally a person who is well known and well respected throughout the community, not necessarily from the tourism industry.
- Often the chair is encouraged to stay in that position for an extended period to ensure continuity of purpose, trust and delivery of the destination strategy.
- There is a small, professional executive led by a director or CEO.
- The work of the board is supported by a network of special-interest or sector-related groups and product clubs.
- There is a strong commitment to regular consultations engaging with all interest groups.
- Representative membership bodies are encouraged; however, these bodies do not get automatic representation on the board of the DMO.
- There is a constant search for fresh and innovative ways of raising capital and revenue funding and other resources.

The need for strong leadership

In this dynamic environment, the need for strong, visionary, responsible leadership of the DMO has become very apparent (UNWTO, 2007; Mendiratta, 2011; Tigu, 2012). Successful DMOs are characterized by a determined, inspiring, chair of a board, who works in unity with the CEO of the executive team. The chair is invariably a well-respected local person whose leadership qualities were well known in the community and whose decisions were regarded as being made for the benefit of the wider community and not just the tourism stakeholders. These leaders have generally held their position for relatively long periods of time thus ensuring continuity and consistency in their approach irrespective of the volatility in the world of local and national politics.

Competitiveness: The role of innovation and creativity

The increasing pressure on DMOs to constantly search for new ideas to maintain their competitive edge to secure differentiation is leading the search for international best practice to reveal the elusive recipe to deliver an environment within the destination that allows innovation and creativity to flourish (Decelle, 2004; Block and Keller, 2008; von Hippel, 1988). Innovation is now widely regarded as one of the most important strategic factors in tourism most likely to assure sustainable growth and community wealth (Peters and Pikkemaat, 2006). For Refai (2011, p. 3), innovation is of fundamental importance:

> Innovation – new and transformative ideas – plays a vital role in tourism's ability to respond to an ever-changing world. Innovative thinking, be it new concepts, products, technologies and even policy, drives competitiveness and opportunity and allows the tourism sector to stay ahead of changing trends. Innovative is particularly important for ensuring that destinations remain sustainable.

Destinations must be in the vanguard in shaping unconventional responses and new experiences because of the dynamic fast-moving consumer and lifestyle trends. Yet all too often tourism is characterized by unimaginative, ordinary, replicated serial reproduction of visitor experiences (Stevens, 2013). Innovation in tourism has not been given the attention it deserves, especially given the urgent need to address the growing concerns among tourists to escape an ever more depressing sea of sameness in many destinations. These symptoms are expressed in popularist writing, such as Martin Cohen's (2006) *No Holiday: 80 Places You Don't Want to Visit* and the mockery of the, Jetlag Travel Guide (2004, 2014) to the fictitious destinations of *Phaic Tăn, Gastronesia,* and *Moustaschistan*. More telling was the comment written in the guest comments book at one of the most treasured visitor attractions in England in July 2017. At Wordsworth's Dove Cottage, a discontent had written, 'We are all bored, we are oh so tired of everything!' (Stevens, 2017b).

Most innovation in tourism emerges from outside the tourism industry (von Hippel, 1988) and even then, according to Sendlinger (2018), personal communication), 'less than 10% of the (tourism) industry follows the 0.1% that really delivers innovation'. Hjalgar's (2015) long-term research (2002–2015) into innovations that have transformed tourism over the past 700 years concludes that none were specifically designed for tourism but have, nevertheless, affected tourism to a significant extent. She pinpoints the need for the tourism industry to continue to respect the influence of innovation in science, life sciences and technology, fearing Stipanuk's (1993) concerns that contemporary tourism researchers will only be concerned with the impacts of information technology. Many studies have concluded that investigations of innovation in tourism have, according to Hjaglar (2015, p. 5), 'often come up with very bleak results' concluding that 'tourism accommodation businesses and restaurants, transportation companies are not particularly innovative', while visitor attraction operators are slightly more innovative (Fuglsang *et al.*, 2011).

Where innovation in tourism occurs, it is as a result of developing networks and being open to external sources of inspiration: from the work of leading DMOs and from other disciplines (Hall and Williams, 2009). According to Torrent (2018), 'interesting things happen in tourism when you allow different disciplines to collide' while for Turner (1982), p. 45), innovation most frequently 'occurs in the interfaces and limina, then becomes legitimated in central sectors'. This implies a direct link with the ideas associated with *communitas* and the core characteristics of the creative clusters – a subject of much needed further research (Filep and Pearce, 2014).

At the heart of this challenge is the quest by DMOs to find the elusive answer to the question: 'Where do good ideas come from?' and rise to the challenge set by Sendlinger (2011, p. 7), who stated that:

> Why not think things that have never been thought of before? This is a time when tourism needs hybrid thinkers to find hybrid solutions. Hybrid thinkers blend various skills and fields of thought to master the most intricate of assignments. They develop innovative concepts. The problem is that those working in tourism are insufficiently innovative or creative to deliver hybrid solutions.

Innovation is a real issue for DMOs where, in the absence of self-developing innovation in their tourism businesses, there is clearly a need to 'step up to the plate' in order to fill the gap, by exploring new ways of working, by exploiting external linkages and by establishing in-destination networks. Where do good ideas come from that can assist DMOs in raising their game? In recent years much has been written on this question, looking at it both as a process and as a phenomenon. It is apparent that there are certain talents and, according to Mlodinow (2018), 'qualities of thought, different aspects of a cognitive style' – which he calls 'elastic thinking' – and systematic new ways of thinking that can help identify and nurture new ideas, such as the capacity to let go of comfortable ideas, becoming accustomed to ambiguity and contradiction, and the willingness to experiment and be tolerant of failure.

Interestingly, recent studies have substantiated what was by many often discarded as a myth; that is, Jarosz *et al.* (2012) have concluded that the average man produces his most creative thinking with an alcohol level of 0.075%. Known as the 'creative peak', it is considered the ideal state for problem solving, inventing and generally coming up with great ideas. For Johnson (2010), however, the source of 'good ideas' is rooted in a more traditional scientific research which he summarizes into a succinct schedule of behavioural change which can be used by DMOs to foster innovation. This is summarized in Table 11.1

Table 11.1. An interpretation of Johnson's (2010) model applied to destination management.

The adjacent possible: Engage those who are sitting next to you
Liquid networks: Involve poets, musicians, anthropologists
Intuition and hunch: be counterintuitive and notional
Serendipity and risk taking: Just do it!
Exaptation: Allow activity to acquire features not originally intended
Hybrid: Be willing to allow a fusion of ideas from different sources

(From Stevens, 2017a)

Competitiveness: the role of design in architecture and the value of tourism in the destination experience

It is now taken for granted that quality is essential in all aspects of tourism service and experience delivery at the destination level. Today we are witnessing a rapidly emerging area in the growth of interest in applying high-quality design to physical spaces and in the architecture of hotels and visitor centres (Stevens, 2017a; Scerri *et al.*, 2019). Indeed, Scerri *et al.* (2016) state that 'it is difficult to discuss tourism without discussing architecture. It is a commodity for tourism consumption, cultural capital and enhances sense of place'. Across the exemplar destinations there is an obvious trend towards the encouragement of investment and the patronage of star architects producing exciting, innovative architecture and quality visual interventions as a means of enhancing the overall visitor experience (Crouch and Lűbbern, 2003; Jencks, 2005; Dempsey, 2006; Nicholson, 2013; Stevens, 2016).

The 'Gehry effect' is evidenced in the Basque Country. Bilbao has embraced leading architects in one form or another to acclaimed success since the city first invited Gehry to design the city's Guggenheim in 1993. San Sebastian, the Basque Country's centre for gastronomy and film, has followed suit, while south of Bilbao, in Elciego, the Marqués de Riscal Winery has created the City of Wine by Gehry, with the chairman of the company stating that the €66 million investment far exceeded their expectations, taking 'no place' to a world class, competitive destination.

Architecture Attracts Visitors by Pla'tou (the Platform for Architecture in Tourism, established in Austria in 2007) promotes the fact that: 'The topics of architecture, culture of construction and design are capturing the attention of the public'. The organization asserts that: 'the innovative potential which lies within contemporary architecture is still hardly used in the tourism industry'. Their research (www.platou.at) shows that success and enhanced profitability were directly correlated to the use of contemporary architecture. Copenhagen, Aarhus, Oslo, Liverpool, Zurich, Bilbao, Barcelona, Montpellier, Abu Dhabi, Dubai, Chicago, Denver, Singapore and Melbourne are taking full advantage of these opportunities.

In rural areas there are two outstanding examples of destinations where contemporary architecture is now driving tourism. Bregenzerwald (Austria), a rural area with a once flagging agricultural economy, has developed a global reputation as a world-class destination by using the power of architecture to give the destination a competitive edge and distinctiveness. The second example is the story of Norway's tourist routes (Guvstasen, 2016), which started in 1994 as a trial project to offer motorists an alternative to main roads that gave them access to stunning architecture along their routes. There are now 18 National Tourism Routes (NTR) in Norway, all of which will be completed by 2025. Running along the coast and fjords, through countryside and mountains, the routes offer world-class scenery and architectural installations.

Co-created experiences

Experiential tourism has evolved rapidly. It is now about deep co-creation: asking and delivering the answer to the question of the tourist: what would you, the visitor, like to do today... what do you want to do and, how can we work together to create the experiences you want? Research shows that the most important memories are experiences where participants had gained personally significant outcomes. The tourist is wanting to become much more in control of the design and management of their own narratives and interests and, hence, their experiences. This is now driving the potential for premium pricing and greater values for all involved.

In this process there is a discernible shift from selling the destination (place) and its assets to the narrative of the people, thus appealing to a tourist in terms of their intellect, spirituality, emotional triggers, senses, feelings of belonging, destinations they love (Roberts, 2005, 2006) and things the visitor cares about, such as: social issues, the environment and its conservation, sustainability and local culture, quality of life of local people as well as themselves. The most ardent emotions are evoked not by landscapes but by people and their stories thus evoking a sense of wonder. It is the role of DMOs to create an environment designed and capable to deliver great experiences in an integrated, joined-up and holistic manner. It is more than publishing lists of experiences; a list of potential experiences is just that – a list! The experiences must continually evolve and demonstrate continuous innovation to deliver the whole story by what Lippard discusses, in her two books about tourism and place, is all about mystifying the mundane, amplifying the exotic, minimising the misery, rationalising the disquietude and romanticising the strange' (Lippard, 1997, 1999).

Talent retention and human resource development

Talent attraction and retention through strategic human resource development initiatives is of central importance to destination competitiveness. A big question for the future of DMOs is how can the general aims of sustainable and transformative tourism development be translated into the specific objects of change including the role of people? These issues have been raised by the President of the WTTC who, in 2015, stated in his closing speech at the WTTC Global Summit in Madrid (World Travel and Tourism Council, 2015), that:

> Tourism is a people industry. We depend on quality people, but we need the right policies, programs, projects and partnerships to ensure that the workforce of the future knows about the opportunities in our sector and the appropriate

skills, talent and knowledge to support the future growth.

Halkier's (2013) conclusions identify the need to increase the capacity and skills of individuals through: the development of networks and collaborative working; knowledge sharing; conditional access to information and training/ advice tourism entrepreneurs; unconditional access to information about marketing and branding and product development; unconditional access to institutional support; conditional access to organizational capacity through network creation; increased emphasis on creative generation of market intelligence; and establishing efficient DMOs.

Most of the human resource initiatives are driven from the 'top down' with the national tourism organization creating the framework and structure for the development of leadership capacity building, providing co-ordination of the activities and overseeing accreditation of programmes with an expectation that delivery of the training will be co-ordinated, and organized, at destination level by the DMO.

Communitas, liminality, flow and knowledge exchange

In finding solutions for DMOs, there is 'no one glove that fits every hand': the approach to the way DMOs are governed, funded and operated varies considerably according to the specific needs, circumstances and stage in the tourism area life cycle of the destination (Butler, 1980, 2011). The recipe for their success involves an eclectic mix of robust evidence-based decision making, hard-nosed business planning infused with lots of intuition, extraordinarily high levels of networking, community engagement and what can only be described as a big dose of 'alchemy' (Albertus, 2000; Ratcliffe and Heath, 2019; Stevens, 2011) – a seemingly magical process of transformation and creation, or a combination derived from the concept of what is widely regarded as the medieval forerunner of chemistry with attempts to convert base metals into gold or to find a universal elixir.

The findings from a new research project presented to SCIENTIA[1] (Stevens, 2017c) supports the notion that there is more to destination management than the traditional business approach. This project examines the flourishing creative and artistic enclaves that gave birth to a generation of niche destinations at various times over the past 100 years in locations around the world suggesting similarities between the way these historic clusters functioned and the way destinations are governed and operate today. The hypothesis states that contemporary interest in cluster theory, as it can be applied to destinations (Nordin, 2003; Estêvão and Ferreira, 2009; Truong, 2016) can inform DMOs by understanding what made the creative clusters a successful phenomenon. The historic clusters include: the art communities in Kirkcudbright (Scotland); Rovinj (Croatia), Skagen (Denmark), Cornwall (England), Montsalvat (Australia) and Sedona (USA); together with the cluster of poets and philosophers who gathered in Swansea in the 1950s; and the rock communities of Woodstock and Laurel Canyon (USA) in the late 1960s.

The early findings from this research have identified four dominant concepts that are common to each of the creative clusters and which seem likely to become increasingly relevant to the future of destination management. They are:

- *Communitas:* inspired fellowship, a group's pleasure in sharing common experiences... a social unit who have something in common...an unstructured (anti-structure) community in which everyone is equal... developing an intimacy among people who experience liminality...existential and spontaneous (Esposito, 2009; Turner, 2012).
- *Liminality:* threshold...at the border, the edge both metaphorically and physically (of place and society)...shifting from normal modes of societal actions...entering a period of scrutiny of central values... normal limits of self-discovery and behaviour are undone...the very structure of society of suspended, de-constructed...anti-structures appear...the fault lines of legitimate and illegitimate (van Gennep, 1909).
- *Flow psychology:* a state in which people are so involved in an activity that nothing else matters... the experience is so enjoyable that people continue to do it even at great cost... the best moments of our lives occur if the

mind and body are stretched to the limit... this is vital to creativity and well-being and is vital to self-actualization (eudaimonia), the antidote to hedonia (Csikszentimihalyi, 1990).

- *Knowledge exchange:* a process that, rather like the triple helix of innovation, brings together different groups and disciplines in order to exchange ideas, evidence and expertise in order to advance understanding and improve the environment that will stimulate success, enterprise, innovation and creativity (Williams, 2017; McTiernan *et al.*, 2018).

Fu(turism)o: The Future Agenda Shaping the Work of DMOs

In 2019, the Museum of Modern and Contemporary Art of Trentino (MART) presented an exhibition called *The Invention of the Modern.* Central to the storyline is a section on Futurism – an artistic movement that masterfully embodied the spirit of the avant-garde, making the meaning of the term going ahead, pushing further on than others and breaking with tradition. This is an appropriate call to arms for DMOs as they face the challenges of destination management over the next 20 years.

There is no shortage of information, advice and evidence about tourism futures and changing market demands (Yeoman, 2012; Arsenhault, 2016), making it relatively straight-forward to predict that DMOs are going to have to carefully consider how they deal with increased demand for customized experiences, the shared economy, increased desire for authentic, ethical and wholesome local experiences and how to harness social media to connect with local stakeholders and the global market place.

What is less easy to forecast is what tourists will want from the destination with expectations that go beyond the basic satisfaction, requiring researchers to revisit models of behaviour as they relate to tourism (Imkova and Holner, 2014; Yousaf *et al.*, 2018). At the same time, the host–guest relationship needs to be re-evaluated in order to gain a better understanding of what the resident wants and expects from its tourism industry and how tourism is managed in respect of their quality of life (Alonso and Nyanjom, 2016; Uysal *et al.*, 2019). In other words, a new agenda for managing destinations is rapidly emerging that will change the way DMOs function and fundamentally challenge how success is measured, where reputation is becoming more important than performance. A recent collection of essays (Lois-González *et al.*, 2014), opens up this discussion by exploring the cultural elements of 21st century tourism, suggesting new product developments that will be linked to reappraising the landscape, heritage, nature and spirituality in a destination. In exposing these scenarios, the authors start questioning tourist typologies, rethinking destination positioning and challenges the roles, responsibilities and, maybe, the very existence of DMOs.

There will be an increasing emphasis upon DMOs to address the wellness and well-being agenda (Hartwell *et al.*, 2018) and to adopt kindness as an aspect of policy (Unwin, 2018). There is evidence to suggest that the importance of kindness must now be taken as a given in the work of DMOs (Carnegie Trust, 2019), resonating with the trend towards what Pritchard *et al.* (2011) refer to as 'Hopeful Tourism', and the trend towards transformational tourism (Tomljenoiç and Ateljevic, 2015). This will require the adoption of a new paradigm where kindness becomes an explicit approach which is embedded in the policies and activities of the DMO (Filep *et al.*, 2017; Haskins *et al.*, 2018). In so doing, the destination can, at the same time, address both the needs of the traveller and provide a positive way of addressing the antitourism, overtourism concerns from within the community (Kuščer and Mihalič, 2019).

Tourism is increasingly about finding larger purpose in our lives and in the world. It is changing mindsets. According to a SKIFT (2019) report:

> Experience development is key and an essential and highly sought-after part of the travel ecosystem, but the concept has spread so widely that experience tourism as we know it today has lost its edge. To compensate we need to start going deeper towards the transformational value and how it helps the traveller become the

> person they aspire to be. It is all about customer centricity delivering personalisation and personal fulfilment.

These trends are creating new types of opportunities for destinations, and we are now on the threshold of the emergence of a new generation of DMOs which must become more 'responsible' organizations, not simply being reactive. Symbolic acts and doing what is right, not just in narrow terms for the tourist and the destination, but for society as a whole will become ever more important (Morrison, 2013). These are big responsibilities indeed for the humble DMO.

Concluding Remarks

The 'responsible' challenge for DMOs has very strong echoes of the optimism of the early luminaries of our tourism experiences, such as Brown (1971) who, in *Islands of Hope*, discussed the importance of the visitor experience and the intersection with responsible environmental management:

> ...to move from (simply) managing the resource itself to managing the people who use it. First and foremost, the experience that the visitor gets (in a destination) is the critical thing – not from the political administration approach, size or the way it is organized but from the dimensions of life that visitors can experience such areas. (Brown, 1971)

The future success of DMOs will require fresh thinking about all aspects of their work. In 2018, S&A identified a new model of successful destination management which represents a significant shift from the traditional reliance on the five Ps of tourism development) to the five Rs (© Stevens & Associates, 2018) comprising:

- **Recognition** – recognizing the importance of tourism to the destination and the rural/urban economy and communities.
- **Relationships** – the building of relationships (vertical and horizontal integration) ensuring a shared vision and strategy for tourism based upon common trust and common values.
- **Relevance** – of products, experiences and the destination to the interests and needs of emerging markets and the host communities.
- **Responsibility** – the development of a sustainable, empathetic, sensitive and responsible approach to tourism development respecting the assets of the destination, its environment, socio-cultural interests of the tourists and capacity of the host community.
- **Respect** – for the interests and well-being of all parties including the host community, business community and the visitors to the destination.

In the pursuit of destination competitiveness, the destinations that excel go their own way with a bespoke, objective mission and vision and clear management strategy. The DMOs that are succeeding are open-minded. They are willing to innovate and to try new ways of working. They adopt the adage, 'change your thought and it will change your mind', and they apply the relevant elements alchemy to their work once the destination leaders understand that the job to be done is doing what is right for the wealth of their community and the visitors.

In summary, therefore, DMOs are about creating the vision, implementing a dynamic strategy and managing the reputation and performance of tourism by delivering the promise across their destination by adopting a clear leadership role, driving collaboration through internal and external networks, normalizing innovation and creativity in all aspects of the work of the organization and delivering exceptional visitor experiences using tourism to create shared values and wealth across the resident community. So, vision, strategy, positioning and co-ordination are then the key, and the key to co-ordination is the establishment of a well-resourced, competently managed destination management organization.

Finally, by far and away the most inspirational destinations that are setting new standards in destination management are, *ipso facto*, some of the best destinations to visit as a tourist. These include: Istria (Croatia), Austin, Flagstaff, Jackson Hole and Park City (USA); Bregenz/Bregenzerwald, Saalfelden-Leogang, Linz, Kitzbühl (Austria); Bled/Julian Alps, Soca Valley and Ljubljana (Slovenia); Muskoka and Toronto (Canada); Copenhagen, Aarhus, Fyn and

Fanoe Island (Denmark); Franschhoek Valley (South Africa); the Bern region in Switzerland; Catalunya/Barcelona, Bilbao and San Sebastian (Spain); and Baiersbronn and Baden-Baden in the Black Forest of Germany.

The most impressive DMOs which are already ahead of the game and delivering best practice are:

i. The Sustainable Destination with the Riverwind Foundation for Jackson Hole, Wyoming;
ii. Saalfelden Leogang Touristik GmbH for Saalfelden – Leogang, Austria;
iii. Linz Turismus for Linz, Austria; and
iv. Turistička zajednica Istre for Istria, Croatia.

DMOs are the lodestars of 21st century tourism. They are comprised of an assembly of alchemists whose most fundamental job is to ensure that the two rules of tourism are never broken or forgotten: Rule one: Tourism is only about great products and experiences; and, Rule two: You never forget Rule one.

Note

[1] A multi-disciplinary research think-tank at Swansea University in May 2018. Unpublished at this stage.

Acknowledgements

I would like to thank the many thousands of tourism professionals that I have had the honour to work with; they are too numerous to mention by name. However, special mention goes to: Stefan Puehringer (the destinations of Kitzbuhel, Saalfelden-Leogang and Kufstein in Austria); K.R. Manfred Grubauer and Georg Steiner (Linz, Upper Austria); Marco Pointner and Hannes Reidlsperger (Saalfelden – Leogang, Austria, Tim O'Donoghue (Jackson Hole, Wyoming), Myles Rademan and Dana Williams (Park City, Utah), Patrick Torrent (Catalunya), Tomislav Popović and Denis Ivosević (Istria, Croatia); the late Jenny Prinsloo (Franschhoek Wine Valley, South Africa); Brian Johnston (Belfast City and Armagh); Harry Connolly (West Belfast); Freda Newton (Loch Ness); Alastair Dobson (Isle of Arran); Eva Stravs-Podolgar, Misa Novak and Maja Pak (Slovenia) and the visionary Claus Sendlinger (Founder of Design Hotels).

References

Adeyinka-Ojo, S.F., Khoo-Lattimore, C. and Nair, V. (2014) A framework for rural tourism destination management and marketing organisations. *Procedia - Social and Behavioral Sciences* 144, 151–163. DOI: 10.1016/j.sbspro.2014.07.284.

Albertus, F. (2000) *The Alchemists' Handbook*. Hay House, London.

Alonso, A.D. and Nyanjom, J. (2016) Tourism, quality of life, and residents of a rural town. *Tourism Analysis* 21(6), 617–629. DOI: 10.3727/108354216X14713487283165.

Anastasiadou, C. and Halkier, H. (2014) Organising tourist destinations in Scotland and Denmark: multilevel governance versus localism. *Paper presented at the 23rd Nordic Symposium Conference*, Copenhagen, 2nd-4th October 2014.

Arsenhault, P. (2016) Top 7 tourism trends of the future. *Tourism Review*. Available at: https://www.tourism-review.com/tourism-trends-of-the-coming-age-news4892 (accessed 6 September 2019).

Australian Government (2016) *A Guide to Best Practice: Destination Management. Department of Resources, Energy and Tourism*. Australian Regional Tourism Networks/AUSTRADE, Canberra.

BCC (2009) *Backing UK Tourism: Destination Recovery*. British Chambers of Commerce, London.

Beritelli, P., Bieger, T. and Laesser, C. (2014) The new frontiers of destination management: applying the variable geometry as a function-based approach. *Journal of Travel Research* 53(4), 403–417.

Beritelli, P., Reinhold, S., Laesser, C. and Bieger, T. (2015) *The St. Gallen Model for Destination Management (SGDM)*. St Gallen University. IMP-HSG, St. Gallen.
Block, F. and Keller, M.R. (2008) *Where do innovations come from?* Transformations of the US National Innovation System 1970 – 2006. Norway. Working Paper in Technology Governance and Economic Dynamics. No. 35. The Other Canon Foundation.
Brooks, R.A. and Forman, M. (2003) *The 25 Immutable Rules of Successful Tourism*. Kendall Hunt, Dubuque, Iowa.
Brown, W.E. (1971) *Islands of Hope*. National Recreation and Park Association, Arlington, Virginia.
Butler, R.W. (1980) The concept of a tourist area cycle of evolution: implications for management of resources. *The Canadian Geographer/Le Géographe canadien* 24(1), 5–12. DOI: 10.1111/j.1541-0064.1980.tb00970.x.
Butler, R.W. (1985) Evolution of tourism in the Scottish highlands. *Annals of Tourism Research* 12(3), 371–391. DOI: 10.1016/0160-7383(85)90005-2.
Butler, R.W. (2011) *Tourism Area Life Cycle: Contemporary Tourism Reviews*. Goodfellow Publishers, Oxford, UK.
Cains, P. (2019) Review. *Sunday Times*, 11 June.
Carnegie Trust UK. (2019) *The Practice of Kindness*. Carnegie Trust UK, Dunfermline, UK.
Cohen, M. (2006) *No Holiday: 80 Places You Don't Want to Visit – A Disinformation Guide*. The Disinformation Company, London.
Cooper, C. and Hall, C.M. (2008) *Contemporary Tourism: An International Approach*. Butterworth Heinemann, London.
Crouch, D. and Lűbbern, N. (2003) *Visual Culture and Tourism*. Berg, New York.
Crouch, G.I. and Ritchie, J.R.B. (1999) Tourism, competitiveness, and societal prosperity. *Journal of Business Research* 44(3), 137–152. DOI: 10.1016/S0148-2963(97)00196-3.
Csikszentimihalyi, M. (1990) *Flow: The Psychology of Optimal Experience*. Harper and Row, London.
Decelle, X. (2004) A Conceptual and Dynamic Approach to Innovation in Tourism. OECD. Available at: https://www.oecd.org/industry/tourism/34267921.pdf (accessed 5 September 2019).
Dempsey, A. (2006) *Destination Art*. Thames & Hudson, London.
Derco, J. (2013) Destination governance in the Czech Republic, Slovakia and Poland. *Tourism Planning & Development* 10(3), 354–364. DOI: 10.1080/21568316.2012.747987.
Esposito, R. (2009) *Communitas: The Origin and Destiny of Communities*. Stanford University Press, Stanford, California.
Estêvão, C.M.S. and Ferreira, J.J. (2009) *The Tourism Clusters Role in Regional Development: Presenting a Competitiveness Conceptual Model*. Tourism Destination Development and Branding, Conference Proceedings, Eilat, Israel, pp. 127–139.
Filep, S. and Pearce, P. (eds) (2014) *Tourist Experience and Fulfilment: Insights from Positive Psychology*. Routledge, Abingdon, UK.
Filep, S., Macnaughton, J. and Glover, T. (2017) Tourism and gratitude: valuing acts of kindness. *Annals of Tourism Research* 66, 26–36. DOI: 10.1016/j.annals.2017.05.015.
Fuglsang, L., Sundbo, J. and Sørensen, F. (2011) Dynamics of experience service innovation: innovation as a guided activity – results from a Danish survey. *The Service Industries Journal* 31(5), 661–677. DOI: 10.1080/02642060902822109.
Getz, D. (1986) Tourism and population change: long-term impacts of tourism in the Badenoch and Strathspey district of the Scottish Highlands. *Scottish Geographical Magazine* 102(2), 113–126. DOI: 10.1080/00369228618736663.
Gretzel, U., Fesenmaier, D., Formica, S. and O'Leary, J. (2006) Searching for the future: challenges faced by DMO. *Journal of Travel Research* 45, 274–289.
Gunn, C.A. (1972) *Vacationscape: Designing Tourist Regions*. Bureau of Business Research. Texas A&M University, College Station, Texas.
Gunn, C.A. (1994) *Tourism Planning*, 3rd edn. Taylor and Francis, New York.
Guvstasen, T.M. (2016) *Nasjonale turistveger: The National Tourist Routes in Norway*. Forlaget Press, Oslo.
Halkier, H. (2013) Innovation and destination governance in Denmark: tourism, policy networks and spatial development. *European Planning Studies* 21(1), 1–12.
Hall, C.M. and Williams, A. (2009) *Tourism and Innovation*. Routledge, London.
Hartwell, H., Fyall, A., Willis, C., Page, S., Ladkin, A. *et al.* (2018) Progress in tourism and destination wellbeing research. *Current Issues in Tourism* 21(16), 1830–1892. DOI: 10.1080/13683500.2016.1223609.
Haskins, G., Thomas, M. and Johri, J. (eds) (2018) *Kindness in Leadership*. Routledge, Abingdon, UK.

HIE (2002) *Cairngorm Vision: International Benchmarking for Aviemore, Badenoch and Strathspey*. Highlands & Islands Enterprise, Inverness, Scotland.

Hjaglar, A.-M. (2015) 100 innovations that transformed tourism. *Journal of Travel Research* 54(1), 3–21.

Holden, A. (2000) Winter tourism and the environment in conflict: the case of Cairngorm, Scotland. *International Journal of Tourism Research* 2(4), 247–260. DOI: 10.1002/1522-1970(200007/08)2:4<247::AID-JTR214>3.0.CO;2-X.

Hristov, D. and Zehrer, A. (2015) The destination paradigm continuum revisited: DMOs serving as leadership networks. *Tourism Review* 70(2), 116–131. DOI: 10.1108/TR-08-2014-0050.

Hudson, S. (2000) *Snow Business: The Study of the International Ski Industry*. Cassell, London.

Hudson, S. (2003) *Sport and Adventure Tourism*. Routledge, London.

Hudson, S., Ritchie, B. and Timur, S. (2004) Measuring destination competitiveness: an empirical study of Canadian ski resorts. *Tourism and Hospitality Planning & Development* 1(1), 79–94. DOI: 10.1080/1479053042000187810.

Imkova, E. and Holner, J. (2014) Motivation of tourism. *Procedia: Social and Behaviour Science* 159, 660–664.

Inskeep, E. (1991) *Tourism Planning: An Integrated and Sustainable Development Approach*. Van Nostrand Rheinhold, New York.

Jarosz, A.F., Colflesh, G.J.H. and Wiley, J. (2012) Uncorking the MUSE: alcohol intoxication facilitates creative problem solving. *Consciousness and Cognition* 21(1), 487–493. DOI: 10.1016/j.concog.2012.01.002.

Jencks, C. (2005) *The Iconic Building: The Power of the Enigma – The Coming of Cosmic Architecture*. Francis Lincoln, London.

Jetlag Travel Guide (2004) *Gastronesia: 200 Islands of the Qic Pu Peninsula*. Hardie Grant, London.

Jetlag Travel Guide (2014) *Surving Moustachistan: Central Asia's Forgotten Jewel*. Quadrille, London.

Johnson, S. (2010) *Where Good Ideas Come From: The Natural History of Innovation*. Allen Lane, London.

Klimek, K. (2017) Destination management organisations (DMOs) and their role in summer product development in selected alpine countries. *European Journal of Service Management* 24(4), 31–38. DOI: 10.18276/ejsm.2017.24-05.

Kozak, M. (2004) Introducing destination benchmarking: a conceptual approach. *Journal of Hospitality & Tourism Research* 28(3), 281–297. DOI: 10.1177/1096348003256603.

Kuščer, K. and Mihalič, T. (2019) Residents' attitudes towards over-tourism from the perspective of tourism impacts and cooperation: the case of Ljubljana. *Sustainability* 11(6), 1823. Available at: www.mdpi.com DOI: 10.3390/su11061823.

Laws, E. (1995) *Tourism Destination Management: Issues, Analysis and Policies*. Routledge, London.

Leiper, N. (1979) The framework of tourism: towards a definition of tourism, tourist, and the tourist industry. *Annals of Tourism Research* 6(4), 390–407.

Lennon, J. (2004) *An Assessment of the Economic Impact of Water-related Recreation and Tourism in the Spey Catchment 1993*. Glasgow Caledonian University for The Spey Catchment Plan Partners, Report IH020365T

Lippard, L. (1997) *The Lure of the Local: Senses of Place in a Multicentred Society*. The New Press, New York.

Lippard, L. (1999) *On the Beaten Track: Tourism, Art and Place*. The New Press, New York.

Lois-González, R., Santos-Solla, X. and Taboada-de-Zuñiga, P. (eds) (2014) *New Tourism in the 21st Century: Culture, The City, Nature and Spirituality*. Cambridge Scholars Publishing, Newcastle upon Tyne, UK.

Luštický, M. and Kincl, T. (2012) Tourism destination benchmarking: evaluation and selection of the benchmarking partners. *Journal of Competitiveness* 4(1), 99–116. DOI: 10.7441/joc.2012.01.08.

MacLellan, R. (2011) The role of destination management organisations: models for sustainable rural destinations in Scotland? In: MacLeod, D. and Gillespie, S. (eds) *Sustainable Tourism in Rural Europe: Approaches to Development*. Routledge, Abingdon, UK, pp. 121–136.

MacLellan, R. and Somerville, H. *et al*. (1998) Tourism and the Environment in Scotland. In: Seaton, A.. (ed.) *Tourism: The State of the Art*. John Wiley & Son, Chichester, UK.

MBIE (2017) Ministry of Business, Enterprise and Innovation, Government of New Zealand. *Learning Lessons from International Examples of Regional Tourism Destination Development*, Stevens & Associates.

McKercher, B. and Prideaux, B. (2014) Academic myths of tourism. *Annals of Tourism Research* 46, 16–28. DOI: 10.1016/j.annals.2014.02.003.

McTiernan, C., Thomas, R. and Jameson, S. (2018) Focusing on knowledge exchange: the role of trust in tourism networks. In: Fayos-Sol, E. and Cooper, C. (eds) *The Future of Tourism: Innovation and Sustainability*. Springer, Cham, Switzerland, pp. 302–314.

Mendiratta, A. (2011) *Come Closer: How Tourism is Shaping the Future of Nations*. Myriad Publications, London.

Mlodinow, L. (2018) *Elastic: Flexible Thinking in a Constantly Changing World*. Allen Lane, London.

Morrison, A.M. (2013) Destination management and marketing: the platform for excellence in tourism destinations. *Tourism Review* 28(1), 6–9.

Morrison, A.M. (2019) *Marketing and Managing Tourism Destinations*. Routledge, New York.

NECSTUoR (2018) *The Barcelona Declaration: Better Places to Live, Better Places to Visit*. Network of European Regions for a Sustainable and Competitive Tourism, Paris.

Neqruşa, A.L. and Coroş, M.N. (2016) Destination management organization roles and performance – a literature review. 2nd *International Scientific Conference*. SAMRO. Romania. October. Available at: http://www.digilibmuscel.ro/wp-content/uploads/2017/11/4_5_Destination-management-organiya-tion-s-Dmos-roles-and-performance_Literature-review.pdf (accessed 5 September 2019).

Nicholson, M. (2013) *The Comedy Carpet, Blackpool*. Booth-Cibberns Editions, London.

Nordin, S. (2003) *Tourism Clustering and Innovation: Paths to Economic Growth and Development*. The European Tourism Research Institute, Sweden.

Pechlaner, H., Hölzl, B. and Tallinucci, V. (2005) Cross-level destination management and the transfer of knowledge. A report of an exploratory study carried out in 2003/2004 in the Tyrol. Private exchange of papers.

Pechlaner, H., Kozak, M. and Volgger, M. (2014) Destination leadership: a new paradigm for tourist destinations? *Tourism Review* 69(1), 1–9. DOI: 10.1108/TR-09-2013-0053.

Peters, M. and Pikkemaat, B. (2006) Innovation in hospitality and tourism. *Journal of Quality Assurance in Hospitality & Tourism* 6(3-4), 1–6.

Pike, S. and Page, S.J. (2014) Destination marketing organizations and destination marketing: A narrative analysis of the literature. *Tourism Management* 41, 202–227. DOI: 10.1016/j.tourman.2013.09.009.

Porter, M. and Kramer, M.R. (2011) The big idea: creating shared value, rethinking capitalism. *Harvard Business Review* 89(1/2), 62–77.

Presenza, A. (2006) El desempeño de un destino turistico. *Quién gestona el destino? Quién realiza el rol auditor? Revisita de Análisis Turistico* 2, 43–55.

Pritchard, A., Morgan, N. and Ateljevic, A. (2011) Hopeful tourism: a new transformational perspective. *Annals of Tourism Research* 38(3), 941–963.

Qirici, E. and Theodhori, O. (2013) Destination management as a way for the competitive tourist destination: case study – south region of Albania. *Romanian Economic Business Review* 8(2), 48–58.

Ratcliffe, J. and Heath, U. (2019) *The Alchemists: The INEOS Story – An Industrial Giant Comes of Age*. Biteback Books, London.

Refai, T. (2011) *Overview. Fields of Vision*. Stevens & Associates, Swansea, UK.

Reinhold, S., Laesser, C. and Beritelli, P. (2015) 2014 St. Gallen consensus on destination management. *Journal of Destination Marketing & Management* 4(2), 137–142. DOI: 10.1016/j.jdmm.2015.03.006.

Ritchie, J.B. and Crouch, G. (2003) *The Competitive Destination: A Sustainable Tourism Perspective*. CAB International, Wallingford, UK.

Roberts, K. (2005) *Lovemarks: The Future Beyond Brand*. Saatchi & Saatchi, London.

Roberts, K. (2006) *The Lovemarks Effect: Winning the Consumer Revolution*. Saatchi & Saatchi, London.

Robinson, K. (2011) Destinations - future performance. *Journal of the Tourism Society* 146(4-5).

Rowan, D. (2013) Editorial: The World in 2013. *Wired*.

Scerri, M., Foley, C. and Edwards, D. (2016) The Value of Architecture to Tourism. *Conference Proceedings. 26th Annual CAUTHE Conference*, Sydney, February.

Scerri, M., Edwards, D. and Foley, C. (2019) Design, architecture and the value to tourism. *Tourism Economics* 25(5), 695–710. DOI: 10.1177/1354816618802107.

Scottish Enterprise (2008) *Tourism: Destination Development Guide*. Scottish Enterprise, Edinburgh.

Sendlinger, C. (2011) *Design Hotels: Exploring Your Senses. Fields of Vision*. Stevens & Associates, Swansea, UK.

Sendlinger, C. (2018) *Personal conversation*. ITB Berlin, March.

Seraphin, H., Gowreensunkar, V. and Ambaye, M. (2016) The Blakeley model applied to improving a tourist destination: an exploratory study. The case of Haiti. *Journal of Destination Marketing & Management* 5(4), 325–332. DOI: 10.1016/j.jdmm.2016.07.004.

Serra, J., Font, X. and Ivanova, M. (2017) Creating shared value in destination management organisations: the case of Turisme de Barcelona. *Journal of Destination Marketing & Management* 6(4), 385–395. DOI: 10.1016/j.jdmm.2016.06.005.

Shikira, E. and Qirici, E. (2013) The Role of Destination Management Organisations in Developing Sustainable Tourism in the Korça Region of Albania. *Economic Science and Practice in The Proceedings of The Second International Conference*, Chitra. Young Scientist Publishers; and discussed by Negrusa, A. et al. (2017) in The Proceedings of the 11th International Management Conference on The Role of Management in the Economic Paradigm of the XXIst Century. Bucharest. November.

SKIFT (2019) The Post-Experience Economy: Travel in an Age of Sameness. Skift. Available at: https://skift.com/sgf-magazine-2018/ (accessed 30 July 2020).

Stevens, T. (2011) *Fields of Vision*. Stevens & Associates, Swansea, UK.

Stevens, T. (2013) Crimes against the consumer: avoiding serial reproduction – Innovation and destination development. *Proceedings of the 2nd International Conference on Tourism in South East Europe*, SSRN.

Stevens, T. (2016) *Landscape Wales*. GRAFFEG, Cardiff, UK.

Stevens, T. (2017a) *Linz, Austria in UNWTO and World Tourism Cities Federation (2017) City Tourism Performance*. UNWTO / WTCF, Madrid.

Stevens, T. (2017b) *Build it and they will come. CLAD Global. Issue 1*. Leisure Media, Hitchin, UK.

Stevens, T. (2017c) Lessons from Creative Clusters of the Past. *Presentation to SCIENTIA*, Swansea University, 23 November.

Stevens, T. (2018) *Transforming Istria. CLAD Global*. Leisure Media, Hitchin, UK.

Stipanuk, D. (1993) Tourism and technology: interactions and implications. *Tourism Management* 14(4), 267–278.

Temperton, J. (2013) Editorial. *Wired*.

The Communications Group (2018) *Placemaking: learning from the best*. London, UK.

Tigu, G. (2012) New challenges for tourism destination management in Romania. In: Murat, K. (ed.) *Strategies for Tourism Industry: Macro and Micro Perspectives*. Inter Open, London, pp. 167–184.

Timur, S. and Getz, D. (2008) A network perspective on managing stakeholders for sustainable urban tourism. *International Journal of Contemporary Hospitality Management* 20(4), 445–461. DOI: 10.1108/09596110810873543.

Tomljenoiç, A. and Ateljevic, A. (2015) TransTourism: operational and typology of transformational travel experiences. Working Paper 21. Zagreb: Institute of Tourism, University of Zagreb. Available at: https://pdfs.semanticscholar.org/6f1b/36fd8a65bc32a023dff91bd4bd96348c556e.pdf (accessed 6 September 2010).

Torrent, P.Q. (2018) *Developing neighbourhood tourism through good destination management*. Felie an Phobail (The Peoples' Festival) Summer School, Belfast.

Truong, H.T. (2016) The industry cluster approach for tourism development in central Vietnam. *International Journal of Business and Management* 11(5), 167–178.

Turner, E. (2012) *Communitas: The Anthology of Collective Joy*. Palgrave Macmillan, Basingstoke, UK.

Turner, V. (1982) *From Ritual to Theatre: The Human Seriousness of Play – The Anthropology of Performance*. PAJ Books, London.

Unwin, J. (2018) *Kindness, Emotion and Human Relations: The Blind Spot in Public Policy*. Dunfermline, Carnegie UK Trust.

UNWTO (2007) *A Practical Guide to Destination Management*. UN World Tourism Organization, Madrid.

UNWTO (2011) *Policy and Practice for Global Tourism*. UN World Tourism Organization, Madrid.

UNWTO (2019) *Concept Notes: International Seminar on Destination Management – New Challenges, New Solutions*. UN World Tourism Organization, Madrid.

UNWTO/UNDP (2017) *Tourism and Sustainable Development Goals – Journey to 2030*. UN Word Tourism Organization, Madrid.

UNWTO (2002) *Destination Management & Quality Programme: Conceptual Framework*. UN World Tourism Organization. Available at: http:/destination.unwto.org/content/conceptual–framework

Uysal, M., Perdue, R. and Sirgy, S. (2019) *Handbook of Tourism and Quality of Life Research*. Springer, London.

van Gennep, A. (1909) *The Rites of Passage*. First published in French and then translated into English and published in 1960 by The Chicago Press.

Vanhove, N. (2005) *The Economics of Tourism Destinations*. Elsevier, Amsterdam.

Varghese, B. and Paul, N. (2014) A literature review on destination management organisations (DMO). *International Journal of Multidisciplinary Research* 4(12), 82–88.
Visit Britain (2010) *Delivering a Golden Legacy: A Growth Strategy for Inbound Tourism 2012-2020*. Visit Britain, London.
Visit Scotland (2013) *Corporate Plan 2013 / 2016, The Winning Years & Beyond*. Visit Scotland, Edinburgh.
Visit Wales (2019) Destination management tool kit. Available at: www.business.wales.gov.uk//dmwales/dmwtoolkit (accessed 28 September 2019).
von Hippel, E. (1988) *The Sources of Innovation*. Oxford University Press, New York.
Williams, A*et al*. (2017) *Trust, Networking and Knowledge Exchange. The Hidden Story: Understanding Knowledge Exchange Partnerships Within the Creative Economy*. Arts and Humanities Research Council, London.
World Travel and Tourism Council (2015) Closing speech by David Scwosil, President at Global Summit, Madrid. Available at: www.wttc.org/news
Yeoman, I. (2012) *2050: Tomorrow's Tourism*. Channel View Publications, Bristol, UK.
Yousaf, A., Amin, I. and C. Santos, J.A. (2018) Tourist's motivations to travel: a theoretical perspective on the existing literature. *Tourism and Hospitality Management* 24(1), 1–15. DOI: 10.20867/thm.24.1.8.
Zlatković, M. (2016) Tourism destination benchmarking analysis. *European Journal of Multidisciplinary Studies* 1(1), 283–293. DOI: 10.26417/ejms.v1i1.

12 Creating an Enabling Environment for Urban Tourism Development: The Future of Bangkok

Louisa Yee-Sum Lee[1]* and Philip L. Pearce[2]
[1]Lee Shau Kee School of Business and Administration, The Open University, Hong Kong; [2]James Cook University, Queensland, Australia

Introduction

Cities are prominent economic powerhouses for most Asian countries. In Thailand, for example, cities collectively contribute over 90% of the country's GDP (World Bank, 2010). At the same time, as cities have come to dominate the economy, so too has tourism emerged as a major income earner in many urban Asian destinations (UNWTO, 2012). The converging forces of globalization, national economic reforms and sophisticated amenities have boosted the attractiveness of cities as tourism destinations and, consequently, the long-term growth of urban tourism looks set to continue for the foreseeable future. Nevertheless, many scholars, together with both governments and businesses, have been concerned for some time about the costs imposed by the tourism industry on both natural and social ecological systems, both generally and in cities in particular (Gore, 2013; Sharpley, 2018; Font and Guix, 2018; Buckley, 2018). In this chapter, contemporary concerns about the future and sustainability will be explored at the level of one pivotal Asian city, namely Bangkok. As a case study, the specific focus of the authors is to explore the foundation and ongoing drivers of development in Bangkok, a process that may also reveal forces of interest operating in other ASEAN cities and shaping their future.

Academics and practitioners have long sought to identify the prerequisites and means to address the challenges of achieving sustainable development, and the idea of creating an enabling environment has emerged as one important dimension in the concerns for sustainability (Brinkerhoff *et al.*, 2007; Gough, 2015). At the same time, burgeoning tourism demand and the growing dominance of cities as drivers of economic development have given rise to questions about their competitive advantages in the marketplace. City competitiveness can be conceptualized as the ability to produce and market a collection of products that present good value in relation to competitors (Choe and Roberts, 2011). The creation of an enabling environment that accommodates the productivity, functionality and livability of cities are the prime roles of local government and together these factors help build competitiveness (Duff, 2009; Meyer and Neethling, 2017).

Currently, the roles of tourism development in shaping Asian cities are not well understood (Novy, 2019) and, hence, a more rigorous understanding of the evolving drivers that shape

*Corresponding author: Yslee@ouhk.edu.hk

 Tourism in Development: Reflective Essays (eds P.U.C. Dieke *et al.*)

DOI: 10.1079/9781789242812.0012

the future of urban tourism development is needed. In this chapter, the authors consider tourism development in Bangkok from the past to the present, and then venture on to examine the city's future. The analysis introduces how the evolution of the city, its urbanization and the overall growth of Thai tourism more generally have shaped the present state of Bangkok. We draw on existing literature augmented by in-depth interviews; specifically, six significant and influential interviewees from both the private and public sectors of Bangkok help reveal how the past and present are shaping the future of tourism in the city.

Bangkok in the Past

Thailand is in the unique position of never having been colonized, yet has a history of receiving immigrants from all over the Asian continent. The economy of Thailand is export-driven, accounting for 60% of the nation's GDP, amounting to US$505 billion as of 2018 (Trading Economics, 2018). Primary products for export are agricultural products including fish and rice, textiles, rubber, automobiles, jewellery and electronic appliances. Some 10% of the country's population lives under the poverty line, although both within the region and globally this figure is relatively low, suggesting that the country can be cast as moderately well developed in economic terms. Thailand is also a very well-known tourism destination. Indeed, 21% of its GDP, or US$109 billion, is generated by travel and tourism (Knoema, 2019). Bangkok, together with Pattaya, Hua Hin, Phuket and Koh Samui, are popular tourism destinations while the country as a whole offers a variety of tourist attractions including historical and religious sites, national parks, wildlife sanctuaries, forests, beaches and local food. Bangkok, the capital of Thailand since 1767, is the country's political, commercial, industrial, educational and entertainment centre. It is also known as the City of Angels.

Urbanization in general is vital in providing economic opportunities and in Thailand in particular, the process is dominated by the Bangkok area (World Bank, 2015). Specifically, rapid urbanization has transformed the face of Bangkok and the lives of its citizens, and posed challenges for urban policy planners and makers. From 1960, the development of infrastructure, including road networks and real estate, along with increasing land values and the growing national economy fuelled the city's rapid urbanization and its burgeoning population (Bangkok Metropolitan Administration, 2007). During the first decade of the new millennium, the population in the Bangkok metropolitan area increased from 7.8 million to 9.6 million (World Bank, 2015) while spatially, the Bangkok urban area grew from 1900 sq. km to 2100 sq. km. In so doing, the city became larger than other Asian megacities including Seoul, Jakarta and Manila; indeed, this growth established Bangkok as the fifth largest urban area in East Asia. In terms of population, the city also ranks 36th among the world's largest cities and 17th among Asian cities (World Atlas, 2019).

Bangkok, with its all-important political base, offers sophisticated financial and producer services. The city became the financial hub of the Baht Zone among Thailand, Laos, Cambodia and Burma, and it also exerts power in international governance and diplomacy in the Asian region. The Thai government has actively welcomed and encouraged various international associations to base their headquarters in Bangkok; these include, for example, the United Nations Economic and Social Commission for the Asia and the Pacific (UNESCAP), United Nations Children's Rights and Emergency Relief Organization (UNICEF), United Nations Educational, Scientific and Cultural Organization (UNESCO), the Asian Institute of Technology and the International Air Transport Organization (IATO). Webster (2004) observes that 66 international organizations and the diplomatic missions of 123 countries are represented in Bangkok. The USA Embassy in Bangkok is the second largest USA diplomatic mission abroad.

The appearance of tourism activity in Thailand was first recorded in the late 1920s while the roots of contemporary mass tourism in the country lie in the late 1950s when tourism was promoted to boost economic development (Meyer, 1988). Despite the absence of a distinct policy for tourism development, the Tourist Organization of Thailand was established in 1959. Thai Airways International commenced operations in the same year as a joint venture with the Scandinavian Airline System

(SAS) (Meyer, 1988). At that time, an extensive world tour undertaken by the Thai royal couple was a key marketing effort aimed at promoting the kingdom abroad (Peleggi, 1996). In the mid-1970s, tourism to Thailand was further boosted by the perception that, as a Buddhist nation, it was the safest country to visit in South-east Asia, particularly in comparison to political instability and the threats and conflicts attached to Islamic fundamentalism in neighbouring states (Peleggi, 1996). Thailand was used during the Vietnam War as a rest and recreation base for American and Australian troops. Images of Bangkok as a leading nightlife destination with a permissive bar scene developed from this period (Cohen, 1982, 1988). The tourism sector also received much more formal support from the government, which has long played an important role by offering incentives to both domestic and foreign investors to establish quality hotel businesses (Pearce and Thanksooks, 2016).

Tourism was officially emphasized by Thailand's government in the Fourth Economic and Social Development Plan from 1977 to 1981 (Peleggi, 1996). The Tourist Organization of Thailand was upgraded to the Tourist Authority of Thailand (TAT) in 1979. As the result of considerable promotional efforts, the country enjoyed annual increases in tourist arrivals at rates that largely surpassed those in other countries in the region (Walton, 1993). Commensurately, tourism receipts grew from 17 million baht in 1980 to 110 million baht in 1990. To a great extent, this growth was fuelled by both increasing air connectivity between Bangkok and many European countries and reductions in air fares (Bowen, 2000). As a consequence, not only did many travellers choose Thailand, and especially Bangkok, either as a destination or as a base for linking to other regional cities, but the city also became a key connection point for travel between Europe and Oceania. Moreover, by the mid-1980s, monuments and historic ruins had been extensively restored throughout Thailand and, hence, celebration of the country's cultural heritage became the iconic element in the capital's tourism promotion. Equally, and importantly, branding the city through cultural tourism was also a deliberate strategy to mitigate the powerful image of this part of Asia as a sex tourism destination (Richter, 1989). The Thai government also advocated mega-projects in the urban area of Bangkok, building new city space with a combination of office, retail, hotel, conference centres and residential land use. All these initiatives added to the extensive offerings of the city.

In more recent years, Bangkok has continued to enjoy increasing popularity as a destination for regional and international visitors while, over the last decade, tourism has made an ever greater contribution to Thailand's economy more generally. Increasing from a base of 15.8% of GDP in 2009 to the present level of 21.6% of GDP in 2019, tourism can clearly be seen as a successful and significant contributor to the national economy (Knoema, 2019). GDP per capita in Bangkok is between two and ten times greater than in the provinces (Thai Websites, 2019) and the economic contribution of tourism is significant. However, specific city-wise data are unavailable on the official website of National Statistical Office of Thailand.

Examining Bangkok's principal tourism products, it is evident that culture and heritage remain the mainstays – The Grand Palace, Wat Arun, Wat Pho and Wat Phra Kaew are the most publicized historical gems (Chinmaneevong, 2015). Thailand also promotes itself as a shopping paradise and Bangkok lies at the heart of this claim. The city accommodates all types of shopper, offering a large number of massive shopping malls to those on a shoestring and to big spenders. While rarely a part of official promotional attempts, the continuing appeal of commercial sex options and precincts in the city attract mostly men from a number of international markets (King, 2017). According to the global shopper index, Bangkok ranks 7th among 25 Asian cities and is 'the champion' for affordability (The Economist Intelligence Unit, 2012). Wellness tourism, embracing massage and spa centres, yoga, dental services, plastic surgery and other medical services, is also an increasingly important sector in Bangkok (Connell, 2007; Cohen, 2008).

The Present: A Review

In order to identify the factors driving the nature and direction of current and future tourism development in Bangkok, six key Bangkok

Table 12.1. Code and corresponding job positions of the Bangkok case interviewees.

Code of interviewees	Sector	Position of interviewees
B1	Private	Hospitality and MICE Entrepreneur (Chain Operator)
B2	Private	Hospitality and MICE Entrepreneur (Chain Operator)
B3	Public	Head of Tourism Authority
B4	Public	Division Head of Public Transport
B5	Public	Head of Monitoring and Evaluation
B6	Public	Head of Department of Tourism

stakeholders from both the private and public sectors were interviewed. The respondents included key players within the hospitality and MICE sectors, along with public sector representatives from the tourism authority, public transport and the government policy monitoring and evaluation unit. Table 12.1 summarizes the professional role of each respondent; all held senior positions within their respective organizations and were well acquainted with tourism, hospitality, aviation or city development. The significant time devoted to building personal connections between the authors and the interviewees facilitated the opportunity to access quality information from senior personnel. They were not easily accessible and so the authors had to be flexible in meeting times and scheduling in the process of eliciting information.

All of those interviewed expressed overall satisfaction with the recent tourism performance and development of Bangkok. They viewed the city as a mature destination that welcomes international tourists and, perhaps unsurprisingly, to be among the leading and most visited tourism cities in the world. Visitor arrivals and tourist receipts were often cited by way of justification relative to other cities in the region.

According to Interviewee B3, the capital city is the first point of entry for approximately 60% of visitors to Thailand, a proportion that reinforces the critical yet significant role of Bangkok in tourism. Interviewees further commented on the source markets. As noted by Interviewee B1, European countries and Japan remain the major source markets of tourism to Bangkok, despite a slight drop in their share of arrivals in recent years. Significantly, the huge influx of visitors from mainland Chinese, coupled with growth in some emerging markets such as Russia and India, has provided a strong base for continuing demand.

Nevertheless, some of the respondents acknowledged that the performance of the tourism sector has not always been optimal. Over the past decade, the global financial crisis of 2008 and more recent political disputes, notably in 2013–14, had an undoubted negative impact on tourism, with political upheavals representing a particularly serious problem that affected and delayed plans for tourism development and promotion. Interestingly, however, some interviewees adopted a 'take-it-for-granted' mindset, assuming that once such problems had been resolved, international tourists would come back anyway. They evidently viewed these variations in arrivals as normal fluctuations in the business cycle. Perhaps because of the unique offerings of Bangkok as a tourism city, interviewees expressed confidence in the resilience of the tourism system and the recovery of tourist numbers within a short period of time.

Nevertheless, the respondents expressed diverging opinions about the current positioning of Bangkok as a tourism city. Those from the public sector believed that tourists came to Bangkok for its modern facilities, natural treasures and variety of entertainment. Indeed, it was suggested that many tourists travel to Bangkok only because of the entertainment opportunities. However, private sector respondents expressed doubts regarding the various forms of entertainment. As Interviewee B2 suggested, 'The Tourism Authority of Thailand has branded Bangkok as a family destination, but I disagree. I would create air quotes, meaning " ", for the kind of entertainment offered to tourists.' Specifically, current entertainment offerings, including the nightlife in general, and pubs, sex tourism and particularly 'lady boy' shows were viewed as inconsistent with the family-market positioning. One interviewee also commented that such offerings have associated Bangkok with a negative and risky destination image. Generally, however, consensus was apparent among all respondents with regards to the promising tourism development and performance. In particular, they emphasized the efforts made by both the public and private sectors over the last decade to enhance the city's appeal. The following section presents the key drivers that appear to have contributed to such success.

The Future

This section draws upon the insights provided by the six key stakeholders who were interviewed. Their views, taken together, characterize the urban tourism outlook for Bangkok. The insights emerged by following the rigorous procedures of the grounded theory approach.

All six interviewees expressed a generic confidence about future tourism prospects. As the capital city of Thailand, Bangkok will be an important tourism city to both international and domestic tourists. Some acknowledged that political instability has affected tourism over the past five years. They were not deeply worried, however, because of the extent of what the city offers. Tourism promotion and advocacy by government and private sector investment in tourism projects, as well as the expansion of infrastructure, are expected to provide a further stimulus for tourism demand. When asked about the future direction of tourism in the city, it was believed that Bangkok has advantageous conditions to brand itself as a destination for heritage tourism, shopping amenities, MICE (meetings, incentives, conference and events), medical tourism and wellness tourism.

Bangkok undoubtedly offers a concentration of state-of-the-art heritage sites, historical buildings and museums. The iconic attraction, the Grand Palace, and the numerous royal temples and shrines are all located in the city. Shopping precincts, favoured by groups of some international tourists, are another opportunity. Major shopping malls have been built in the city alongside the rail line or adjacent to the tourism attractions. These malls feature thematic and innovative designs, and a sense of novelty through their chic images. For example, the shopping mall Terminal 21 has adopted airport themes. Another example is the embracing of multimedia innovations by the Siam Centre, thereby enhancing the mall's interior decoration and functionality. In 2013, the Siam Paragon shopping mall, a massive and major shopping complex, became one of the top two shopping locations identified on Instagram (Instagram, 2014). Meanwhile, a number of other shopping complexes are being planned. All of these projects are mega-malls requiring considerable investment. Furthermore, the concentration of company regional headquarters will continue to make Bangkok a popular MICE destination.

The capital city has been performing well in medical tourism. Interviewees reported the value-for-money, professional and high-quality services of medical tourism. One of the two famous private hospitals in town, the Bumrungrad International Hospital, treats more than half a million international patients annually (Bumrungrad International Hospital, 2016). This figure represented over half of the patients treated in the city.

Interviewee B1 suggested that, 'Tourists combine a holiday with medical check-ups in Bangkok. We have got huge business from the Middle East countries, especially the Emirates.' According to B1, civil servants in Middle Eastern countries are entitled to healthcare coverage in overseas countries. Bangkok is a popular medical tourism destination and will continue to benefit from solid demand from these countries. All other interviewees echoed the theme of a prosperous future for medical tourism, despite the concern of rising costs and the potential loss of niche procedures for some source markets. Interviewee B6 proposed the potential for developing health and wellness tourism, noting that,

> Bangkok is popular for medical tourism. But we have to think carefully. We are popular among the Middle East countries. The European prefers going to India for medical services while the Chinese mainly come for dental services. I think we should promote health and wellness tourism instead. Health and wellness tourism covers spa, massage, herbal massage and plastic surgery.

In terms of source markets, tourists from ASEAN countries were seen as generating significant tourist numbers to Bangkok owing to the relaxation of visas for ASEAN citizens. Visitors from mainland China and from other emerging markets, such as Russia, will also continue to support Bangkok. Some private-sector businesses, especially small-to-medium-sized enterprises, were seen as not well prepared to welcome tourists from these sources markets. Furthermore, interviewees expressed the view that the sector does not know the preferences or needs of some of these emerging tourist markets.

The six interviewees also addressed the key driving forces changing the future development of urban tourism. Their thoughts align with the quadruple bottom line of sustainability – social, environmental, economic and governance bottom lines – that were advocated by Wheeller (2015).

Towards Social Sustainability

Thainess

Thainess was the first driver for success that was mentioned by the interviewees. They were in agreement that Thainess is the distinctive smiling and accepting culture of the country. The term has an inherently abstract quality and is manifest in the way of life, rules for behaviour and cultural products (Naghizadeh and Shahzadeh, 2013). The interviewees suggested that Thainess is expressed in the arts and architecture, and is embedded in the cultural rituals and practices of everyday behaviour.

Interviewee B6 emphasized that 'Culture is our most outstanding product of Bangkok'.

The government also recognized the subtle role of showcasing Thai culture. According to interviewee B3, it is likely that future promotions will focus on the country's culture and heritage. The new Thai boxing stadium and the new Thai textile museum featuring Thai silk will be the signature features. The lifestyle of Thais is indeed an intangible reflection of the Thai culture. Part of the professional duty of Interviewee B6 was to ensure that the tourism policy formulated by the Department of Tourism was aligned with international standards. UNESCO assigned her and her team a project about the conversion of Thai culture into a tourism product. The challenges here included the view that the Thai lifestyle is an intangible reflection of Thai culture; it is not a staged culture that is simply performed for tourists. In the academic literature this is captured by the concept of subjective embodied authenticity (Rickly and Vidon, 2018). The beauty of Thai lifestyle is the contemporary behaviour of Thai people, including etiquette, such as the way they greet one another and the ritual of hosting guests. These behaviours are distinct. It is believed that the Thai lifestyle will be the drawcard in differentiating Bangkok with other urban tourism destinations in the future.

Interviewee B6 stated that,

> Lotus (the flower) can mean hello, sorry and thank you in our culture. The garland is indeed not just for decoration. It is used to worship our Buddha... And then there is the wai ...When a younger fellow meets a senior person, he or she has to ...use a wai... close the palms, lift it up to the nose, and say hello in a soft tone of voice.

Compared with other ASEAN countries, the Thai lifestyle is more delicate and gentle. Respondent B6 argued that the strategy of attracting tourists through products and experiences that were not aligned to Thai culture would be ineffective. Sophisticated international tourists would like to be informed and experience cultural insights and the Thai lifestyle. It was suggested, however, that the public and private sectors were not always aware of the significant role of, or the way to incorporate, the Thai lifestyle into their offerings. The training of manpower and the issuing of licences for tour guides were the duties of her department. When designing the training curriculum, she stressed the importance of reinforcing Thai cultural identity and educating practitioners how to showcase Thai lifestyles to tourists. The key driver identified here is that the Thai culture and lifestyle will be the drawcard for both new visitors and for those who have visited before.

Workforce

The Thai culture is well known for its warm hospitality and friendliness. With the advent of the ASEAN Economic Community multilateral trade agreement, interviewees generally believed that workers from neighbouring countries would seek to join Bangkok's tourism workforce. A plausible reason for the immigration of workers was the search for better living conditions and opportunities for employment, given that Bangkok was a relatively developed and mature tourism city. Interviewee B6 reflected on the worries of the Thai workers. The local frontline practitioners expressed concerns about their own language limitations. Owing to the colonial and historical contacts, immigrant workers from several other ASEAN countries had better English literacy compared to Thais.

Interviewee B6 explained that, 'Only a small percentage of the Thai tourism workers, like street vendors and retailers, can speak English. The others have to use body language.'

It was reported that the workers are understandably concerned about losing their jobs. Interviewee B6 further commented that the newly arrived workers may possess better English speaking skills but lack insights into Thai culture. The native workforce's cultural insights are their one enduring comparative advantage. Other interviewees expressed concern about the loss of Thai hospitality. Interviewee B2 worried about the loss of Thai quality by frontline staff owing to the mobilities generated by the working visas that are now available to those from neighbouring countries. The capacity of Bangkok hotels to deliver Thai hospitality continuously is becoming a challenge. Interviewee B1 emphasized that, 'International tourists keep coming to Bangkok because of the Thai hospitality. They cannot tell whether the staff are from Thailand or other ASEAN countries. They just expect the same style and level of service. But the new immigrants do not know the Thai style.'

Changes to the ethnic and cultural background of frontline workers may dilute Thai-style service standards and cause disappointment. Despite the potential impacts on local labour, interviewee B1 perceived that it may not be bad for the overall labour market. The relaxation of business visas may provide Bangkok with a solid supply of labour with better English skills. The viewpoint indeed pointed to the need for language training for Thais to accommodate the future growth.

Towards Economic Sustainability

ASEAN Economic Community (AEC)

The ASEAN Economic Community will play an important role in steering the economic and tourism future of the ten ASEAN countries. As one of the ten nations, Thailand will undoubtedly be influenced by the AEC. Interviewees from the public and private sectors expressed divergent views about the likely impacts of AEC. The private-sector personnel were inclined to see competition among the ten territories; in contrast, public-sector respondents perceived AEC to be one destination and stressed the opportunities for co-operation among the ten ASEAN countries.

Private-sector interviewees generally viewed the multilateral economic co-operation among the ASEAN countries as a form of competition. Interviewee B2 remarked that,'As we are going into the AEC, the game would become even more critical. We have to make sure that we deliver just like other major cities in our competition sets, such as Kuala Lumpur and Singapore, less so Jakarta.' Interviewee B2 also explained that the more neighbourhood countries open up to tourism, the fiercer will be the competition. Some cities compete strongly and have more effective mechanisms than Bangkok. The capital city has competitive advantages, but so do other Asian cities. The opening up of other countries through, for example, the relaxation of visa requirements, has had the effect of eliminating the conventional barriers for international tourists. As a consequence, Bangkok is expected to lose part of its share. The worry of interviewee B1 centred on the workforce issue that was mentioned in the previous section. By way of contrast, the public-sector interviewees generally welcomed the economic integration with Asian neighbours.

Interviewee B6 noted that, 'The real goal of the AEC is to strengthen the tourism co-operation in our region. If tourists come to Cambodia, people there will recommend them coming to Thailand. We will do the same.' She emphasized the aim of the AEC and the value in bringing along long-haul tourists to ASEAN countries as a single and unified destination. The unity of the ten countries should be the focus. Statements from Interviewees B3 and B5 supported this stance, the former claiming that 'The AEC will lead to a major influx of tourists, business travellers and convention delegates from our neighbours. Many new businesses are being started to tap the opportunities of this new market.' Interviewee B5 evaluated the opportunities brought by the AEC and empathized with the worries of the private sector, noting the homogenous tourism offerings of all tourism cities. However, the sector should be free from anxiety, she suggested, owing to its possession of its competitive advantage – Thai hospitality in the land of smiles. This was the reason for Bangkok's popularity as a tourism city with repeat visitation. Nevertheless, the common concern of all interviewees was the uncertainty for tourism surrounding ASEAN integration. The likely positive and negative impacts were based on their assumptions. The actual implementation and effectiveness of the multilateral co-operation remains in doubt.

Mobility in city

It was admitted by all six interviewees that Bangkok has been suffering from traffic congestion, a significant problem that hinders the effective flow of information and people. In particular, the movement between the two airports, between the central business district and suburban areas are issues suppressing economic potential. As Interviewee B4 stated, 'In terms of city tourism, the first thing came to my mind was the traffic between the domestic and international airports'. According to him, the only mode of transportation connecting the two airports was ground transport and, hence, the lack of rail connectivity was a concern. Though the authorities provide a shuttle bus for transiting tourists, he had seen it caught up in traffic during peak hours. He revealed the plan of the Skytrain Company to extend the railway by adding a new route to connect the two airports, which would resolve the congestion problem in the long run. Additionally, the movement from the international airport to the city centre was also of concern to interviewee B4. Although the city centre and the international airport were connected by a railway, the design of the compartments in the trains and the station were not suited to carrying luggage. He therefore emphasized the importance of understanding railway users and to develop specific actions that would cater to their needs. There is a sense of urgency about this issue because of the increasing arrivals. He also mentioned the railway expansion plan to more suburban areas in order to enhance the fluidity of movement within the tourism city. Accessibility within Bangkok represents an inhibitor for the future appeal of Bangkok, although potential improvements to the existing systems may help to reduce congestion and inconvenience.

Toward Environmental Sustainability

Liveability

The interviewees all stressed the significance of making Bangkok liveable in the pursuit of sustainable city tourism development. They generally saw local residents as the primary city users and thought that the needs of locals should be prioritized. For example, Interviewee B3 suggested that 'It is more important to first improve the quality of life for the local people that in turn will lead to an improved quality of experience for visitors. Both go hand in hand.' going on to observe that Bangkok is home to millions of urban dwellers, including up to three quarters of a million expatriates. Interviewee 5 admitted there is conflict between foreign tourists and locals regarding the use of transportation facilities, particularly during the rush hour. No foreseeable plan has been launched by government to address this issue. She further expressed her worries over the problem becoming more critical when coupled with the increasing number of tourists, suggesting that 'We heard the voice of the local citizens complaining about the congested space on public infrastructure. The BTS is always packed with tourists, the ground transport is full of tour coaches.'

Interviewee B1 adopted the same stance, arguing that the quality of life and standard of living for Bangkok dwellers should be recognized and improved. He observed how well New Zealand has performed in promoting responsible behaviour for locals and tourists, claiming that,

'If New Zealand is about responsible tourism, Bangkok is indeed about irresponsible tourism. Bangkok is a dirty city. No government initiatives exist to recycle and keep Bangkok a green city. The only green initiatives are advocated by private organizations.'

Several interviewees pointed to the excessive use of plastic water bottles by locals and tourists alike. According to them, making Bangkok liveable should not only be the agenda of the city government, but also requires the co-operation of industry to embrace green initiatives. The central point of livability rests on issues of waste management. Interviewee B3 admitted that the promising growth in tourist numbers may intensify the problems and stated the government's determination to work on education, regulatory controls and proper enforcement mechanisms. The clean-up of the city's legendary canal system will also be on the agenda.

Towards Sustainable Governance

Against corruption

Some interviewees made reference to corruption in Thailand. According to the benchmark values provided by the corruption perception index, Thailand was ranked 38, below the world average of 43 in 2014 (Transparency International, 2015). Overall, Thailand ranked 85th on this list of 175 countries or territories. Public-sector interviewees stated that anti-corruption has always been a priority for the country. The government of Thailand has, however, been accused of using such declarations as a smokescreen to discredit its political opponents prior to elections (Macan-Markar, 2004). Mr Thaksin Shinawatra, the first elected prime minister of Thailand, ruled the country for five years. In 2006, he was ousted owing to corruption and the abuse of power. Two years later, he was sentenced to two years imprisonment over a corrupt land deal (MacKinnon, 2008), having been found to have violated the conflict of interest laws in helping his wife to buy land from a state agency at a reduced price. Since then, he has been in self-imposed exile in order to escape a jail sentence. The case was seen as a typical example of corruption by policy, a term referring to the collusion among politicians, state bureaucrats and private business. As has been reported by Fernquest (2010), the police, parliament and government construction projects have been especially corrupt. Corruption by policy has often involved big projects and a tremendous amount of graft – estimated at 25 billion Thai baht per year. Interviewee B5 described corruption by policy as a rampant problem in the Thai public sector; each new government promises to change the situation, but the suspicion persists that this is an endemic problem and one that is difficult to resolve. Interviewee B5 emphasized that 'Corruption is the problem all over Thailand, not just Bangkok. Under the martial law, the new government is very tight on the rules.'

The government is keen on gaining public confidence, earning credibility and complying with international obligations (The National Anti-Corruption Commission of Thailand, 2015). According to Interviewee B5, the country has had anti-corruption regulation in place but with weak enforcement. The upper class has benefitted while the poor have suffered. She further expressed the view that corruption was part of the culture and that it was not easy to change such a mindset, a view echoed by Interviewee B1: 'Bangkok cannot afford to be dented by this image'.

New destinations are emerging both within and beyond the country. Although the locals feel generally safe in Bangkok, the sensational photos of the coup in the press during the middle of the last decade soured the image of the city and shocked overseas tourists. Attending to the causes of corruption and managing this image may be pivotal for Bangkok tourism to grow in the long run.

Conclusions

The six interviewees all expressed viewpoints about the major drivers shaping the future of tourism in Bangkok. As displayed in Fig. 12.1, six drivers were revealed. The two drivers under the social sustainability bottom line were Thai lifestyle and workforce. The ASEAN Economic Community and mobility within the city were framed under the economic sustainability

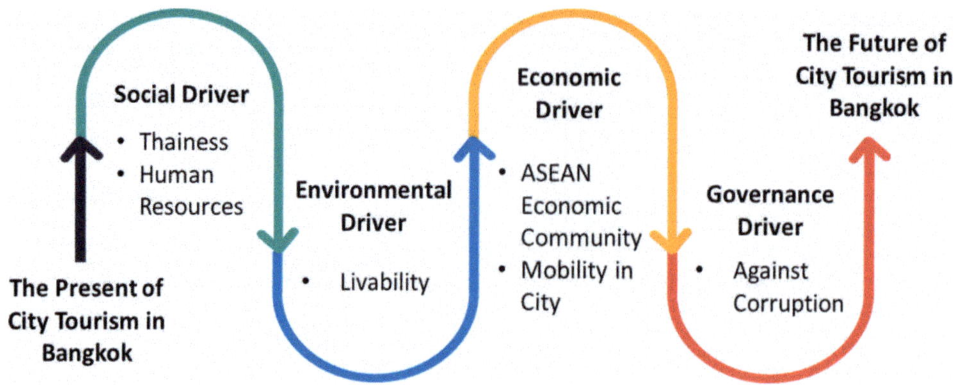

Fig. 12.1. The quadruple wheels driving the future of Bangkok.

bottom line. Livability was conceived as being the environmental sustainable driver, while eliminating corruption was the force for achieving governance sustainability.

The outcomes of the research discussed in this chapter can to some extent be generalized to other major Asian tourism cities. The manner in which these cities develop their tourism sectors will largely depend on how well they can meet global and domestic demands from businesses and investors, embrace the complexity brought by dynamic external factors and provide locals with a good quality of life (Milne and Ateljevic, 2001). The identified drivers and their management will affect the ability of cities such as Bangkok to evolve sustainably.

References

Bangkok Metropolitan Administration (2007) Bangkok Info. Available at: http://www.bangkok.go.th/main/page.php?2-Bangkok%20Info (accessed 14 February 2015).

Bowen, J. (2000) Airline hubs in Southeast Asia: national economic development and nodal accessibility. *Journal of Transport Geography* 8(1), 25–41. DOI: 10.1016/S0966-6923(99)00030-7.

Brinkerhoff, D., Smith, S. and Teegen, H. (eds) (2007) *NGOs and the Millennium Development Goals: Citizen Action to Reduce Poverty*. Palgrave Macmillan, New York.

Buckley, R. (2018) Managing the natural environment for tourism. In: Cooper, C., Volo, S., Gartner, W.C. and Scott, N. (eds) *The Sage Handbook of Tourism Management: Applications of Theories and Concepts to Tourism*. Sage Publications, London, pp. 479–493.

Bumrungrad International Hospital (2016) Corporate information. Available at: https://www.bumrungrad.com/en/about-us/overview (accessed 10 March 2016).

Chinmaneevong, C. (2015) Movenpick will double Thai hotels. *Bangkok Post*, 20 May. Available at: http://www.bangkokpost.com/business/tourism/566695/movenpick-will-double-thai-hotels (accessed 1 August 2019).

Choe, K. and Roberts, B. (2011) *Competitive Cities in the 21stCentury: Cluster-based Local Economic Development*. Asian Development Bank, Mandaluyong City, Philippines.

Cohen, E. (1982) Thai girls and Farang men the edge of ambiguity. *Annals of Tourism Research* 9(3), 403–428. DOI: 10.1016/0160-7383(82)90021-4.

Cohen, E. (1988) Tourism and AIDS in Thailand. *Annals of Tourism Research* 15(4), 467–486. DOI: 10.1016/0160-7383(88)90044-8.

Cohen, E. (2008) Medical tourism in Thailand. In: Cohen, E. (ed.) *Explorations in Thai Tourism: Collected Case Studies*. Emerald, Bingley, UK, pp. 225–255.

Connell, J. (2007) Tummy tucks and the Taj Mahal? Medical tourism and the globalization of health care. In: Woodside, A.G. and Martin, D. (eds) *Tourism Management: Analysis, Behaviour and Strategy*. CAB International, Wallingford, UK, pp. 232–244.

Duff, C. (2009) The drifting city: the role of affect and repair in the development of 'enabling environments'. *International Journal of Drug Policy* 20(3), 202–208. DOI: 10.1016/j.drugpo.2008.08.002.

Fernquest, J. (2010) The many dimensions of corruption. *Bangkok Post*, 10 November. Available at: http://www.bangkokpost.com/learning/learning-from-news/205721/the-many (accessed 10 March 2016).

Font, X. and Guix, M. (2018) Tourism and corporate social responsibility. In: Cooper, C., Volo, S., Gartner, W.C. and Scott, N. (eds) *The Sage Handbook of Tourism Management: Applications of Theories and Concepts to Tourism*. Sage Publications, London, pp. 567–580.

Gore, A. (2013) *The Future*. Random House, New York.

Gough, M.Z. (2015) Reconciling livability and sustainability: conceptual and practical implications for planning. *Journal of Planning Education and Research* 35(2), 145–160.

Instagram (2014) Top locations on Instagram in 2013. Available at: http://blog.instagram.com/post/69877035043/top-locations-2013 (accessed 21 August 2015).

King, R. (2017) An accidental Bangkok theme park. *Tourism Geographies* 19(5), 717–733. DOI: 10.1080/14616688.2017.1320583.

Knoema (2019) Thailand: contribution of travel and tourism to GDP in current prices. Available at: https://knoema.com/atlas/Thailand/topics/Tourism/Travel-and-Tourism-Total-Contribution-to-GDP/Contribution-of-travel-and-tourism-to-GDP (accessed 31 July 2019).

Macan-Markar, M. (2004) Corruption-Thailand: Thaksin's anti-graft drive challenged. Inter Press Service News Agency, September 23. Available at: http://www.ipsnews.net/2004/09/corruption-thailand-thaksins-anti-graft-drive-challenged/ (accessed 10 March 2016).

MacKinnon, I. (2008) Former Thai PM Thaksin found guilty of corruption. *The Guardian*. Available at: http://www.theguardian.com/world/2008/oct/21/thaksin-thailand-corruption (accessed 10 March 2016).

Meyer, W. (1988) *Beyond the Mask: Toward a Transdisciplinary Approach of Selected Social Problems Related to the Evolution and Context of International Tourism in Thailand*. Verlag Breitenbach, Saarbrucken, Germany.

Meyer, D.F. and Neethling, J.R. (2017) Measurement of the enabling developmental environment: a comparative study in a developing region. *Economics & Sociology* 10(4), 67–82. DOI: 10.14254/2071-789X.2017/10-4/6.

Milne, S. and Ateljevic, I. (2001) Tourism, economic development and the global-local nexus: theory embracing complexity. *Tourism Geographies* 3(4), 369–393. DOI: 10.1080/146166800110070478.

Naghizadeh, Y. and Shahzadeh, H. (2013) The role of media in progressing the people's culture. *Journal of Current Research in Science* 1(6), 559–562.

Novy, J. (2019) Urban tourism as a bone of contention: four explanatory hypotheses and a caveat. *International Journal of Tourism Cities* 5(1), 63–74. DOI: 10.1108/IJTC-01-2018-0011.

Pearce, P.L. and Thanksooks, J. (2016) Towards the future of the domestic hotel in Thailand: a timeline approach. *Journal of Vacation Marketing* 22(3), 199–211.

Peleggi, M. (1996) National heritage and global tourism in Thailand. *Annals of Tourism Research* 23(2), 432–448. DOI: 10.1016/0160-7383(95)00071-2.

Richter, L.K. (1989) *The Politics of Tourism in Asia*. University of Hawaii Press, Honolulu.

Rickly, J. and Vidon, M. (2018) *Authenticity and Tourism: Productive Debates, Creative Discourses*. Emerald, Bingley, UK.

Sharpley, R. (2018) The host community: perceptions of and responses to tourism. In: Cooper, C., Volo, S., Gartner, W.C. and Scott, N. (eds) *The Sage Handbook of Tourism Management Applications of Theories and Concepts to Tourism*. Sage Publications, London, pp. 537–552.

Thai Websites (2019) GDP per capita and population data for the provinces of Thailand. Available at: http://www.thaiwebsites.com/provinces-GDP.asp (accessed 1 August 2019).

The Economist Intelligence Unit (2012) The Global Shopping Index. Available at: http://globeshopperindex.com/ (accessed 28 February 2016).

The National Anti-Corruption Commission of Thailand (2015). International Anti-corruption Day 2015. Available at: https://www.nacc.go.th/download/article/article_20160108110806.pdf (accessed 10 March 2016).

Trading Economics (2018) GDP Thailand. Available at: https://tradingeconomics.com/thailand/gdp (accessed accessed 31 July 2019).

Transparency International (2015) Corruption perceptions index 2014. Available at: https://www.transparency.org/cpi2014/results (accessed 10 March 2019).

UNWTO (2012) *Global Report on City Tourism*. UN World Tourism Organization, Madrid.

Walton, J. (1993) Tourism and economic development in ASEAN. In: Hitchcock, M., King, V.T. and Parnwell, M.J.G. (eds) *Tourism in South-East Asia*. Routledge, London, pp. 214–233.

Webster, D. (2004) Bangkok: evolution and adaptation under stress. In: Gugler, J. (ed.) *World Cities beyond the West: Globalization, Development and Inequality*. Cambridge University Press, Cambridge, UK, pp. 82–118.

Wheeller, B. (2015) Sustainable tourism: milestone or millstone? In: Singh, T.V. (ed.) *Challenges in Tourism Research*. Channel View Publications, Bristol, UK, pp. 253–263.

World Atlas (2019) The 150 largest cities in the world. Available at: https://www.worldatlas.com/citypops.htm (accessed 1 August 2019).

World Bank (2010) *Thailand Bangkok Metropolitan Administration Urban Transformation Project*. World Bank, Washington, DC.

World Bank (2015) *East Asia's Changing Urban Landscape: Measuring a Decade of Spatial Growth*. The World Bank Group, Washington, DC.

13 Smart Tourism Development

Ulrike Gretzel*
University of Southern California, Los Angeles, California, USA

Introduction

The beginning of the 21st century has seen an acceleration of tourism development that in many ways has been fuelled by information and communication technology (ICT) (Xiang, 2018). Such technology underlies the value propositions of new mass tourism industry players such as Airbnb, Ryanair, Tripadvisor and Booking.com, supports critical tourist behaviours, including information search, booking/payment and en route navigation, and facilitates government actions such as electronic visa programs and security screenings at airports or cruise ship terminals. At the same time, there is also, paradoxically, increasing trust in the same technology to help mitigate the many negative impacts of tourism around the world. Social media, for instance, are simultaneously blamed for creating overtourism and praised for their abilities to support community and employee activism as well as their potential to encourage environmentally friendly behaviours (Gretzel, 2019). Another way in which ICT is believed to both stimulate growth and create a more sustainable future for tourism is through its advanced use in generating and managing tourism experiences. Such efforts are usually labelled as 'smart tourism', which – from an instrumental perspective – is defined as integrated efforts to collect, aggregate, transform and distribute data using smart technologies with the goal of achieving efficiency, sustainability and experience enrichment (Gretzel *et al.*, 2015a).

While the relationship between tourism and ICT has been an intricate one for some time (Werthner and Klein, 1999), there seems to be an ever-greater emphasis on technological development in the context of tourism planning and development. ICT has moved to the bottom of the tourism development hierarchy of needs, paving the way for tourism by making many essential aspects of tourism possible, from efficient transportation to the destination and effective provision of food and shelter to the creation of safe environments for tourists. Places outside technological connectivity are now conceptualized as too remote but for the most adventurous tourists (Gretzel, 2010). Milne *et al.* (2004) describe the role of ICT in tourism development as a critical one for creating tourism destinations in a global marketplace, ensuring the survival of small and medium tourism enterprises and enabling greater public participation and, therefore, community buy-in. Thus, tourism development without necessary investments in ICT development has become unimaginable. Smart tourism development takes this trend one step further by making ICT the central focus as well as the predominant means of development.

This increasing reliance on ICT is also visible in relation to international development efforts, with ICT for development (ICT4D) shifting the focus from pure economic growth goals

*gretzel@usc.edu

 Tourism in Development: Reflective Essays (eds P.U.C. Dieke *et al.*)
DOI: 10.1079/9781789242812.0013

to technological empowerment of marginalized people (Unwin and Unwin, 2009). Heeks (2008) summarizes the core ICT4D issues as: (i) having policies and infrastructure for ICT deployment in place (readiness); (ii) supplying ICTs to those in need (availability); (iii) ensuring the usefulness of ICT (uptake); and, (iv) using ICT to make the greatest development impact (impact). Rega and Inversini (2016) highlight the importance of ICT investments, especially for tourism development in developing country contexts, but in general there has been very little emphasis on ICT4D in tourism theory and practice.

In contrast, there has been a tremendous boom in smart tourism-related initiatives over the last decade (mostly in developed countries) (Gretzel, 2018). This is also reflected in the growing academic literature on smart tourism (Rafael, 2020; Mehraliyev *et al.*, 2020). Smart tourism efforts emphasize the use of ICT in the form of intelligent systems (Gretzel, 2011) to achieve a variety of tourism development goals. They require substantial infrastructure investment as well as mindset shifts in terms of tourism governance. This chapter reflects on the principles of smart tourism development, charts its conceptual and practical evolution and discusses its role in future tourism development around the globe. Specifically, it argues for a shift away from using smart tourism to advance a purely technological agenda and highlights the need to develop smart tourism ideas into frameworks that can guide tourism development research and practice.

Smart Development

Originally derived from the United Nations Sustainable Development Goals (SDGs), smart development has become a ubiquitous strategy that nowadays often supersedes sustainability when describing development efforts (Joss *et al.*, 2019). Yigitcanlar *et al.* (2018), however, note that the original focus on sustainability typically gets lost in practice, with initiatives centred on technological infrastructure and innovation management to enable economic benefits. It is also important to mention that smart development has been almost exclusively applied to urban areas, with the realization of the 'smart city' being the pinnacle of most contemporary smart development projects. This emphasis emerged as a direct response to increased global urbanization and mounting urban sustainability challenges (Bibri and Krogstie, 2017). Consequently, smart development has so far mostly focused on understanding and addressing urban problems and challenges, predominantly by providing data-led solutions (Grace *et al.*, 2020).

Smart city definitions

Reviews of the smart city literature present a multitude of definitions but also identify common elements (Mora *et al.*, 2017; Bibri and Krogstie, 2017; Yigitcanlar *et al.*, 2018). Overall, the smart city concept describes a utopian vision of efficiency and sustainability in terms of managing and controlling city assets, from public transport, water supply and energy consumption to waste management, law enforcement and schools (Colding *et al.*, 2020). Smart technologies, such as sensors, the Internet of Things (IoT), cloud computing and big-data analytics tools provide the technological backbone of the smart city. Kitchin (2014) conceives of the smart city as one that implements pervasive ICT systems to monitor and integrate all its critical infrastructures, optimize its resources, plan its activities and maximize services. Bibri (2018) defines smart cities as combining sustainable planning principles with big data and context-aware applications embedded into critical city infrastructure (for example, smart energy grids and smart mobility services). According to Bibri (2018), this combination enables new models, processes, methods, techniques and practices to help the smart city function constructively, safely, efficiently and sustainably. Similarly, Yigitcanlar *et al.* (2018, p. 156) posit that this advanced technology embedded into every aspect of the city acts like an 'artificial nervous system' that enables smart cities 'to behave in intelligently co-ordinated ways'. In sum, smart cities are developed with the assumption that ICT is critical for achieving prosperity, effectiveness and competitiveness and is central to urban operations, functions, services and designs (Bibri and Krogstie, 2017).

In addition to the ICT-centric definitions of smart cities, there are also people-oriented conceptualizations that envision smart development as a way to improve the quality of life for city inhabitants. For instance, Lara *et al.* (2016) stress that a smart city is a community that systematically promotes the overall well-being for all its members and strives to proactively and sustainably become an increasingly better place to live, work and play. Comprehensive smart city frameworks recognize the importance of evaluating smart development efforts from a quadruple bottom-line perspective, applying environmental, social, economic and governance indicators to gauge sustainability (Teriman *et al.*, 2009). From such a perspective, smart development can be defined as 'investments in human and social capital and traditional (transport) and modern (ICT) communication infrastructure [that] fuel sustainable economic growth and a high quality of life, with a wise management of natural resources, through participatory governance' (Caragliu *et al.*, 2011, p. 6).

In line with these frameworks, Giffinger and Haindlmaier (2010) use six distinct dimensions to rank the 'smartness' of cities: (i) smart mobility; (ii) smart environment; (iii) smart living; (iv) smart people; (v) smart economy; and (vi) governance. More recently, Yigitcanlar *et al.* (2018) presented a smart city framework based on a comprehensive review of the literature that identifies city assets as the inputs, policy, technology and community as the drivers of smart city asset development and management, and productivity and innovation (economic development), liveability and well-being (socio-cultural development), sustainability and accessibility (spatial development), and governance and planning (institutional development) as the desired outcomes (Fig. 13.1). However, they also stress that attention of theory and practice so far has been mostly on the smart city drivers, with desired outcomes, and especially governance aspects, being mostly neglected.

Smart governance

Fernandez-Anez *et al.* (2018) highlight the key role of governance, alongside stakeholder involvement and a comprehensive vision, in successful smart city development. Zheng *et al.* (2020) reveal a corresponding emphasis on smart city governance and strategic planning in their review of smart city literature. Meijer and Bolívar (2016) suggest that smart city governance is not just about good administration of smart cities but involves a substantial transformation of governing bodies and governance processes. The need for innovation in governance is also emphasized by Bolívar (2018), who advises that smart development processes exceed the capacities, capabilities and reaches of traditional institutions. To address governance issues related to smart development, Deakin (2014) proposes a triple helix model approach built on public–private–academia partnerships. In this model, universities engage in capacity building and government institutions monitor, evaluate and facilitate, while industry co-designs solutions with both. In addition to the value placed on such innovative partnerships, transparency, openness, accountability, collaboration,

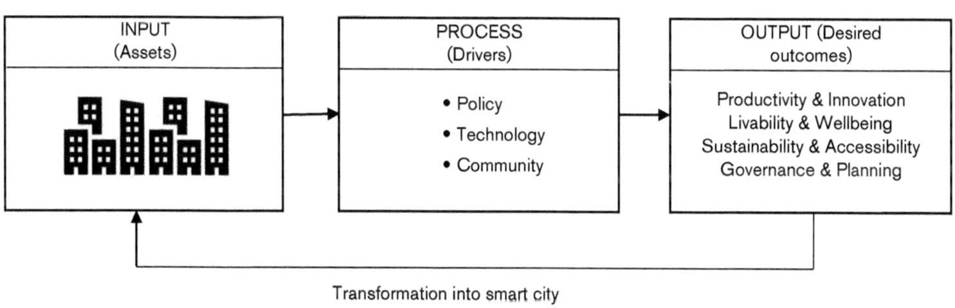

Fig. 13.1. Smart city development framework.

(Adapted from Yigitcanlar *et al.*, 2018)

innovation and efficiency are common keywords used when smart city governance processes are described (Chourabi *et al.*, 2012). There also seems to be a growing recognition of the need for cross-city governance (Zheng *et al.*, 2020), recognizing that smart cities are systems within systems, such as smart regions and smart nations.

Participatory governance is especially highlighted in the literature. There is a general consensus that current North-east Asian models of top-down smart city governance, although effective in the short term, will not enable the achievement of smart development goals in the long run (Yigitcanlar *et al.*, 2018). Thus, Eggers and Skowron (2018) assert that smart cities of the future need a shift in governance 'from "doing things" to enabling participation in civic innovation' and promote the idea of the city-as-a-platform where governments: (i) cultivate ecosystems of partners; (ii) build governance infrastructure that enables residents to become proactive, connected and collaborative participants in the smart city movement; and (iii) provide funding that encourages businesses to align with smart city development goals. Importantly, no matter how it is currently defined, practised or envisioned, the central role of governance for successful smart development is undeniable.

Smart Tourism

Smart tourism emerged from the smart city idea, with smart destinations representing special cases of smart cities (Buhalis and Amaranggana, 2014). Koo *et al.* (2020) have recently introduced the term 'smart tourism cities' to highlight this overlap between smart cities and smart destinations as they share space and essential infrastructure, intersect in terms of data needs and pursue similar goals, from mobility and accessibility to sustainability and liveability/visitability. Using the case of Seoul, for instance, Gretzel *et al.* (2018) illustrate how smart city infrastructure, such as free wireless networks and smart bike sharing, add to the touristic value of the city and, consequently, positively influence the attractiveness and functioning of the smart destination. However, while there is convergence between the two concepts, there is also a lot of uniqueness in smart tourism that warrants special consideration. In addition to improving the quality of life of residents at tourism destinations, smart tourism development pays special attention to offering enhanced experiences to tourists (Gretzel *et al.*, 2015a; Romão *et al.*, 2018). These experiences are more concentrated in temporal and spatial terms and are functionally as well as emotionally different from work and leisure. Further, tourism governance is often separated from overall city governance, with tourists lacking veto powers and usually also the motivation for involvement in designing the long-term future of the destination. Thus, a conceptualization of smart tourism and smart destinations separate from general smart development is warranted.

Similar to the smart city development literature, recent papers stress that smart tourism overall remains a rather elusive concept (for example, Molinillo *et al.*, 2019). Rafael (2020) demonstrates that data, technology and innovation are at the centre of smart tourism conceptualizations in the academic literature; sustainability is mentioned but remains insufficiently addressed, although it is theoretically inseparable from the smart tourism concept (Perles Ribes and Ivars-Baidal, 2018). Similarly, Mehraliyev *et al.* (2020) show that the scientific literature on smart tourism tends to focus on single technological applications and their adoption with only a few papers dealing with smart tourism planning and management. There is indeed danger in equating smart tourism with smartphone use in tourism (for example, Dorcic *et al.*, 2019) and in focusing on the tourist perspective (for example, Femenia-Serra *et al.*, 2019a), as the value of smart tourism clearly lies in providing a comprehensive and normative framework to guide tourism development.

Smart tourism definitions

As indicated in the introduction to this chapter, smart tourism has been defined in the literature as:

> tourism supported by integrated efforts at a destination to collect and aggregate data derived from physical infrastructure, social connections, government/organizational sources and human

> bodies/minds in combination with the use of advanced technologies to transform that data into on-site experiences and business value-propositions with a clear focus on efficiency, sustainability and experience enrichment. (Gretzel *et al.*, 2015a, p. 181)

From a structural perspective, Gretzel and Scarpino-Johns (2018) describe smart tourism as involving a composite of six layers (physical, technological, data, governance, business and experience) that work together to realize what are commonly referred to as pillars or axes of smart tourism in the literature, namely: (i) effective use of advanced technology; (ii) mobility/accessibility; (iii) sustainability; and (iv) knowledge development/innovation (Gretzel, 2018). As can be seen from these, smart tourism conceptualizations clearly emphasize technology and innovation aspects. Technology-based smart environments form the basis of smart tourism but are not enough to realize its goals. Boes *et al.* (2016) conceptualize the technological infrastructure as 'hard smartness' and illustrate that smart tourism also requires 'soft smartness', meaning investments in human and social capital. Del Chiappa and Baggio (2015) further highlight the need for co-ordination and knowledge transfer among the often very diverse tourism entities to achieve smartness. Interestingly, all existing frameworks and definitions only apply to well-defined, spatially limited smart destinations and have not been scaled up to the level of smart tourism regions (Gretzel, 2018).

While Femenia-Serra *et al.* (2019b) have called for an integration of the tourist perspective into smart tourism frameworks, it seems to be clear that the notion of the smart tourist is mostly irrelevant if the focus of smart tourism development does not go beyond the creation of smart end-user applications. Although co-creation matters for the experience layer, this is just the tip of the iceberg of smart tourism development efforts. Active participation of tourists is typically not required to realize the fundamental tenets of smart tourism. However, goals regarding the kinds of businesses and touristic experiences smart tourism wants to support, the kinds of interactions between tourists and residents it seeks to foster, and the types of processes it plans to put in place to realize and evaluate its development need to be articulated, as they should shape the foundations of smart tourism development but remain absent from the current literature.

Smart tourism and smart destination definitions are also created and communicated by practitioners. For example, the Spanish government agency SEGITTUR (2015, p. 32) defines the smart destination as:

> an innovative space, accessible for all, established on a cutting edge technology infrastructure which guarantees sustainable development of the land, facilitates the interaction and integration of the visitor with the surroundings and increases the quality of their experience in the destination, as well as the quality of life of residents.

Looking at existing smart tourism brands, Gretzel and Collier de Mendonça (2019) find that the definitional uncertainty found in the academic literature is also reflected in the communications of smart tourism practitioners. Smart tourism brands seem to be aimed at attracting investments and mobilizing key stakeholders rather than achieving fundamental changes in tourism development and governance. Gretzel and Collier de Mendonça (2019) further stress that industry and government discourse is firmly rooted in techtopian ideology and shaped by the rhetoric of powerful corporations, while ignoring social and environmental sustainability gorsals. Ivars-Baidal *et al.*'s (2019b) research confirms the centrality of technology and data-driven scenarios for the perceptions of smart tourism among practitioners. They clearly identify governance as the predominant challenge for smart tourism development.

Smart tourism governance

Smart tourism is seen as vital to increasing the long-term competitiveness of destinations by allowing them to improve their market offerings and image, implement resource stewardship and increase the effectiveness of their management (Koo *et al.*, 2016). While a healthy business ecosystem is seen as crucial for delivering the technological solutions needed by smart tourism (Gretzel *et al.*, 2015b), there is also a deep recognition (at least in theory) of the need for government intervention to provide for basic infrastructure, essential regulation and necessary co-ordination (Shafiee *et al.*, 2019). The kind of government involvement envisioned in the smart tourism literature corresponds to the active government involvement in tourism development described by Jenkins and Henry (1982). Indeed, strong destination management organizations are perceived to be critical for smart tourism development (Gretzel and Scarpino-Johns, 2018; Ivars-Baidal *et al.*, 2019a).

Perles Ribes and Ivars-Baidal (2018) posit that smart tourism development requires destination management at applied, instrumental and strategic/relational levels. The applied level focuses on smart solutions, the instrumental level puts key infrastructure and information systems in place, while the strategic level develops policy and enables the kind of co-operations a smart tourism ecosystem requires (Gretzel *et al.*, 2015b). Perles Ribes and Ivars-Baidal (2018) specifically mention long-term planning, scenario building, efficient resource use, real-time management, monitoring, transparent and participatory governance, public–private co-operation and open innovation as aspects of smart tourism management. Lalicic and Önder (2018) discuss participatory decision making in the context of smart destination management. However, a comprehensive smart tourism governance framework is currently missing from the academic literature as well as industry documents. As a result, the policies that should be put in place, the processes that need to be enabled and the strategic visions that need to be developed to enable smart tourism remain obscure, and most of the development and discussion therefore happens at the instrumental level.

Smart Tourism in Development

While smart city development literature and practice have had 30 years to evolve (Zheng *et al.*, 2020), smart tourism has only existed as a concept for about ten years. Thus, it is not surprising that, both in theory and practice, smart tourism development lags behind smart city development. One of the most striking differences is that the smart city literature seems to be much more interdisciplinary, a lot more planning-focused, as well as considerably more critical. Indeed, the critical smart city literature is growing especially fast (Zheng *et al.*, 2020). For instance, Colding *et al.* (2020) warn of the danger of smart cities running into unmanageable complexity issues. In addition, Kitchin (2014) voices concerns regarding the politics of data, technocratic governance, corporatization of city management, system vulnerabilities (rendering cities hackable) and ethical issues regarding privacy and discrimination. Söderström *et al.* (2014) and Wiig (2016) question the motives of smart development efforts and their ability to address the mounting urban challenges because they seem overwhelmingly focused on advancing the agendas of big, global corporations.

In contrast, most of the smart tourism discourse happens at the instrumental level (as evidenced by the review conducted by Mehraliyev *et al.*, 2020) and does not question the technological agendas promoted by the bulk of smart tourism initiatives. Gretzel and Jamal (2020) have also identified a huge gap in smart tourism theory and practice regarding justice considerations. Fairness, equity and ethics are not even mentioned in most smart tourism documents despite the sustainability and accessibility issues smart tourism is supposed to be concerned with. Gretzel and Jamal (2020) provide initial guidelines for ethical smart destination management, including:

- accountability for and regulation of local-global ‘virtual spaces’;
- special care with respect to technology design and use to protect against privacy violations and algorithmic discrimination;
- awareness of new levels and forms of dependency on non-local technology providers;
- identification/monitoring, co-ordination and distribution of wealth as well as responsibility among a dynamically emerging set of stakeholders beyond traditional tourism realms; and
- consideration of the typically invisible environmental, social and health consequences of technology use at the destination and elsewhere.

However, a lot more critical work is needed to guide smart tourism development over the next decade with a clear emphasis on justice.

Smart tourism development of course co-evolves with general tourism development. The last few years have seen tremendous growth in mass tourism, especially to urban areas. Smart tourism quickly became a panacea for overtourism-related issues due to its promise of equipping destinations with heightened abilities to monitor visitor flows and redirect demand (Ivars-Baidal *et al.*, 2019b). This created tremendous opportunities for funding and general support for smart tourism development agendas. This trend came to an abrupt halt in early 2020 when the coronavirus pandemic paralyzed tourism, and tourism-related movements as well as economic activities came to a standstill. As the focus now shifts to crisis management and resilience, smart tourism development has the chance to yet again be identified as a necessary step in equipping the tourism industry and destination management organizations with important capabilities. Gretzel and Scarpino-Johns (2018) describe the resilient smart destination as one that not only has a smart tourism mindset but can also rely on essential sensing, opening (of organizational silos), sharing, governing and innovating skills. Given the likely return of pandemics and other potential threats to tourism in the near future, using smart tourism concepts to build resilience will likely become a priority for destinations. In this context, the smart tourism city also gains in importance as it underscores the advantages of flexibly switching between touristic and recreational offers during travel bans.

Klein and Peet (2008) warn of disaster capitalism in the wake of events like the coronavirus crisis – that is, the increased adoption of liberal economic policies that voters would not likely accept under normal circumstances. There is also fear that governments will abuse their powers to extend their control via modern surveillance structures. Paradoxically, such steps will make smart tourism development simultaneously more needed and more feasible. The infrastructure investments that governments are likely to make to foster work-from-home and create more flexible gig economy jobs, such as through faster internet connections, and to track individuals/virus outbreaks through, for example, mobile phone applications or drone patrols, will put in place technologies, capabilities and co-operations that are critical for smart tourism management or could be repurposed to serve tourism needs once the crisis is over. The drastic economic shortfalls resulting from the crisis also create a new appreciation for tourism as an economic engine and source of wellbeing. With tourism in the news, smart tourism development plans will likely receive more attention. However, while such initiatives will promote smart tourism in the short term, they may challenge smart tourism principles of sustainability, liveability and well-being in the long run. Thus, while the present moment provides a unique opportunity to push for smart tourism development, its strategic orientation beyond technological development needs to be strengthened to secure its success.

Conclusion

Whereas for several decades it looked like technology was undermining tourism planning efforts and hijacking destination management, smart tourism development provides a new-found opportunity to attract investments in tourism infrastructure, secure a place for tourism agencies at urban planning tables and expand tourism-related goals beyond the short-term economic gains typically promoted by traditional tourism stakeholders. Smart tourism elevates the role of destination marketing organizations as enablers of a kind of tourism that ultimately seeks sustainability. Its intricate link and interdependence with broader smart development plans promise more opportunities for and more recognition of tourism. However, the discussion of current smart tourism definitions and frameworks in this chapter has clearly revealed the shortcomings in conceptualizing smart tourism development as a strategic mindset rather than an applications-focused promotional tactic to establish short-term competitive advantages.

Further, the above elaborations have underscored the need to ask difficult questions regarding smart tourism's ultimate goals, its desired processes and its beneficiaries. This is especially critical for instances where smart tourism is used as an international development tool. What is needed is tourism that is both smart and just. There also seems very little co-ordination across smart destinations beyond the smart destination network established by the Spanish government (see https://www.destinosinteligentes.es/). Moreover, the triple- helix model of partnerships established for smart city development has not been successfully implemented in tourism, leading to a rift between academic work and practical implementation projects. Most importantly, smart tourism discussions have been largely initiated by computer science and tourism management scholars. The voices of tourism planning and development experts need to be urgently brought into the conversation. It is hoped that this chapter will provide an important impetus for more planning and development-focused smart tourism research.

References

Bibri, S.E. (2018) *Smart Sustainable Cities of the Future*. Springer, Heidelberg, Germany.

Bibri, S.E. and Krogstie, J. (2017) Smart sustainable cities of the future: an extensive interdisciplinary literature review. *Sustainable Cities and Society* 31, 183–212. DOI: 10.1016/j.scs.2017.02.016.

Boes, K., Buhalis, D. and Inversini, A. (2016) Smart tourism destinations: ecosystems for tourism destination competitiveness. *International Journal of Tourism Cities* 2(2), 108–124. DOI: 10.1108/IJTC-12-2015-0032.

Bolívar, M.P.R. (2018) *Smart Technologies for Smart Governments*. Springer, Cham, Switzerland.

Buhalis, D. and Amaranggana, A. (2014) Smart tourism destinations. In: Xiang, Z. and Tussyadiah, I. (eds) *Information and Communication Technologies in Tourism 2014*. Springer, Cham, Switzerland, pp. 553–564.

Caragliu, A., Del Bo, C. and Nijkamp, P. (2011) Smart cities in Europe. *Journal of Urban Technology* 18(2), 65–82. DOI: 10.1080/10630732.2011.601117.

Chourabi, H., Nam, T., Walker, S., Gil-Garcia, J.R., Mellouli, S. *et al.* (2012) Understanding smart cities: an integrative framework. In: *45th Hawaii International Conference on System Science (HICSS)*. IEEE, pp. 2289–2297.

Colding, J., Colding, M. and Barthel, S. (2020) The smart city model: a new panacea for urban sustainability or unmanageable complexity? *Environment and Planning B: Urban Analytics and City Science* 47(1), 179–187. DOI: 10.1177/2399808318763164.

Deakin, M. (2014) Smart cities: the state-of-the-art and governance challenge. *Triple Helix* 1(1), 1–16. DOI: 10.1186/s40604-014-0007-9.

Del Chiappa, G. and Baggio, R. (2015) Knowledge transfer in smart tourism destinations: analyzing the effects of a network structure. *Journal of Destination Marketing & Management* 4(3), 145–150. DOI: 10.1016/j.jdmm.2015.02.001.

Dorcic, J., Komsic, J. and Markovic, S. (2019) Mobile technologies and applications towards smart tourism – state of the art. *Tourism Review* 74(1), 82–103. DOI: 10.1108/TR-07-2017-0121.

Eggers, W.D. and Skowron, J. (2018) Forces of change: smart cities. Deloitte Center for Government Insights Report. Available at: https://www2.deloitte.com/us/en/insights/focus/smart-city/overview.html (accessed 1 March 2020).

Femenia-Serra, F., Neuhofer, B. and Ivars-Baidal, J.A. (2019a) Towards a conceptualisation of smart tourists and their role within the smart destination scenario. *The Service Industries Journal* 39(2), 109–133. DOI: 10.1080/02642069.2018.1508458.

Femenia-Serra, F., Perles-Ribes, J.F. and Ivars-Baidal, J.A. (2019b) Smart destinations and tech-savvy millennial tourists: hype versus reality. *Tourism Review* 74(1), 63–81. DOI: 10.1108/TR-02-2018-0018.

Fernandez-Anez, V., Fernández-Güell, J.M. and Giffinger, R. (2018) Smart City implementation and discourses: an integrated conceptual model. The case of Vienna. *Cities* 78, 4–16. DOI: 10.1016/j.cities.2017.12.004.

Giffinger, R. and Haindlmaier, G. (2010) Smart cities ranking: an effective instrument for the positioning of the cities? *ACE: Architecture, City and Environment* 4(12), 7–26.

Grace, M., Scott, A.J., Sadler, J.P., Proverbs, D.G. and Grayson, N. (2020) Exploring the smart-natural city interface; re-imagining and re-integrating urban planning and governance. *Emerald Open Research* 2(2), 7. DOI: 10.35241/emeraldopenres.13226.1.

Gretzel, U. (2010) Travel in the network: redirected gazes, ubiquitous connections and new frontiers. In: Levina, M. and Kien, G. (eds) *Post-global Network and Everyday Life*. Peter Lang, New York, pp. 41–58.

Gretzel, U. (2011) Intelligent systems in tourism: a social science perspective. *Annals of Tourism Research* 38(3), 757–779.

Gretzel, U. (2018) From smart destinations to smart tourism regions. *Journal of Regional Research* 42, 171–184.

Gretzel, U. (2019) The role of social media in creating and addressing overtourism. In: Dodds, R. and Butler, R.R. (eds) *Overtourism: Issues, Realities and Solutions*. de Gruyter, Boston, Massachusetts, pp. 62–75.

Gretzel, U. and Collier de Mendonça, M. (2019) Smart destination brands: semiotic analysis of visual and verbal signs. *International Journal of Tourism Cities* 5(4), 560–580. DOI: 10.1108/IJTC-09-2019-0159.

Gretzel, U. and Jamal, T. (2020) Guiding principles for good governance of the smart destination. In: Paris, C.M. and Benjamin, S. (eds) *Proceedings of the 51st Annual Travel and Tourism Research Association International Conference, forthcoming*. TTRA, Lapeer, Michigan.

Gretzel, U. and Scarpino-Johns, M. (2018) Destination resilience and smart tourism destinations. *Tourism Review International* 22(3), 263–276. DOI: 10.3727/154427218X15369305779065.

Gretzel, U., Sigala, M., Xiang, Z. and Koo, C. (2015a) Smart tourism: foundations and developments. *Electronic Markets* 25(3), 179–188. DOI: 10.1007/s12525-015-0196-8.

Gretzel, U., Werthner, H., Koo, C. and Lamsfus, C. (2015b) Conceptual foundations for understanding smart tourism ecosystems. *Computers in Human Behavior* 50, 558–563. DOI: 10.1016/j.chb.2015.03.043.

Gretzel, U., Ham, J. and Koo, C. (2018) Creating the city destination of the future: the case of smart Seoul. In: Wang, Y., Shakeela, A., Kwek, A. and Khoo-Lattimore, C. (eds) *Managing Asian Destinations*. Springer, Cham, Switzerland, pp. 199–214.

Heeks, R. (2008) ICT4D 2.0: the next phase of applying ICT for international development. *Computer* 41(6), 26–33. DOI: 10.1109/MC.2008.192.

Ivars-Baidal, J.A., Celdrán-Bernabeu, M.A., Mazón, J.-N. and Perles-Ivars, Á.F. (2019a) Smart destinations and the evolution of ICTs: a new scenario for destination management? *Current Issues in Tourism* 22(13), 1581–1600. DOI: 10.1080/13683500.2017.1388771.

Ivars-Baidal, J.A., Hernández, M.G. and de Miguel, S.M. (2019b) Integrating overtourism in the smart tourism cities agenda. *E-review of Tourism Research* 17(2), 122–139.

Jenkins, C.L. and Henry, B.M. (1982) Government involvement in tourism in developing countries. *Annals of Tourism Research* 9(4), 499–521. DOI: 10.1016/0160-7383(82)90070-6.

Joss, S., Sengers, F., Schraven, D., Caprotti, F. and Dayot, Y. (2019) The smart City as global discourse: Storylines and critical junctures across 27 cities. *Journal of Urban Technology* 26(1), 3–34. DOI: 10.1080/10630732.2018.1558387.

Kitchin, R. (2014) The real-time city? Big data and smart urbanism. *GeoJournal* 79(1), 1–14. DOI: 10.1007/s10708-013-9516-8.

Klein, N. and Peet, R. (2008) Book review: The Shock Doctrine: The Rise of Disaster Capitalism. *Human Geography* 1(2), 130–133. DOI: 10.1177/194277860800100215.

Koo, C., Shin, S., Gretzel, U., Hunter, W.C. and Chung, N. (2016) Conceptualization of smart tourism destination competitiveness. *Asia Pacific Journal of Information Systems* 26(4), 561–576. DOI: 10.14329/apjis.2016.26.4.561.

Koo, C., Gretzel, U. and Donnellan, B. (2020) Editorial - special issue on smart tourism cities. *Asia Pacific Journal of Tourism Research.* Forthcoming.

Lalicic, L. and Önder, I. (2018) Residents' involvement in urban tourism planning: opportunities from a smart city perspective. *Sustainability* 10(6), 1852–1867. DOI: 10.3390/su10061852.

Lara, A.P., Moreira Da Costa, E., Furlani, T.Z. and Yigitcanlar, T. (2016) Smartness that matters: towards a comprehensive and human-centred characterisation of smart cities. *Journal of Open Innovation: Technology, Market, and Complexity* 2(1), 8p.. DOI: 10.1186/s40852-016-0034-z.

Mehraliyev, F., Chan, I.C.C., Choi, Y., Koseoglu, M.A. and Law, R. (2020) A state-of-the-art review of smart tourism research. *Journal of Travel & Tourism Marketing* 37(1), 78–91. DOI: 10.1080/10548408.2020.1712309.

Meijer, A. and Bolívar, M.P.R. (2016) Governing the smart city: a review of the literature on smart urban governance. *International Review of Administrative Sciences* 82(2), 392–408. DOI: 10.1177/0020852314564308.

Milne, S., Mason, D. and Hasse, J. (2004) Tourism, information technology, and development: revolution or reinforcement? In: Lew, A., Hall, C.M. and Williams, A. (eds) *A Companion to Tourism*. Blackwell Publishing, Malden, Massachusetts, pp. 184–194.

Molinillo, S., Anaya-Sánchez, R., Morrison, A.M. and Coca-Stefaniak, J.A. (2019) Smart city communication via social media: analysing residents' and visitors' engagement. *Cities* 94, 247–255. DOI: 10.1016/j.cities.2019.06.003.

Mora, L., Bolici, R. and Deakin, M. (2017) The first two decades of smart-city research: a bibliometric analysis. *Journal of Urban Technology* 24(1), 3–27. DOI: 10.1080/10630732.2017.1285123.

Perles Ribes, J.F. and Ivars-Baidal, J. (2018) Smart sustainability: a new perspective in the sustainable tourism debate. *Journal of Regional Research* 42, 151–170.

Rafael, C. (2020) Analysis of scientific production: smart tourism destination, technology and sustainability. In: Rocha, R., Abreu, A., Vidal de Cavalho, J., Liberato, D., Gonzalez, E., *et al.* (eds) *Advances in Tourism, Technology and Smart Systems*. Springer, Singapore, pp. 599–613.

Rega, I. and Inversini, A. (2016) eTourism for development (eT4D): the missing piece in the ICT4D research agenda. *Information Technologies & International Development* 12(3), 19–24.

Romão, J., Kourtit, K., Neuts, B. and Nijkamp, P. (2018) The smart city as a common place for tourists and residents: a structural analysis of the determinants of urban attractiveness. *Cities* 78, 67–75. DOI: 10.1016/j.cities.2017.11.007.

SEGITTUR (2015) *Smart Destinations Report: Building the Future*. SEGITTUR, Madrid.

Shafiee, S., Rajabzadeh Ghatari, A., Hasanzadeh, A. and Jahanyan, S. (2019) Developing a model for sustainable smart tourism destinations: a systematic review. *Tourism Management Perspectives* 31, 287–300. DOI: 10.1016/j.tmp.2019.06.002.

Söderström, O., Paasche, T. and Klauser, F. (2014) Smart cities as corporate storytelling. *City* 18(3), 307–320. DOI: 10.1080/13604813.2014.906716.

Teriman, S., Yigitcanlar, T. and Mayere, S. (2009) Sustainable urban development: a quadruple bottom line assessment framework. In: Jaiswal, S. and Wang, L.L. (eds) *The Second Infrastructure Theme Postgraduate Conference: Conference Proceedings*. Queensland University of Technology, Faculty of Built Environment and Engineering, Brisbane, Australia, pp. 228–238.

Unwin, P.T.H. and Unwin, T. (eds) (2009) *ICT4D: Information and Communication Technology for Development*. Cambridge University Press, Cambridge, UK.

Werthner, H. and Klein, S. (1999) *Information Technology and Tourism: A Challenging Relationship*. Springer-Verlag Wien.

Wiig, A. (2016) The empty rhetoric of the smart city: from digital inclusion to economic promotion in Philadelphia. *Urban Geography* 37(4), 535–553. DOI: 10.1080/02723638.2015.1065686.

Xiang, Z. (2018) From digitization to the age of acceleration: on information technology and tourism. *Tourism Management Perspectives* 25, 147–150. DOI: 10.1016/j.tmp.2017.11.023.

Yigitcanlar, T., Kamruzzaman, M., Buys, L., Ioppolo, G., Sabatini-Marques, J. *et al.* (2018) Understanding 'smart cities': Intertwining development drivers with desired outcomes in a multidimensional framework. *Cities* 81, 145–160. DOI: 10.1016/j.cities.2018.04.003.

Zheng, C., Yuan, J., Zhu, L., Zhang, Y. and Shao, Q. (2020) From digital to sustainable: a scientometric review of smart city literature between 1990 and 2019. *Journal of Cleaner Production* 258, 120689. DOI: 10.1016/j.jclepro.2020.120689.

14 The Past, Present and Future of Tourism in the Academy

David Airey*
University of Surrey, Guildford, Surrey, UK

Introduction

It is now nearly 50 years since the first handful of students graduated from the tourism programme at the University of Strathclyde. This, along with the programme offered by the University of Surrey, was the first of two such programmes in the UK and one of the very first in the world. The author was one of the graduates of this first programme in 1973. Since then, tourism, as a field of study in the academy, has developed in so many ways. It has grown almost beyond recognition, such that most countries of the world now have a well-developed tourism provision at university level. It has broadened and deepened both in its teaching and research in a way that has allowed it to take its place alongside other social sciences. And it is supported by a well-developed infrastructure of teachers, researchers, journals, conferences, organizations and alumni, all of which contribute to its standing. Along with this, those involved in tourism studies have responded to the changing needs and issues surrounding tourism; for example, the early focus on the economic benefits of tourism and the proliferation of multiplier studies have been replaced by a variety of foci, from studies of artificial intelligence in tourism to the consequences and responses to so-called 'overtourism'. In brief, in many ways tourism in the academy has come of age.

Yet at the same time, it is still plagued by uncertainties and puzzles. Has the subject really matured as a part of the academy? Can it really compete with other social sciences for students and funding? Has it developed an easy and valued relationship with the tourism sector itself? And, as universities become more competitive, what does it offer that will secure its future in the academy? This chapter explores the past, present and future of tourism in the academy, starting with the early days but focusing particularly on what tourism has to offer in the modern academy. In doing so, it also reflects on the role of the academy in making a mark on the development of this major world activity. The chapter draws on a range of references including earlier work by the author, and it rests heavily on the experience of the UK.

The Early Days

There have been a number of attempts to plot the development of tourism as a field of study. Jafari (2001), for example, recorded the development in what he called four platforms: *advocacy* in the 1960s, dominated by economists extolling the virtues of tourism; *cautionary* in the 1970s, emphasizing both the negative and positive effects of tourism, particularly in relation to the

*D.Airey1@btinternet.com

 Tourism in Development: Reflective Essays (eds P.U.C. Dieke *et al.*)
DOI: 10.1079/9781789242812.0014

environment; *adaptancy* in the 1980s, drawing attention to alternatives to mass tourism; and *knowledge* in the 1990s, where a more comprehensive appreciation of tourism is achieved. Airey (2008b) used these four platforms to identify four stages in the development of education for tourism: *an industrial stage*, with the emphasis on business and vocationalism; *a fragmented stage*, brought about by the application of different disciplines to the study of tourism, and the consequential fragmentation and uncertainty about the curriculum; *a benchmark stage*, marked by a general agreement about the range and scope of the curriculum for tourism as a multidisciplinary field of study; finally arriving at a *mature stage?* where tourism could be considered as taking its place alongside other social sciences. However, it is noticeable that this maturity stage is accompanied by a question mark as an indicator that he identified that there were still uncertainties about tourism in the academy. These uncertainties are especially in relation to tourism's theoretical coherence and the broad acceptance and influence of tourism knowledge.

From the perspective of the second decade of the 21st century, these earlier attempts to plot the development provide interesting and insightful perspectives into how this new subject area developed. After all, before the 1960s there was no education for tourism in the way in which we understand it today or as Medlik (1965, p. 5) initially described it, as courses 'dealing more or less comprehensively with the tourism phenomenon as such'. Although Medlik, in the same source, identified four universities with tourism programmes in Europe before the end of the Second World War, he could only locate ten European universities with tourism programmes in the early 1960s and, of these, only two provided degree programmes in which students could specialize in tourism, with only three having professors of tourism (Airey, 2018). This contrasts starkly with the tens of thousands of tourism programmes worldwide today, with more than 1000 programmes and half a million students in China alone (Yong and Song, 2011), together with hundreds of professors worldwide and, at a recent count (McKercher and Tung, 2015), nearly 300 academic journals in tourism and related areas, 150 of which are in English.

So the origins of tourism in the academy are really found in the 1960s. Airey (2018) identified three major drivers for its emergence and early growth. The first is the growth of tourism itself both in terms of tourist numbers, tourism operations and government recognition of tourism. It is certainly not accidental that the General Assembly of the United Nations designated 1967 as 'International Tourist Year' or that in 1969 the UK government passed its first piece of legislation specifically aimed at tourism, 'The Development of Tourism Act' (Burkart and Medlik, 1974). Tourism had become an important focus of attention. The second and third drivers come from the education sector itself. The recognition of a link between, especially, higher education and economic prosperity (Ayikouru *et al.*, 2009) encouraged both expansion in the higher education sector as well as a recognition of the graduate employment possibilities. But more than this, the developments also included the creation of new universities. The latter played a particularly important role because being less restricted by tradition, the new institutions felt freer to respond to new needs for education and new areas of study. For them, tourism was an attractive option. In the UK, the first two tourism programmes started in 1972, both in newly created universities. Since then, of the 100 or so universities that now offer tourism programmes in the UK, virtually all were established in new universities.

The point has already been made that these first expressions of tourism in the academy were primarily vocational and with a strong business and economics focus. The first programmes, as Airey (2008b, p. 28) has put it, 'sought to .. prepare prospective employees for the tourism sector' and early textbooks (Peters, 1969; McIntosh, 1972; Burkart and Medlik, 1974) and early research focused very much on the business and market aspects of tourism. At the same time, many of the early studies of tourism destinations, influenced by academia and often carried out by those associated with academia, focused on the economic aspects. Prominent among these, as already noted, were the many multiplier studies which allowed one of the main researchers (Archer, 1977, p. ix) in this area, by 1977 'to bring together and summarize [the] considerable body of literature' about tourism multipliers.

The Road to Maturity

After the early days, and against a background of significant development of research and scholarship across the academy in general, tourism, along with other areas of study, significantly developed and strengthened its provision (Airey, 2015). This was noticeable in the programmes of study, in the research endeavours where, according to Botterill and Gale (2005), PhD completions in tourism in the UK increased eight-fold between 1990 and 2002, and in tourism's support mechanisms. As it grew and attracted scholars from a range of different disciplines, so its programmes broadened to include, for example, environmental, social and political issues with greater emphasis on critical approaches, which is the hallmark of higher education. As the research activities of the scholars increased, so too the contents of the programmes were deepened to provide critical insights into the issues raised by tourism. These developments, in turn, were supported by a range of external developments (Airey *et al.*, 2015; Airey, 2015), such as: growth in the number of academic journals dealing with tourism, from two in the 1960s to 150 or more, in English, today; the creation of national and international tourism research and educational associations, from one created in the 1940s to many hundreds today; the provision of hundreds of national and international conferences; and the many thousands of books devoted to various aspects of tourism. It is on this basis that a number of writers have spoken of tourism in the academy as 'coming of age' (Council for National Academic Awards, 1993), a point that Airey (2008a, p. 102) summarized:

> If the following are indicators of maturity: a community of scholars with the support structures of dedicated journals, other publications and scholarly conferences: a curriculum which is not constrained to a particular territory and about which there is a measure of agreement; an appreciation that teaching is as much about encouraging students to challenge and think, as it is about passing on knowledge; a recognition in common with other social sciences that tourism has taken a cultural turn ....; an ability for scholars to be self-critical and aware of the nature of the truths of their work.....; then .....tourism has reached a point of maturity.

However, notwithstanding these developments in moving towards maturity, there remained some important uncertainties about tourism in the academy. In connection with this, Airey (2008b, p. 31) queried, among other uncertainties, whether tourism showed 'the kind of independence and theoretical coherence associated with more traditional disciplines' and that it was 'still a long way from truly informing debate and development in the wider world'. The latter point in some ways is curious because, almost by definition, tourism as a field of study is intimately connected with a real-world activity. Yet it is criticized for not engaging fully with that world. As Cooper (2006, p. 47) expressed it, there existed a 'hostile knowledge adoption environment' for tourism studies, suggesting that the work on tourism in the academy had little bearing or influence on the development of tourism itself.

Neo-liberal Setbacks

The neo-liberalism introduced with the elections of Ronald Reagan in the United States in 1981 and Margaret Thatcher in the UK in 1979 took a long time before it really began to affect higher education. While obvious neo-liberal measures introducing 'smaller government' (Sharma and Gupta, 2006) in tourism, such as the deregulation of air transport, were well under way by the early 1980s, it was not until well into the 1990s that equivalent moves relating to education were made. When they arrived, they typically took the form of passing the financing of education from the state to the student with the immediate consequence of opening up a commercial market for demand for university places where universities competed with each other for funding from the students. This has also been complemented with the introduction of private providers of higher education for tourism and related areas. These developments served to increase awareness by universities of their reputations on the basis that reputation is a key influence in attracting student applications. The development of performance league tables in their turn gave an added push to the awareness of competitive reputations. Against this background, the key focus of universities was turned to reputation and income. As Airey *et al.* (2015,

p. 147) put it: 'Above all, institutional managers have two main concerns: incomes and reputation'. Barnett (2013) expressed it perhaps more brutally with the comment: 'the university's gaze has shrunk to a concern with its fortunes in the immediate world'.

These developments presented particular dangers for tourism in the academy, an outcome explored by Airey *et al.* (2015) in their study of 'The Managerial Gaze' (see also Chapter 17). Here, they reviewed the performance of tourism against income and reputation, as seen by institutional managers, under three headings, namely, teaching, research and impacts, and in three countries, Australia, China and the UK. They found tourism wanting in a number of directions: notably, the quality of the students recruited to tourism programmes, the research income generated, the standing of the published work in tourism and the wider impact of tourism research. Based on existing metrics of these, they suggested that the performance of tourism was often seen as being lower than that of many other subject areas. What the authors described as a 'long tail' of relatively poor performance led them to suggest that the future of tourism studies was uncertain and, indeed, they pointed to a number of tourism programmes and departments being closed or subsumed into larger units, notably business schools.

At the same time, a number of other associated changes brought pressure to bear. For example, the neo-liberal agenda brought with it increased globalization that has enhanced an international dimension to competition. Government interest in the performance of education has also brought a range of performance assessment devices from research assessment, to student satisfaction surveys and to impact assessments – all of which, in their turn, are used in the construction of the league tables that serve to keep the universities' attention firmly fixed on their place in a highly competitive world.

The response to these pressures for tourism in the academy has typically focused on hard metrics, especially those concerned with research income and output, but also with student satisfaction and with the impact of academic endeavours. Of these, the initial and key driver has related to research, and particularly research that can achieve publication in top-ranked academic journals. The effect of this has been a significant move away from the original vocationalism of tourism in the academy toward an emphasis on high-level academic output. In many ways, success in these endeavours at some institutions has helped to secure tourism's place in the academy, notwithstanding the long tail elsewhere. But it has also had the effect of positioning tourism research efforts clearly in an almost exclusively academic context with the result that, as Airey (2018) comments, 'Securing links between academic research [in tourism] and the wider tourism world, including the economic sector of tourism, however, still remains a challenge'. In other words, there has almost been a growing divide between the academic and practical worlds of tourism. At the same time, this emphasis on research has created a real tension between the teaching and research roles of institutions and individuals, with individual academics viewing their career success as being most closely related to their research success and within this, the academic nature of the research takes priority over the practical. In this regard tourism in the academy has often become remote from the practical world of tourism.

Against this, the introduction of student satisfaction surveys and, more recently, impact assessment exercises have served to shift some of the focus, at least at an institutional level, to ensure that the teaching is of high quality and that attention is paid to the impact of research. In many ways, as explored later in this chapter, tourism has some in-built advantages in relation to these drivers. As a subject area it lends itself well to inventive teaching experiences and its obvious links with a large and important world activity provide a good setting for making an impact. So, while the period of neo-liberalism has brought challenges for tourism in the academy it has also served to encourage institutions to address some of the weaknesses in research and, just as important, it has brought into perspective some of the important strengths of tourism in the academy that need to be brought to the fore.

Present Dilemmas

Tourism in the academy at present faces a number of dilemmas. At their heart lies the dilemma of competing with other subject areas in academic terms which, as explained above, relates strongly to research performance, while at the same time maintaining and enhancing links and relevance to the sector which was the original *raison d'être*. Over the past few years, the route for many of the universities offering tourism has been to enhance their research performance. With opportunities for career development as well as financial reward offered to successful researchers, it is certainly not a surprise that, for individual academics, priority is given to research efforts. Partly as a result, the various indices of research quality have shown important increases in the output performance of tourism researchers. This provides the important basis for tourism to maintain its foothold in the increasingly competitive academic world. But whether this alone provides a secure basis for its future against more traditional areas of study is questionable. When, for example, Nobel prizes are open to economists, physicists and chemists, does tourism really offer a home for the truly top-ranked researchers? Similarly, notwithstanding the overall strengthening of the research base, when one of the journal databases (SJCR, 2019) gives an impact factor for the top tourism journal of about 3 compared with similar scores for social sciences more generally of 34, business, management and accounting 18, economics, econometrics and finance 29 and medicine 61, can tourism feel secure if impact factors are dominant in the managerial gaze?

The other side of this dilemma relates to the link with the practical world of tourism. Here, the increased importance of the impact of research on the wider world is certainly nudging tourism in the academy to pay increased attention to its wider world. Some of the case studies presented in the most recent research assessment exercise in the UK (REF, 2014) provide evidence that this is being accomplished with topics specifically related to tourism including Modelling and Forecasting, Reducing Social Exclusion, Modelling Economic Impact, Redesigning Tourism Policy and Practices in Africa, and Food Tourism and Sustainable Development. However, the list of tourism cases among the 50 or so universities included in this group of impact statements related to tourism is very short and few receive the top ranking.

Perhaps more important in terms of links with the wider world is the potential that comes from teaching provision made by universities. This is another dimension of the dilemma prompted by the focus on research: the shifting balance between teaching and research. Reference has already been made to the tension between teaching and research both for institutions and for individuals and also reference has been made to the very origins of tourism in the academy prompted by the growth of the tourism sector. As Medlik (1965, p. 42) argued, university-level programmes were needed that could 'provide a suitable education and training for the future tourist executive'. Arguably, more than research, it is this dimension of tourism that offers a distinctive role for tourism in the academy. After all, the boundaries of tourism in the academy are set by tourism in practice and the subject content in the academy reflects developments in the outside world. Issues such as overtourism, responsible tourism, tourism and poverty alleviation, or the use of virtual reality in tourism all set an agenda for the changing content of tourism programmes such that, in many ways, it can have a head start over other subject areas. The fact that tourism programmes typically offer direct links with the wider world of tourism through case studies, field visits, visiting faculty and professional placements all potentially put tourism in a strong position to exploit its advantages in being in the vanguard of industry links.

Playing to Strengths

A recent survey (Ross, 2018) of 200 leaders of top-ranked world universities reports on the major changes in and influences on universities in the period up to 2030. Perhaps not surprisingly, they expect their universities to thrive, but they also point to some important changes. Notable among these are the importance of links with industry (applied research, utilitarian, applied teaching); the importance of science, technology, engineering, maths and medicine (STEMM) subjects; the development of shorter degrees and other programmes; the reduction of public expenditure on teaching (not value for money); greater attention given to mature students and lifelong learning; and, the increasing casualization of the workforce. Clearly some of these, like the focus on STEMM subjects, point to difficulties ahead for subject areas like tourism, and the reduction in public expenditure and casualization of the workforce highlight problems and tensions for the university sector generally. But there are other pointers to possibilities for tourism.

One of the themes that runs through this chapter is that of the link between tourism in the academy and tourism in the world at large. The needs of the tourism sector have been identified as the starting point for tourism in the academy but the suggestion has also been made that this link has not always been well developed or exploited. In some ways, tourism in the academy has been reticent in responding to the needs of industry and for its part industry has not developed an easy relationship with its academic counterparts – the so-called 'hostile knowledge adoption environment' (Cooper, 2006, p. 47).

It is, perhaps, this industry link that contains the strongest feature for tourism in the academy and one that provides a uniqueness for tourism. This is not only because there is an obvious platform of common interest on which to build, but also tourism itself is dynamic and presents so many different facets of the issues that need to be dealt with in what Barnett refers to as a world of supercomplexity (Barnett, 1999). On this theme, Barnett suggests that to meet this supercomplexity, it is not so much knowledge and skills that are required but, as he puts it, 'certain kinds of human qualities... carefulness, thoughtfulness, humility, criticality, receptiveness, resilience, courage, stillness' (Barnett, 2004). From this, Airey (2018) draws attention to Austin's (2012, p. 252) comments on Barnett's work where she writes that 'the primary responsibility of those who teach in higher education institutions is to prepare students for an uncertain and changing world'. Taking this as a starting point and drawing on the work of Belhassen and Caton (2011), Airey (2018) further suggests that: 'in an age of supercomplexity, higher education cannot be solely about leading students to deal with one set of relatively predictable challenges. It needs to prepare them to deal with the unexpected and uncertain world in which employment patterns and sectors as well as social and leisure environments will constantly change.' Seen from this perspective, tourism itself provides an ideal testing ground for examining the modern and changing world. It presents so many examples of the challenges and changes of the modern world, from the environment and global warming to social impacts and land-use conflicts, to the effects of changing technology, and to the need for poverty alleviation and food security. In this sense, tourism is an ideal world for exploration by the academy and it is one that tourism scholars could make exclusively their own.

At the same time, the way in which tourism has developed in the academy as a multidisciplinary and multi-method field of study that combines both vocational and liberal education makes it very well placed to meet modern demands in this complex world. As Dredge *et al.* (2015) comment, such education in tourism 'provides a balance between liberal education and vocational aspects [allowing for the emergence of] mindful, world-making practitioners'. This draws on Tribe's (2002) earlier work about the philosophic practitioner combining both a critical academic grounding with practical applications. As Airey (2018) puts it, 'those working in tourism need to be conscious of their role not only in meeting income targets or providing tourists with rewarding experiences but also influencing a world activity with major social, cultural, environmental, political and other implications'. The important issue here is that tourism as a subject of study permits great diversity in its content, methods and orientation, and tourism in practice needs this diversity if it is to develop fully and successfully. And this applies not just to the teaching role of universities. Research can also play an important role. As Airey (2013, p. 15) has written, 'Tourism research now regularly draws on a wide range of disciplines, and it is as likely to take an interpretivist or critical stance as it is to take a positivist one and qualitative methods are just as likely to be found as quantitative'. The end result, as Airey (2015) has put it, is that 'as a relatively new field of study with imprecise boundaries that span a range of issues as well as employment possibilities and drawing upon many disciplines, it [tourism] provides a rich context for preparation for a post-industrial age, both in terms of research as well as in terms of teaching'. With a close link between the academy and practice, tourism can play to its real strengths and compete strongly with other subject areas in what has become the competitive world of higher education.

Where from Here?

From the comments made in this chapter, there are two big challenges for tourism in the 21st century. The first and, in many ways, the most difficult is to raise its research outputs (and associated incomes) to the level of competing subject areas in the academy; and, to a large extent, this means all subjects, since all are competing for the attention of senior managers and all are contributors to the key drivers of reputation and income. This is hard for tourism in a world where STEMM subjects seem to come first, at least in the thoughts of institutional leaders, and where, even in the social sciences, tourism is rarely seen as a high-flier. This is not to suggest that attention should be taken away from excellence in research because this is part of the legitimate currency of higher education and is vital for the development of any subject area. But it should not become the dominant challenge because the other challenge, that of building on its link with its natural practical base, not only needs much further development but also potentially provides a much more secure basis for a unique place in the academy.

So, the other big challenge is for tourism in the academy is to be fully cognisant of the world of tourism in practice. In many ways, as already noted, tourism has an in-built advantage in achieving this because it has an external world whose boundaries match those of the academic world. Issues and developments in tourism in practice are of immediate relevance to the teaching and research in the academy. This presents the ongoing challenge for teachers to ensure that the tourism that is explored in the classroom and in research is generally aligned with the tourism in the world at large, ideally in a critical sense. The development of classes, research and publications related to the development and implications of the sharing economy provide a good recent example of this. The edited collection by Dolnicar (2018) is rich with cases demonstrating how the academy has absorbed and articulated this important development in tourism. Similarly, the internships offered to students, as well as guest lectures and study visits, all serve to bring the wider tourism world to the academy and those institutions where such activities are well developed will almost inevitably secure this external link to their work.

At the same time, of course, what happens in the academy also needs to be brought to the wider tourism world. For tourism, there seems to be more work to be done in securing this. There is a regular criticism that the findings reported in academic journals stay within the academy, that there are relatively few examples where academic findings have brought significant change in practice, and that the leaders of tourism, whether public sector or private sector, seem to consult far more rarely than in other subject areas with academia before making major decisions. The topic of change and innovation, or lack of it, has been well explored in the tourism sector (Hall and Williams, 2008; Shaw and Williams, 2009; Thomas, 2012; Thomas and Wood, 2015), with references to employment and industry characteristics, absorptive capacity and the nature of knowledge all thought to have an influence on the extent to which the work of academia finds its way into the world of practice. Whatever the cause, the fact remains that this transfer remains an issue both for academia as well as for industry.

So, over nearly 50 years, tourism in the academy has come a long way. Fifty years ago, to admit to teaching or researching in tourism would have been met with complete surprise and scepticism. Something of that still remains, but what cannot be denied is that tourism is now found in well-established centres of learning across world. Whole generations of students have come through the education system and progressed into long and highly successful careers in tourism. Our knowledge and understanding of tourism is immeasurably greater and, in brief, tourism is established in the academy. But one of the themes that comes through this chapter on the last 50 years is that it cannot take its place for granted. It has developed and responded to the changing macro environment, to the general changes in education and to the developments in tourism itself. It has also taken its place among the social sciences as a legitimate areas of academic scholarship. But throughout, the thing that distinguishes it above all is its link with an important world activity. This is what brought it into existence in the first place and it needs always to recall that it is this that will nurture it for the future.

References

Airey, D. (2008a) In search of a mature subject? *The Journal of Hospitality Leisure Sport and Tourism* 7(2), 101–103. DOI: 10.3794/johlste.72.per.

Airey, D. (2008b) Tourism education: life begins at 40. *Teoros* 27(1), 27–32.

Airey, D. (2013) Forty years of tourism education and research. *Poznan University of Economics Review* 13(4), 11–19.

Airey, D. (2015) 40 years of tourism studies – a remarkable story. *Tourism Recreation Research* 40(1), 6–15. DOI: 10.1080/02508281.2015.1007633.

Airey, D. (2018) Tourism education and scholarship. In: Cooper, C., S.Volo., Gartner, W. and Scott, N. (eds) *The Sage Handbook of Tourism Management*. Sage Publications Ltd, London, pp. 49–60.

Airey, D., Tribe, J., Benckendorff, P. and Xiao, H. (2015) The managerial gaze: the long tail of tourism education and research. *Journal of Travel Research* 54(2), 139–151.

Archer, B.H. (1977) *Tourism Multipliers: The State of the Art*. University of Wales Press, Cardiff, UK.

Austin, A.E. (2012) Challenges and visions for higher education in a complex world: commentary on Barnett and Barrie. *Higher Education Research & Development* 31(1), 57–64. DOI: 10.1080/07294360.2012.642840.

Ayikouru, M., Tribe, J. and Airey, D. (2009) Reading tourism education: neoliberalism unveiled. *Annals of Tourism Research* 36, 191–221.

Barnett, R. (1999) *Realizing the University in an Age of Supercomplexity*. SRHE and Open University Press, Buckingham, UK.

Barnett, R. (2004) Learning for an unknown future. *Higher Education Research & Development* 23(3), 247–260. DOI: 10.1080/0729436042000235382.

Barnett, R. (2013) *Imagining the University*. Routledge, Abingdon, UK.

Belhassen, Y. and Caton, K. (2011) On the need for critical pedagogy in tourism education. *Tourism Management* 32(6), 1389–1396. DOI: 10.1016/j.tourman.2011.01.014.

Botterill, D. and Gale, T. (2005) Postgraduate and PhD education. In: Airey, D. and Tribe, J. (eds) *An International Handbook of Tourism Education*. Elsevier, Oxford, UK, pp. 469–480.

Burkart, A.J. and Medlik, S. (1974) *Tourism Past Present and Future*. Heinemann, London.

Cooper, C. (2006) Knowledge management and tourism. *Annals of Tourism Research* 33(1), 47–64. DOI: 10.1016/j.annals.2005.04.005.

Council for National Academic Awards (1993) *Review of Tourism Studies Degree Courses*. Council for National Academic Awards, London.

Dolnicar, S. (ed.) (2018) *Peer to Peer Accommodation Networks: Pushing the Boundaries*. Goodfellow Publishers, Oxford, UK.

Dredge, D., Airey, D. and Gross, M.J. (2015) *The Routledge Handbook of Tourism and Hospitaity Education*. Routledge, Abingdon, UK.

Hall, C.M. and Williams, A. (2008) *Tourism and Innovation*. Routledge, Abingdon, UK.

Jafari, J. (2001) The scientification of tourism. In: Smith, V. and Brent, M.M. (eds) *Hosts and Guests Revisited*. Cognizant Communications Corporation, New York.

McIntosh, R.W. (1972) *Tourism Principles, Practices and Philosphies*. Grid, Ohio.

McKercher, B. and Tung, V. (2015) Publishing in tourism and hospitality journals: Is the past a prelude to the future? *Tourism Management* 50, 306–315. DOI: 10.1016/j.tourman.2015.03.008.

Medlik, S. (1965) *Higher Education and Research in Tourism in Western Europe*. University of Surrey, Guildford, UK.

Peters, M. (1969) *International Tourism: The Economics and Development of the International Tourist Trade*. Hutchinson, London.

REF (2014) *Research Excellence Framework 2014* [online]. Available at: https://results.ref.ac.uk/(S(h5crreew05v25n44b0lprcqa))/Results/ByUoa/26/Impact (accessed 19 March 2019).

Ross, J. (2018) *THE University Leaders Survey 2018* [online]. London: THE. Available at: https://www.timeshighereducation.com/features/forward-thinking-university-leaders-survey-2018 (accessed 18 March 2019).

Sharma, A. and Gupta, A. (eds) (2006) *The Anthropology of the State*. Blackwell, Oxford, UK.

Shaw, G. and Williams, A. (2009) Knowledge transfer and management in tourism organisations: an emerging research agenda. *Tourism Management* 30(3), 325–335. DOI: 10.1016/j.tourman.2008.02.023.

SJCR (2019) *Scimago Journal and Country Rank* [online]. Available at: https://www.scimagojr.com/journalrank.php?category=1409 (accessed 16 March 2019).

Thomas, R. (2012) Business elites, universities and knowledge transfer in tourism. *Tourism Management* 33(3), 553–561. DOI: 10.1016/j.tourman.2011.06.009.

Thomas, R. and Wood, E. (2015) The absorptive capacity of tourism organisations. *Annals of Tourism Research* 54, 84–99. DOI: 10.1016/j.annals.2015.06.004.

Tribe, J. (2002) The philosophic practitioner. *Annals of Tourism Research* 29(2), 338–357. DOI: 10.1016/S0160-7383(01)00038-X.

Yong, J. and Song, H. (2011) Tourism education programmes in mainland China. *AngloHigher, The Magazine of Global English Speaking Higher Education* 2(6), 8–9.

15 Achieving a Balance between Education, Research and Practice

Geoffrey Wall*
Department of Geography and Environmental Management, University of Waterloo, Waterloo, Ontario, Canada

Introduction

Universities have multiple roles in society and, accordingly, these roles are reflected in the activities of those who engage with them, including their employees. Although the emphasis varies from institution to institution and may change for individuals as their careers mature, most academics are expected to teach, undertake research and assume roles in administration. They may also be expected to provide community service, although the nature of the communities that they are expected to serve may be unclear and unspecified. At the same time, the time available to undertake these tasks is not unlimited so that difficult choices must be made concerning the allocation of time and effort, guided by personal preference, institutional policies and the nature of the reward system, as well as other factors such as family commitments and leisure choices.

Furthermore, in recent years and for various reasons including reputation, information exchange and financial benefits associated with access to new markets, there has been a growing interest in internationalization which is now enshrined in the goals and policy documents of many universities and colleges. Whatever this term may mean, it is has implications for the educational, research, administration and service roles identified above. It implies greater interaction with distant places, different peoples, disparate languages and cultures, foreign local, national and international organizations, new ways of doing things and more, with substantial implications for universities and their employees, what they do and how they do it. It requires commitments both at home and abroad and, in the latter case, both directly and indirectly, may require involvement in development initiatives elsewhere. These things are occurring at a time when class sizes are increasing and budgets are tight.

There is never enough time and money to do everything that is desirable and, in a situation of resource constraints, it is necessary to find efficiencies and set priorities. Tensions may exist between the time available for and allocated to the various roles, as well as the extent to which the focus is on global or local places and problems. I am aware that, in my own career, the question has been asked, more by neighbours than academic colleagues, 'How can he teach his students if he is always away?'. I will state in defence that while I have made many international journeys, these were never the cause of a cancelled lecture or a missed meeting. In this chapter, then, I will address a related question,

*gwall@uwaterloo.ca

DOI: 10.1079/9781789242812.0015

'How can the tensions that often exist among the various academic roles be managed such that synergies can be created among them?', enhancing the performance of the individual academic, the experiences of those with whom they engage and serve, while providing funds to cover many of the costs, thereby meeting the goals and objectives of the institution and enhancing its reputation. This contribution, then, examines relationships among teaching, research, administration and service in the belief that, with careful management, they can be mutually reinforcing activities rather than competitors, as they are often portrayed to be.

Fig. 15.1. Exchanging ideas at the 'The Wanch', Hong Kong.

Context

The purpose of this publication is, in part, to honour Professor Carson (Kit) Jenkins. Professor Jenkins is an individual who has had notable success in managing the trade-offs that have been introduced above, to the extent that he has enviable achievements in all four roles. He has been, and still is, a successful teacher at both undergraduate and graduate levels, and has a well-respected research and publication record. He has assumed major administrative tasks, notably as Head of the Scottish Hotel School at the University of Strathclyde, and has a stellar range of consulting experiences related to tourism, particularly tourism planning, in many parts of the world that is second to none. His is a remarkable record that spans decades, continents and encompasses both pure and applied research, and academic and practical accomplishments. Thus, he is an individual who has managed the tensions with considerable success.

Although I was previously aware of his writings, I first met Professor Jenkins at a conference in Yogyakarta, Java, Indonesia, in 1992. He was a consultant providing inputs into the national tourism plan for Indonesia and I was working on a sustainable development strategy for Bali. We were both participants in an international tourism conference. We arranged to meet at the Borobudur bar where we soon discovered that our common interests were not confined to tourism and beer. We are of a similar age and cultural background, having both been born in Wales. We are devotees of rugby, which is a sport that is certainly a core element of Welsh culture where it has a status that causes it to be followed with almost religious fervour. We both studied at Hull University in England. Also, much of our academic activity has involved engagements with Asia, notably in Indonesia and China. We also share a number of friends in academia.

Since our initial meeting we have kept in contact, have visited each other's homes, and have participated together in a number of international events and other activities. Although now both technically retired, we have continued to be active academics, continue to visit Asia and communicate regularly regarding both academic and personal matters. Although based in different continents, and there are differences between us (for example, one was trained as an economist and the other as a geographer), we have much in common and often find ourselves thinking in similar ways and can have friendly exchanges when we do not. We have created opportunities to exchange ideas in social settings (Figs 15.1–15.3). Thus, in reporting on my own challenges in developing and managing an academic career, I am cognisant that both Professor Jenkins and I have striven to achieve a balance among various university expectations and have reacted in broadly similar ways.

Evidence

The evidence upon which this communication is based is derived entirely from personal

Fig. 15.2. A casual moment outside 'The Wanch', Hong Kong.

Fig. 15.3. Geoff Wall, Han Shan, Wantanee Suntikul, Dick Butler and Kit Jenkins at 'The Wanch', Hong Kong.

experience. No rigorous attempt has been made to canvas relevant literature although there are specific literatures that do address many of the topics that will be touched upon here. However, conclusions can only be justified by the provision of evidence, the strength of conclusions being a reflection of the pertinence of the evidence on which they are based. For this reason, it is appropriate to introduce relevant aspects of my own career.

With respect to teaching, I long had the responsibility for delivery of a large introductory course in human geography, which initially involved both lectures and laboratory sessions and was later also made available by correspondence. The course was taken by many students in the Faculty of Environment, where I was based, as well as from elsewhere in the university. I also taught upper-level undergraduate courses in resources management, and tourism and recreation, as well as graduate courses on the latter topics. I supervised more than 100 graduate theses, about one third of which were PhD theses.

From the outset, my research emphasis has been on various aspects of tourism and recreation. Although in the early years I was enabled to undertake research on hazards (which became particularly useful as climate change became a concern), the main focus was always on tourism, although this interest evolved from an emphasis on travel patterns, to impacts, to sustainability, sustainable livelihoods and heritage, embracing economic, environmental, socio-economic and management dimensions, and incorporating many specific topics, such as ecotourism, climate change, indigenous tourism, and community well-being. International dimensions to this work, through reading, existed from the beginning but came to the fore in the late 1980s when the opportunity was taken to provide inputs into the formulation of a sustainable development strategy for Bali, Indonesia. Since then, research activities have focused especially upon Asia, particularly Indonesia and China, and have been supported through both academic funding and consulting opportunities.

Administratively, shortly after joining the University of Waterloo, I became undergraduate officer in the Department of Geography. This meant that I was automatically a member of many other committees at both faculty and university levels. Since the Department of Geography had programmes in two faculties, I had many meetings in both the Faculties of Environmental Studies (as it was then called) and Arts. The Department of Recreation and Leisure Studies, which included several prominent tourism researchers, was in the Faculty of Applied Health Studies. Thus I was introduced very quickly to the university's administrative structure.

For 11 years, in two stints, I was the Associate Dean for Graduate Studies and Research in the Faculty of Environment. Although there were opportunities to serve in other administrative roles, this was my preferred position because there was only a limited requirement to evaluate colleagues (compared with a chair or dean) and only a small budget to administer, most of which was allocated and gratefully received by successful graduate students. This situation changed radically when I became the director and principal investigator of large multi-year international development projects, involving several university and government partners.

With respect to service, although I sat on a number of local committees and undertook a small number of projects for local and provincial government agencies, my local service inputs were not particularly prominent. However, my involvements with the national government, particularly with respect to climate change issues, were substantial and involved committee work, conferences and research activities. I also undertook all of the tasks expected of an active academic, such as reviewing manuscripts, evaluating proposals and promotion cases, and giving guest lectures.

Throughout my career, I have always regarded myself as an academic and have always taken my teaching at both undergraduate and graduate levels seriously. In spite of my many international visits, the number of lectures that have been cancelled can be counted on the fingers of one hand, and I took pride in reviewing the work of my graduate students thoroughly and in a timely fashion. As will be revealed, the administration of large international projects was essential for the support of research programmes and graduate students, and these projects also had considerable research components.

As an academic, I have regarded my main task as to try to understand and share such understandings with others. Of course, if one is acting as a consultant, one is obliged to try to resolve the problems of one's clients. My interests have focused upon tourism rather than hospitality, and my goal has been to promote education broadly conceived rather than to train people for specific jobs. A key to this is the fostering of critical thinking. This involves providing opportunities to learn to create logical arguments, how to gather and present information, to communicate orally and in writing, to read critically, to prepare literature reviews and to write reports. In these activities, tourism is the subject matter but the skills are generic.

Teaching

When teaching introductory courses, I regarded the main task as laying down a foundation on which others could build. This required bringing students from disparate backgrounds to a common understanding by exposing them to a set of concepts and examples, rather than through pursuing a particular topic in depth. The latter could be done in later courses, with intermediate courses being the most difficult to teach because they are commonly open-ended at both the start and finish, not knowing exactly what students will bring to the course or what can be left to subsequent learning opportunities. Key elements of the approach to introductory courses that I adopted were coverage of a wide range of topics so that there would be something novel for all participants, linked through an overarching framework to promote coherence rather than fragmentation of ideas. Upper-level and graduate courses were designed to challenge students to question interpretations and the effectiveness of research methods, and course content was often decided through interaction with students, to discover why they registered for the class and what they actually wanted to learn, and in responding to their interests as expressed in the initial session. Regardless of level, research experiences commonly provided compelling teaching materials, students invariably responding positively to the sharing of personal experiences. Academic concepts can be illustrated, and often qualified, with real-world examples taken from the field. Students have more confidence in instructors who can share more than book learning and use readings to support the course content, rather than slavishly following a text created by others.

The graduate programmes in which I was involved invariably had a major research component: a thesis that was defended in an oral examination. Education in research often occurred through exploring together. When working in foreign areas and with limited language skills, I often had to rely heavily on my graduate students and learn from them regarding the meaning and significance of what I observed. I always expected them to select their own research topics but they always knew my interests so that it was usually not difficult to find something that attracted our mutual curiosity. My attitude to supervision was to meet regularly, often in casual settings, to share insights and progress, and to try to achieve a balance between challenging students and supporting them. I believe that the main tasks of a supervisor are to ensure that an interesting and worthwhile research topic is selected and placed in context, defined so that there is the prospect of finishing in a reasonable time, and that appropriate research methods are selected and can be justified. In almost all cases, I expected the student soon to know more about the specifics of the topic than I did.

If instructors are to devote maximum effort to the preparation of courses, assurance should be provided where possible on their teaching responsibilities with the expectation that they will have ongoing responsibility for the same suite of courses. Thus, I taught the same courses for many years, although content changed over time. The preparation of a new course is as demanding as writing a book, so, if the preparation of new courses is continually required, then this takes away from the time for other things, reducing the possibility of creating efficiencies and synergies. Therefore, there should be continuity from year to year in the courses for which one has responsibility. It is difficult to give attention to other things if one is continually preparing new courses. Although course evaluations may provide a different perspective, I sensed that courses are often best on their third or fourth iterations when ideas are still relatively new and fresh but there have been opportunities to learn from experience.

Scheduling is also important at other temporal scales. Eventually, I was allowed to schedule some courses for four hours per week rather than the normal three to allow for the cancellation of classes in mid-term if project activities required this. The schedule was spelled out in the course outline and not imposed at the last minute and students liked the fact that they had a pre-arranged break in one course, often at a time in the term when pressures were building and they had forgotten about this gap in the class schedule. Of course, it would not be possible for all instructors to do this, but I was blessed with a supportive dean who even offered to teach an occasional class as a substitute to help out if desired, but this was never necessary.

The University of Waterloo, where I spent most of my career, has the largest co-op programme in the world and this necessitates that a range of courses is offered in the summer. I only taught in the summer once so, for me, this became the major time of the year for international activities. Other than a one-week break in classes in February, the only other time in the year when one could get away from campus without disrupting classes was December. Yet it is essential to maintain personal contacts in an international project: without meetings or the prospect of meetings, interest flags and less is achieved. December, thus, became the busiest time of the year, with marking and grade submission required before the end of the calendar year and outlines to be prepared in time for the new term starting in January. This needed to be done at a time when one often needed to be abroad on project activities, staff were in a holiday mood, and the university closed for Christmas and New Year.

Research

As indicated above, ideally one's teaching and research should be closely related in order to make efficient use of information. Research production is usually evaluated under three headings: research output as indicated primarily in publications; research funding; and the production of highly qualified personnel. I have touched upon the latter in the comments above on graduate education and supervision. It is sufficient to say at this point that, rather than teaching graduate students, I always believed that we learned together. Also, if one wants to be a successful professor, one should be involved with capable graduate students. I will concentrate in this section on commenting on the other two points.

Involvement in international projects exposes one to new environments, new problems and new ideas. For example, I found myself involved in tropical rather than accustomed temperate environments, in developing rather than developed countries, with very different cultures from my own. I became curious about the informal economic sector which scarcely existed in my home society. I learned that governments operated in different ways in different places and this caused me to question the veracity of much of the literature on tourism planning. The texts were written mostly by western tourism academics, which did not fit well the situations that I was experiencing. I did my best to become culturally sensitive by observing carefully and reading novels in translation, although I believe that one can never fully understand another culture unless one speaks the language. Thus, my research topics and experiences became informed by my international activities, both specifically regarding the locations and problems specified in projects, but also by exposure to new ways of thinking. I believe that I would be less well informed and less curious had I concentrated my attention solely, or even primarily, on my own local situation.

It takes time to undertake research in a distant location, to collect and interpret data, to become familiar with a new place, and to become sufficiently confident that one has something worthwhile to say to an international audience that will also make sense in the communities with which one has been working. Thus, there is a time lag between the onset of a project and the publication of research findings that can be particularly long in international research. I found myself being asked, 'What are you actually doing in Bali?', and later, 'When will you have something to say about Hainan?', by colleagues from other institutions who were aware of my international involvements but had seen few outputs. Administrators responsible for evaluating research productivity should be aware that it takes time to produce quality research, particularly when one is working on the other side of the world. Furthermore, depending upon the funding source, data and outputs may be proprietary and this may prevent wide dissemination of results. Researchers need to clarify ownership of data at the outset of a project if they hope to publish the results.

Most research requires money, and international projects may or may not provide a supplementary personal income through consultancy, but they will usually cover the costs of transportation and subsistence. Depending upon the project, they may also provide the resources to fund graduate students and research assistants. The acquisition of external funding is usually welcomed by universities and, by combining internal and external sources, it can be possible to attract and fund very strong graduate students who comprise a large proportion of the research personnel at most universities.

Unlike some who bid on contracts, thereby competing with consultants and, if successful, often by bidding low, find themselves with more pressure from tight deadlines, I designed my own large projects and persuaded governments to fund them, often in competitive situations. Once these projects were up and running, other opportunities often followed. The preparation of research proposals is a skill that is not often taught but should be. Also, applicants should put in significant time and effort commensurate with the amount of money requested when preparing proposals. Otherwise they are wasting their own time as well as that of reviewers and evaluation committees.

In my career there have been two main sources of funding: large multi-year aid projects and smaller research grants made available through academic research councils. Both are highly competitive to obtain. They differ somewhat in their content and format and, to be successful, one must ensure that one addresses clearly all of the objectives of the funding agency. This may appear to be an obvious statement but I have reviewed many proposals that do not meet this requirement. The large proposals were not specifically on tourism but were on broader topics, such as sustainable development, coastal zone management, ecoplanning, and biodiversity, of which tourism may be a part. These projects and the funding agencies tend to be fairly inflexible and require the delivery of what is promised in the proposal. In contrast, my applications to research councils have been more narrowly focused on aspects of tourism (with tourism placed within a broader context), but with greater flexibility in their execution. These proposals were often submitted to pursue interests and opportunities identified through the former larger projects but were on the edge or outside of their boundaries. For example, a large project on coastal management in Hainan, which incidentally identified the role of indigenous people, eventually led to a project concerning tourism and indigenous people in Taiwan, another tropical island with a dominant Chinese culture but with very different forms of government.

Adminstration

Unlike Professor Jenkins, who long held the reigns of the tourism and hospitality programme at Strathclyde University, my university administrative experiences, by choice, have been more modest. However, as director of multi-year international projects, the management of these projects, even with the involvement of capable administrative assistants, was a time-consuming task. In addition to a substantial proposal and a final report, there was a responsibility to provide quarterly activity and financial reports to tight deadlines. There was never enough time or money to complete these tasks in a leisurely fashion. I accept the need for accountability and essentially regard these requirements as the price that had to be paid to acquire the opportunities to work internationally on projects that I had designed (with the assistance of others).

However, I admit that I learned from these administrative experiences. For example, the rigorous reporting procedures, which sometimes changed in mid-project, made me understand the differences between outputs and outcomes, a distinction which has been useful in my research activities. Also, the demand to provide quantitative indicators of progress, while often frustrating, provoked creative thinking.

Discussion

While some have argued that there are tensions between teaching and research, it has been argued that it should not be difficult to develop synergies between them in terms of content, for the latter can enliven course content by providing case examples that ground abstract ideas in reality, and at the same time can be used to test more abstract ideas. The tensions exist primarily in the allocation of time. Thus, time management is critical. As indicated above, course scheduling is a key aspect of time management for it influences the number of new course preparations, specifies the times when it is essential to be on campus and, as a corollary, decide when it is feasible to be away. Ideally, these time commitments should be known well in advance.

At the same time, and somewhat paradoxically, it necessary to be flexible, for unexpected demands on one's time will arise at home and abroad. Foreign collaborators will have their own agendas, schedules and activities, and will proffer invitations to participate in these. At least some of these requests for input must be accepted in order to show support. These are actually great opportunities to learn but they sometimes come at short notice and are not always easy to fit in. At the same time, international visitors will arrive at inconvenient times and their interests must be accommodated.

Time management and flexibility are ultimately an individual responsibility but they can be facilitated by institutional culture and the administrators to whom one reports. Senior administrators may espouse internationalization but, if they have limited international experience themselves, they may not be aware of what is actually involved. They may recognize the importance of a strong international reputation for their institutions but not fully understand the international engagements that underpin this.

It is often forgotten that successful international collaboration should involve benefits for all participants. It is easy to see the personal benefits of visiting new places, the novel experiences that may be gained and the ideas that may be generated. However, unless there is also something in it for one's foreign collaborators, their enthusiasm can easily wane. They can visit their local sites and undertake research on local problems in the absence of international visitors whose arrival can complicate their lives. Reciprocity is important. It should not be necessary to state that opportunities need to be created for foreign partners and their interests also need to be met. Unfortunately, these things can be easily neglected. One should try to be a good host as well as a grateful guest.

I have always regarded it as a privilege to be able visit new places and engage with different cultures. Although I know that I will always be a visitor and that I can never fully understand a culture that is not my own, especially given my foreign-language limitations, I have appreciated the opportunity to try new things and always endeavour to respect my international partners and behave in a culturally sensitive manner. I am sure that I have made many mistakes but, if one hangs back, one is likely to be advised about what to do in any given circumstance. If one has a smile on one's face, one is more likely to be forgiven for indiscretions.

While, like Professor Jenkins, I have always found my international experiences to be both challenging and enriching, it is not everyone who is able to perform at their best in unfamiliar circumstances. Unfortunately, on one occasion it was necessary to request the withdrawal of a respected and otherwise competent colleague because they could not function adequately in an unaccustomed setting. They were unable to eat the food, use the utensils, drink the local beverages, find a way to set up meetings or exchange information with partners. Our hosts recognized what was happening and the culture shock of my colleague was a cause of discomfort for both him and them. They asked what they could do to resolve the problem but it was not their fault. If one cannot relate to one's partners and bridge the cultural divide, then there is a lower chance of one's recommendations being considered seriously. I sensed their feeling that 'If he does not understand us, how can he help us?'. Clearly, not everyone is suited to undertake research in international settings or able to collaborate well as a team member with others who are different from themselves. Ironically, the individual to whom I refer has published papers on community involvement. Thus, it is not everyone who is suited to international activities and can find such activities to be both intellectually exciting and socially rewarding.

Finally, while both Professor Jenkins and I have spent our academic lives focusing on tourism, it is important to recognize that most project activities have a wider remit, particularly if one is trying to create a tourism plan. Tourism competes with other sectors for the scarce resources of land, water, capital, labour, waste assimilation capacity and so on, so it is insufficient to be well informed about tourism if one is unable to relate one's knowledge to other areas of interest. As mentioned above, most of the research funding I received has been for broader topics such as sustainable development, climate change, coastal zone management, biodiversity, community livelihoods and the like, in which tourism is important but only a part of the subject matter to be explored. Thus, I espouse a broad perspective on tourism and the importance of being able to place tourism in context. This has many implications for tourism education, which I believe has become excessively narrow, but that is another topic and it is beyond the remit of the present contribution.

Conclusions

The academic profession usually requires involvement in teaching, research, administration and service. It is often argued that since time is not unlimited, more spent on any one of these activities leaves less time for the others, creating tensions in their performance. However, I have suggested that synergies can occur among these activities that result in both efficiencies and enhancements. However, this does not occur automatically but requires good time management, careful planning, flexibility and a supportive administrative setting. Where international activities are involved, success is facilitated by respect for one's collaborators and cultural sensitivity. International involvements can ground both teaching and research in real situations and events, while greatly enriching the life experiences of those who follow this path. It provides them with many stories to share.

Finally, Professor Jenkins has had a stellar career in education, research, administration and consulting. He has been able to balance the demands that emanate from these sources and manage the pressures that they involve. He needs few teaching aids because he can command an audience with his gentle Welsh lilt and the ideas that he conveys are grounded in a wealth of international experiences. He is a good listener but he also has many stories to tell. Many of his graduate students, benefitting from his mentorship, have become accomplished in their own careers. His research is highly regarded and he assumed important administrative leadership roles, particularly at the University of Strathclyde. His wise counsel has been sought by international agencies and governments throughout the world. I have heard him say on several occasions, usually with a glass in his hand, 'Those who worry about respect do not have it'. Kit need not worry for he has earned high respect for his many accomplishments. I am proud to call him my friend.

16 The Development of Tourism Higher Education in China

Hanqin Qiu* and Dora Dongzhi Chen
Nankai University, Tianjin, P.R.China

Introduction

China did not open to the world until 1978, when Deng Xiaoping's so-called 'Open Door Policy' gave access to international businesses wishing to set up in the country for the first time. This was a historical turning point in China's development history, since significant political and economic reform has occurred subsequently, accompanied by remarkable economic growth; between 1989 and 2019, China's GDP grew by an annual average of 9.39% (Trading Economics, 2020). Within this context, China's tourism industry has also developed rapidly. Not only has the country become the world's largest and arguably most valuable outbound tourism market – in 2018, almost 150 million overseas trips were made by Chinese tourists (Travel China Guide, 2019) who 'exhibit the highest international spending per trip among global travelers' (Dichter *et al.*, 2018) – but it has also become a major international tourism destination. In 2018, 141.2 million international tourist arrivals were recorded (notably, over 78% of these were from Hong Kong, Macau and Taiwan), generating US$127.1 billion. In addition, some 5.54 billion domestic tourist trips were made that year (Travel China Guide, 2019).

During the same period, the provision of tourism higher education in China has both witnessed significant expansion and also played an important role in promoting the development of China's tourism industry, achieving considerable success in not only the number of colleges and students, but also the development of different levels and mechanism of programmes (Zhang and Fan, 2006). Many scholars may, therefore, be curious as to how China's tourism higher education has been developed over the past 40 years and what challenges the sector currently faces. In addition, as tourism higher education confronts an arguably uncertain future in a number of predominantly western countries, it may be asked what its future development trends are likely to be in China, particularly in the emerging 5G digital era.

The purpose of this chapter is to address these questions, exploring in particular the relationship between higher education and tourism and development. It briefly traces the historical development of tourism higher education in China and identifies contemporary opportunities and challenges before going on to consider potential future trends that, in particular, exploit opportunities afforded by digital technology and 5G development.

An Historical Perspective: 1978–2019

The history of China's tourism higher education development can be broadly divided into three

*Corresponding author: qiuhanqin@nankai.edu.cn

 Tourism in Development: Reflective Essays (eds P.U.C. Dieke *et al.*)
DOI: 10.1079/9781789242812.0016

identifiable stages, each of which is discussed in the following sections.

The first stage: the exploratory development period (1979–1988)

Coinciding with the opening up of the country, the establishment of Shanghai Institute of Tourism College in 1979 marked the beginning of tourism higher education in China (Du, 2003). A few other vocational-level colleges, such as the Shanghai Institute of Tourism, were established in locations including Tianjin, Nanjing and Guilin. Subsequently, seven other higher education institutions, including Nankai University, Beijing International Studies University, Hangzhou University and North-west University, began to set up undergraduate-level tourism education programmes (Bao and Zhu, 2008). Interestingly, at the outset, each of these colleges started with less than 100 students. Following the instigation of economic reforms by Deng Xiaoping in 1978, China started to develop its tourism industry. Initially, the focus was on promoting only political and educational visits. However, from the early 1980s, tourism was recognized as an important source of foreign exchange earnings and, hence, an emphasis was placed on increasing inbound tourism more generally. As a result, the number of international tourists, both from regional markets, including Hong Kong, Taiwan and Macao, and from further afield, began to increase – in 1978, for example, roughly 1.8 million international arrivals were recorded, a figure that grew to about 33 million by 1991 (Lim and Pan, 2003). Consequently, the Chinese government realized the necessity of appropriate education and training of personnel to meet the needs of the burgeoning tourism sector, reflected in tourism education in China entering a developmental stage when a relatively small number of universities and colleges began to offer tourism programmes. By the end of 1989, there were 69 colleges offering tourism programmes attended by 8551 students (Lam and Xiao, 2000).

The second stage: the large-scale development period (1989–2000)

During this second stage, the scale of tourism higher education in China began to expand significantly, commensurate with the rapid development and growth of China's tourism industry based on the country's political stability and wider economic growth. Indeed, both the demand for and supply of tourism services developed much faster than might have been expected. For example, all of the most prominent international brand hotels which had entered the China market started building hotels in major cities such as Beijing, Shanghai and Guangzhou. At the same time, local government also started to transform traditional guesthouses into hotels. As a consequence, the shortage of hospitality human resources with the appropriate skills and knowledge was acknowledged by both the federal and provincial governments, as well as by the tourism industry. The number of tourism colleges and schools expanded at all levels, ranging from vocational and undergraduate to the postgraduate level across the whole of China. By 2000, 252 tourism colleges and universities were in existence, with a total 73,580 students enrolled on tourism programmes (China National Tourism Administration, 2011).

The third stage: the standardization development period (2000–present)

Following the rapid growth in the number of tourism higher education programmes and the successful establishment over the past two decades of many tourism 'Academies', 'Departments' and 'Research Institutes' within universities, increasing attention has been paid by both the Ministry of Education and China National Tourism Administration (CNTA) to issues of standards and quality with regard to teaching materials, teachers and the educational methods being employed. CNTA has recently been merged with the Ministry of Culture, which has subsequently been renamed as the Ministry of Culture and Tourism of the People's Republic of China. Rather than focusing on the continuing expansion of higher education in tourism, increasing attention has been given to managing its scale and, particularly, to maintaining and enhancing the quality of existing programmes. At the same time, the level of tourism higher education has broadened during this period and has come to embrace

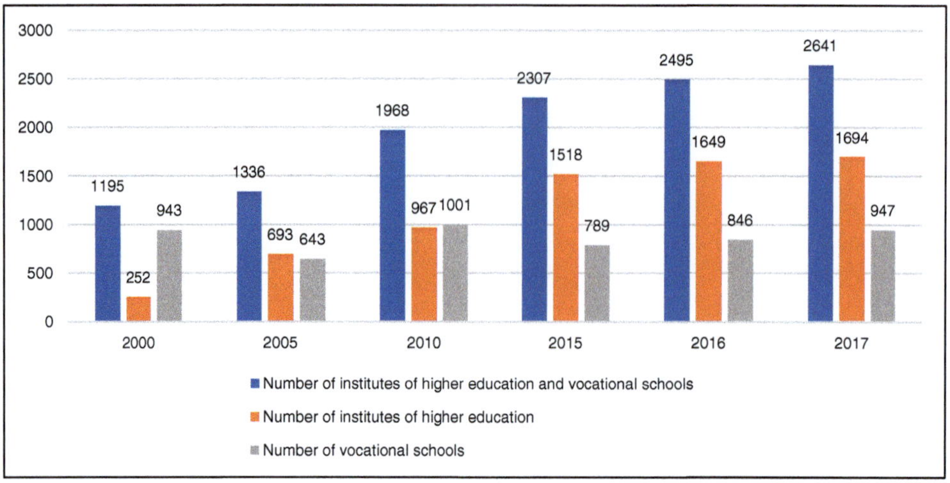

Fig. 16.1. Number of institutes of higher education and vocational schools offering tourism programmes (2000–2017).

(Adapted from China Tourism Statistics Yearbook, 2000–2017)

not only vocational and undergraduate programmes, but also postgraduate programmes including Master's and PhDs. There are course and research dissertation requirements for both Master's and PhD-level courses which combine elements of the British and American systems.

Figure 16.1 presents the number of tourism schools offering Bachelor and Master's degrees at both college and university level, as well as PhD-level education provision. The number of vocational schools offering sub-degree tourism education is also indicated. It is clear that the

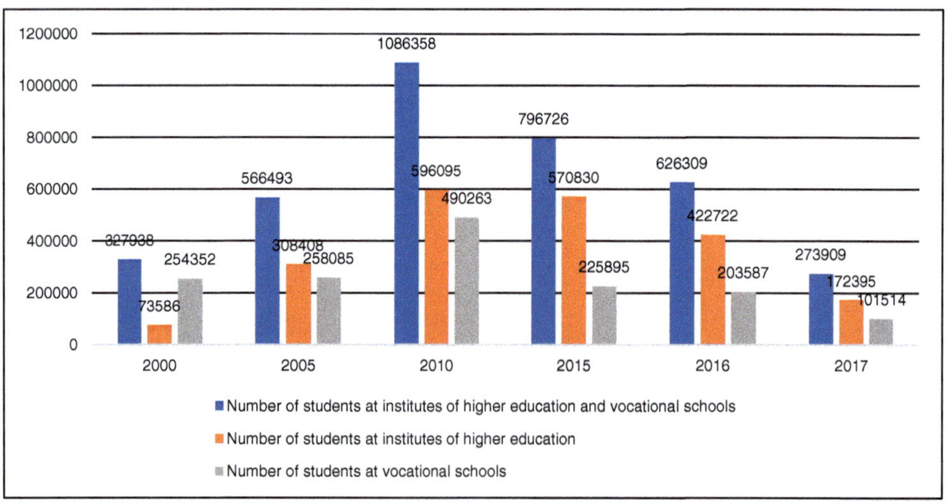

Fig. 16.2. Number of students on tourism programmes at institutes of higher education and vocational schools.

(Adapted from China Tourism Statistics Yearbook, 2000–2017)

number of schools has increased considerably since 2000. In contrast, Fig. 16.2 indicates the trends in the number of students attending tourism programmes at both college and vocational level over the period 2000–2015. It is clear that the number of students attending tourism programmes at all types of institution has been in decline since 2010. This suggests, of course, that secondary school leavers are becoming less interested in studying tourism and related areas at the higher educational level.

It is important to note, then, that while tourism higher education provision continues to increase, the number of students attending tourism programmes is decreasing. Inevitably, this is leading to an imbalance between demand and supply; if this continues, it is likely that there will be some institutions that are unable to attract sufficient students and, consequently, will face various challenges.

Unsurprisingly, perhaps, what has happened in China with regards to tourism education over the past 40-year period has also been experienced by other countries, and reflects Jafari's (1989) four 'platforms' of tourism development. According to Jafari (1989), during the first advocacy stage or platform (1960s), tourism was supported, often uncritically, for its perceived economic benefits, and mostly supported by economists. In the secondary, cautionary platform in the 1970s, tourism development was regarded more critically and recognized as being both beneficial and problematic in many circumstances while subsequently, the third, adaptancy platform in the 1980s was defined by the emergence of a general consensus on tourism and on what a tourism curriculum should include. Finally, since the 1990s, the knowledge platform has been in evidence whereby growing numbers of tourism courses, teachers and students have been supported by an increasing volume of publications and have been better informed by wider research into and understanding of tourism. Building on these platforms, Airey (2008) suggested four stages in the evolution and development of tourism education. First was what he described as the 'industrial stage', which was characterized by a concentration on business aspects of tourism and its vocational nature. Second was the 'fragmentation stage', where the growing number of courses in tourism and difference in backgrounds of the initiators made it difficult to agree on what might be an appropriate curriculum for tourism, and what should be core components. The third stage was described as a 'benchmark stage' which saw the emergence of a general agreement on the content and scope of these courses. The fourth stage was described as a 'mature stage' where tourism was sufficiently developed to rank alongside other social science courses. There is little doubt that the provision of tourism education has been facilitated at many levels in China over the past 40 years. It now appears to have reached a 'mature' stage, as defined by Airey (2008), who summarized this as follows:

> If the following are indicators of maturity; a community of scholars with the support standing of dedicated journals; other publications and scholarly conferences; a curriculum which is not constrained to a particular territory and about which there is a measure of agreement; an appreciation that teaching is as much about encouraging students to challenge and think, as it is about passing on knowledge; a recognition in common with other social sciences that tourism has taken a cultural turn...; an ability for scholars to be self-critical and aware of the nature of the truth of their work; then...tourism has reached a point of maturity.

Challenges and Opportunities

After 40 years of development, higher education in tourism in China has now reached a mature stage with a range of undergraduate and postgraduate programmes being offered by both comprehensive universities, such as Nankai and Zhongshan, and applied colleges and schools, such as Beijing International Studies University, Beijing United University and Guilin Institute of Tourism. However, just as the rapid development of tourism in general in China has resulted in a number of issues and challenges, so has the development of tourism education is particular. More specifically, in comparison to the development of tourism higher education in countries such as the USA, the UK and Australia, China's tourism education sector has faced a number of problems such as a relatively short but rapid history of development, and a lack of appropriate governance structures and of a clear development strategy, while the process has primarily been one of adopting or applying the models and experiences of tourism higher education schools in the USA or the UK. Moreover, owing to the uneven economic and tourism development of the eastern and western regions of China, tourism education in the latter is facing a serious shortage of qualified staff, affordable campus facilities, and affordability for students who are struggling to meet the ever-increasing fees as a result of inflation.

Some problems remain. Elite educational institutes (including those with a tourism programme) generally attain prestige from their research output, sometimes at the cost of teaching quality. Hence, more attention should be given to secure a balance between research and teaching where this is discerned as a problem. It is through excellence in teaching that students can be stimulated to adopt a critical perspective of their subject. Developing a critical perspective should take precedence over memorizing information and material. These issues can be addressed within a drive to internationalizing Chinese educational institutions and providing exposure to both Chinese and foreign academics to writing and publications in their subject areas.

Rather than being about 'Why do we do this?', the proposed approach will require clarifying the use and development of tourism knowledge, before moving to the next stage of development – enhancing the knowledge base through original research. The research dimension is now very important as publications reflect prestige on both the individual and the institution, particularly in international journals and media. As the elite tourism institutions grow in stature and reputation, they still must battle to attract the best students.

Nevertheless, there remains a paradox in providing tourism education, particularly at degree and postgraduate levels. It is regarded by many academics as a 'second-tier' subject, lacking the necessary basic theory and development to be regarded as 'first-tier'. However, this is arguably an invalid perspective because most social science subjects are not 'pure' in the way that one may regard physics, chemistry or mathematics, all with long histories of development in both research and application. For example, business studies is a subject area within the social sciences which attracts many top-quality students, presumably because over the years it has become more research-oriented and is seen to provide the business-related tools applicable to jobs in the business world. Yet many of the courses and subjects in business degrees are widely used in tourism courses, such as economics, statistics, mathematics, psychology, sociology and, more recently, Big Data and Artificial Intelligence (AI). Why then does tourism studies continue to be regarded in this light?

There are several reasons why this continues to be a problem in many countries, including China. First, from an academic perspective, many colleagues do not know or understand the comprehensive scope of tourism. It is sometimes viewed as a focus on travel industry management – essentially an applied, 'how to' teaching programme. Yet in China and in many developed countries, tourism as a sector contributes a substantial input to Gross National Product, to earnings of foreign exchange, creation of employment, contributions to government revenues and regional development. For example, according to statistics from the Ministry of Culture and Tourism, in 2018, China's domestic tourism and inbound tourism revenue totalled US$5.97 trillion, and the overall contribution of the national tourism industry to GDP was US$9.94 trillion, accounting for 11.04% of total GDP (Wu and Gao, 2019). The measurement of all these impacts requires specific training skills as part of a tourism degree, many identical to those taught in business programmes.

A second reason may be both the potential student's lack of knowledge of what the study of tourism includes and the job opportunities available. China is very much influenced by traditional values, and many parents have minimal knowledge of tourism (other than as holidaymakers). They would prefer their children to opt for recognized and better-known degree courses such as medicine, accountancy or economics.

A third reason is related to cultural attitudes. In Dubai, a city-state Emirate which has emerged as a world-class tourist destination over four decades, in excess of 90% of those employed in the tourism industry are non-Emirati. The benefits of job creation have been experienced by foreigners rather than locals. The reason here is cultural, where Emiratis see service work as being a form of servility and both parents and children are reluctant to work in tourism. In other Muslim countries, families do not want or permit females to work in tourism where they would interact with foreigners and particularly men. Cultural attitudes to employment are both sensitive and complex and outside the scope of this chapter.

A fundamental and continuing problem is a lack of understanding of words such as tourism, hospitality, travel industry and what travel might mean. The mission statement of the world-leading School of Travel Industry Management at the University of Hawaii is comprehensive: 'To develop and disseminate hospitality, tourism and transportation concepts, knowledge and skills through global leadership in teaching, research, training and outreach. These activities contribute to economic, social-cultural, and environmental sustainability including support and enriching local cultures (University of Hawaii, 2018). As a long-established school, it can afford to adopt a comprehensive mission statement. However, elite Chinese universities offering tourism programmes should give greater attention to establishing thorough research and teaching across a specialist curriculum.

A further continuing problem is the fragility of the link between tourism programmes and the needs of the tourism industry. This sometimes leads to a situation of what might be called 'classroom tourism' where the content of a tourism course often reflects the special interests of the teacher and with little relevance to the real-world situation. The attempts to overcome this problem, often through the formation of academic–industry consultative committees have often had mixed results. Jenkins (1999) identified this as a 'great divide between tourism academics and tourism practitioners. Much later, Airey (2018, p. 54) commented that 'securing links between academics and the wider tourism world, including the economic sector, still remains a challenge'. In the developing countries, these problems can be major barriers to development of tourism and to a robust tourism curriculum (Jenkins, 2015). A final observation by Tribe (2002) is appropriate as he describes his concept of the 'philosophic practitioner' as the person who can take their knowledge of tourism and apply it to various circumstances and conditions in the world, with the emphasis not only on the knowledge base but also on the human qualities to work successfully in often complex situations.

With the emphasis now on quality assurance in all university programmes across China as a result of the national strategy to build 'Double First Class', meaning top-quality schools and programmes, it is perhaps inevitable that a 'Chinese model' will emerge. This will likely draw upon China's political philosophy and educational structure and the predominance of the so-called 'socialist market economy'. Although borrowing from tourism education systems and practices in the developing countries, the model in China will be strongly based on outreach techniques such as MOOCs. This approach will offer better prospects of distributing tourism knowledge and educational opportunities to all parts of the country and to as big an audience as possible. Educational specialism in the form of designated universities and vocational colleges will develop new knowledge and 'trickle down' the benefits of their teaching and research to lower levels. Recognition of tourism by the government of China as a pillar industry will help to sustain this dynamic sector. Unlike many developed countries where government support for tourism is often, at best, ambivalent, the government in China has supported the development of the tourism sector both financially and politically. The Belt and Road Initiative, initiated in 2013, will have major implications for the development of infrastructure and connectivity between and within countries and should be a stimulus for tourism.

Future Perspectives

With the development of the internet and information communication technology (ICT), modes of learning and methods of assessment have changed. In particular, there has been an increasing emphasis on student-centred learning as a new educational method. ICT has been widely used in tourism higher education. Recognizing the strengths and weaknesses of traditional teaching and learning in established classroom settings, constant ICT developments have transformed student learning behaviours and call for a completely different set of learning and teaching approaches. Students are looking for a more interactive, friendly and accessible learning environment and resources. In responding to such changes and student demands, a more innovative student-centred learning and teaching approach needs to be developed as a response to such dynamic changes in today's ICT-driven world.

The outbreak of the coronavirus in Wuhan, China, which (at the time of writing) is now spreading across the world, provides an opportunity to speed up the adoption of blended online approaches to learning and teaching. Recently the Ministry of Education in China has issued a policy named 'schools are stopped, but learning and teaching must continue' with the support of online teaching and learning. All staff must start teaching as scheduled, using different online methods which can be as simple as (i) SPOC (Small Private Online Courses), (ii) apps-supported learning and teaching named Rain Classroom, an app developed by the XuetangX organization, or (iii) blended learning by using well-established MOOCs plus online tutorial discussions. The 5G digital era represents a significant opportunity to promote reform in the country's tourism higher education, particularly as online education has entered a relatively mature stage (Huang, 2016). At high transmission rate, 'AR/VR (Augmented Reality/Virtue Reality)+Education' will be activated again. First, it can achieve teaching with virtual hotel or theme part scene presented. Second, it can simulate many high-cost, high-risk teaching and training opportunities such as food and beverage laboratory. Third, it can restore history or other three-dimensional scenes. Fourth, it can simulate a live dialogue. With the popularization of 5G technology, the application of artificial intelligence technology in education will be 'smarter'. As a result, the '4A' model (anyone at any place and anytime can learn with any device) is one of the ways to transform the learning and teaching, Anyone can learn the courses developed by leading institutions in their respective fields. In the future, blended learning methods with interdisciplinary and multi-disciplinary learning contexts will be the focus for tourism higher education. More cooperative learning through social media will be used.

Massive Open Online Courses

Massive Open Online Courses (MOOCs) are, arguably, the most significant extension of open learning and teaching in the 5G digital age (Murphy *et al.*, 2014). By the end of 2018, more than 900 universities around the world had announced or launched a total of around 11,400 MOOCs (Shah, 2019). On the one hand, MOOCs break down barriers to tourism higher education knowledge, promote the sharing and exchange of educational resources and reflect the concept of global co-operation and resource sharing (Miller, 2012). On the other hand, they also establish communication platforms for universities and enterprises, connect the enterprise demand seamlessly with the university training, and innovate the university–enterprise co-operation teaching method.

Table 16.1 lists the tourism courses provided through four MOOCs platforms, namely, FutureLearn, Edx, XuetangX and Wisdomtree. From the content of these courses and their providers, the practice of knowledge dissemination and practitioner training in tourism and service industries in mainland China is still in the wait-and-see stage (Lin, 2016). Well-known colleges and universities in China lack practice in using MOOCs for tourism training, and the overall quality of the MOOCs in tourism education is not yet high.

Blended Learning Model Developed by Nankai University

Considering the ever-changing learning and teaching behaviour as a result of technological advances, the need for continued teaching during crises, such as the coronavirus outbreak, and maybe even more unpredictable future events, the College of Tourism and Service Management at Nankai University will accelerate the development of online learning through creating more blended

Table 16.1. MOOCs on the theme of tourism and service offered by the four MOOCs platforms.

Platforms	Courses	Universities
FutureLearn	Introduction to the Travel & Tourism Industry: Passport to the World	GTTP (The Global Travel & Tourism Partnership)
	Global Tourism Development Trends	Nankai University
	Tourism Policy and Planning	Nankai University
	Research Methods in Tourism Studies	Nankai University
	Culture and Tourism	Nankai University
	Understanding Data in the Tourism Industry	Edinburgh Napier University
	How to Perform a Hotel Market Analysis and Valuation	Glion Institute of Higher Education
	Hotel Revenue Management: The Strategy and Tactics of Hotel Room Pricing	Glion Institute of Higher Education
EdX	Tourism and Travel Management	The University of Queensland
	Sustainable Tourism: Society & Environmental Aspects	Wageningen University & Research
	Introduction to Hospitality and Tourism Industry	University of Maryland Eastern Shore
	Sustainable Tourism	Wageningen University & Research
	Sustainable Tourism: Rethinking the Future	Wageningen University & Research
	Hospitality and Tourism Technology and Innovation	The Hong Kong Polytechnic University
	Managing Marketing in the Hospitality and Tourism Industry	The Hong Kong Polytechnic University
	Managing Human Resources in the Hospitality and Tourism Industry	The Hong Kong Polytechnic University
	Luxury Management	The Hong Kong Polytechnic University
	Hospitality and Tourism in China: A Global Perspective	The Hong Kong Polytechnic University
	International Hospitality Management	The Hong Kong Polytechnic University
XuetangX	Global Tourism Development Trends	Nankai University
	Tourism Policy and Planning	Nankai University
	How to Become a Good Guide?	Ningbo Tourism Training Center
	Situational English Speaking for Tour Guide	Ningbo City College of Vocational Technology
	Hotel Housekeeping Operation and Management	South China University of Technology

Continued

Table 16.1. Continued

Platforms	Courses	Universities
Wisdomtree	Tourism Route Geography in China	Qingdao Technical College
	Airborne Etiquette	Sanya Aviation and Tourism College
	Ice and Snow - Ice and Snow Tourism Resources and Culture in North-east, China	Heilongjiang Agricultural Economy Vocational College
	Tourism Etiquette	Shaanxi Vocational and Technical College
	Rural Tourism Development and Management	Hainan Vocational and Technical College
	Tourism Landscape Culture and Appreciation	Harbin University of Commerce
	Travel English (Hainan Union)	Hainan Normal University
	Introduction to Tourism Economics	Changchun Normal University
	Tourism Studies	Shaanxi Business College
	Landscape and Tourism Planning	North-east Agricultural University
	Tourism Resources	Shan Dong University
	Destination Management	Shandong Youth University of Political Science
	China's Role in Global Tourism Development	Nankai University
	Tourism Policy and Planning	Nankai University

(Assembled from authors' personal contacts)

learning courses. In this way about 50% of the learning can take place online and the remaining 50% of the teaching and learning will take place in the classroom for thought-provoking discussion and brainstorming. The very first pilot micro-credential project, called Culture and Tourism Development, was first developed by the College of Tourism and Service Management. This project includes four courses: China's Role in Global Tourism Development, Tourism Policy and Planning, Culture and Tourism, and Research Methods. This project invites well-known tourism scholars from all over the world to contribute and uses Chinese and English subtitles for learners to choose to achieve personalized education. Students can choose to study all these four courses in English through FutureLearn. For example, the first course, China's Role in Global Tourism Development, has been launched on FutureLearn and has attracted more than 2000 learners from 135 countries. For those students within China can study in Chinese on XuetangX or Wisdomtree.

Future Development Strategies

To enhance staff development in tourism education

The future development of tourism higher education depends on the quality of staff provided to each level of education institution, ranging from vocational to undergraduate and postgraduate. Collaboration across schools, regions and countries is one of the strategies to be adopted. In this way, staff from different cultural backgrounds can work together to promote teaching and research excellence. Tourism is a multi-disciplinary discipline which is essential for creating an arena that nurtures future talents with skills such as critical thinking, global outlook, and multi-cultural communication through modern technology.

Establishing college and industry collaboration mechanisms

Tourism is an applied field and needs to pay close and regular attention to industry developments. It will be essential for tourism education providers to work closely with industry partners. Such partnerships can ensure that what students learn can be applied through industry practice. Schools need to develop strategic partnerships with industry that address emerging development issues and to identify the staff who should be developed to serve the ever-changing human talent needs of the tourism sector. To achieve this, it is essential for schools to build regular communication channels, involving industry leaders in the curriculum design and update, jointly hosting research seminars and workshops to ensure that the students get the most updated industry development information and schools can research and feed the industry with the most up-to-date research outputs.

Creating a smart learning environment

With 5G being developed and implemented in all aspects of the living environment, creating a smart learning environment and mechanism is essential to the success of future higher education. Students need innovative and interactive learning and teaching environments instead of the traditional large classes and long hours of teacher-centred learning style. With the MOOCs provided by leading universities such as MIT, Harvard, Cambridge and Oxford, students can learn some basic knowledge easily online without being physically present in schools and classrooms. What students really need and benefit from is the amount of quality time spent with the lecturers and professors to discuss and debate some critical issues derived from their learning process. The role of class teachers will be transformed from lecturing to facilitating and supporting students learning and critical thinking. Smart learning devices can facilitate students' learning, discussion, and practice online and offline. All assessment methods are made available online to reduce classroom lecturing and tutoring time and effort. Students and learners are calling for a different set of learning approaches and these need to be addressed by leaders, educators and administrators from across the higher education sector.

Traditional modes of teaching and learning have been transformed as a result of internet advances, and heavy use of social media platforms such as Facebook, Twitter and WeChat. Traditional teaching and learning have shifted from a physical learning world to an e-learning environment. The transformation process needs a different approach and mindset to address teaching and learning pedagogy; the need for staff development and training; student assessment; and all Quality Assurance matters. Innovative methods and transforming online learning such as Massive Open Online Courses (MOOCs) have been developed at both undergraduate and postgraduate levels in most universities in the world. Students from all over the world can take courses offered by leading universities at any time and at their own pace without coming to the classroom. The learning market includes students from many countries of the world studying online, which provides completely different experiences and group dynamics of learning among students. All these developments have revolutionary implications for university admission policies, student development strategies, staff and resource deployment, and ongoing quality assurance which have been in place for a few decades.

Conclusions

In summary, we may observe that tourism higher education in China has undergone 40 years of development, addressing the human talent needs in tourism. This has been well received, both domestically and internationally. It is recognized that there have been limitations – system reforms, faster-than-expected industry development, and unequal development across eastern and western China. However, the development of tourism education from virtually nothing to its current status with 1086 universities and schools and about 160,000 students (including university and vocational level) must be viewed as a remarkable success. Looking to the future, the tourism education sector, including government, educators and industry stakeholders, needs to act jointly and strategically to sustain future development. It will need to address the ever-changing industry needs, the demand for innovative learning approaches from learners across different levels of education, and government calls for a highly qualified and responsible younger generation. China has become one of the fastest-developing and most successful countries generating the largest outbound tourism demand. This places China in an important role in the global tourism industry. The future development of talent will be essential if China is to sustain its leadership position and promote its model of tourism industry development.

References

Airey, D (2008) In search of a mature subject? *The Journal of Hospitality Leisure Sport and Tourism* 7(2), 101–103. DOI: 10.3794/johlste.72.per.

Airey, D. (2018) Tourism education and scholarship. In: Cooper, C., Volo, S., Gartner, W. and Scott, N. (eds) *The Sage Handbook of Tourism Management*. Sage Publications Ltd, London, pp. 49–60.

Bao, J. and Zhu, F. (2008) On the shrinking of China's tourism undergraduate education and way out: thoughts about thirty years of development of tourism higher education. *Tourism Tribune* 23(5), 13–18.

China National Tourism Administration (2011) *The Yearbook of China Tourism Statistics*. China Travel & Tourism Press (in Chinese), Beijing.

Dichter, A., Chen, G., Saxon, S., Yu, J. and Suo, P. (2018) Chinese Tourists: Dispelling the Myths. McKinsey & Company. Available at: https://www.mckinsey.com/~/media/McKinsey/Industries/Travel%20Transport%20and%20Logistics/Our%20Insights/Huanying%20to%20the%20new%20Chinese%20traveler/Chinese-tourists-Dispelling-the-myths.ashx (accessed 15 January 2020).

Du, J. (2003) Reforms and development of higher tourism education in China. *Journal of Teaching in Travel & Tourism* 3(1), 103–113. DOI: 10.1300/J172v03n01_07.

Huang, L. (2016) Thinking about the development of network education in colleges and universities under the 'Belt and Road' strategy. *Science and Education Wenhui* 9, 138–139.

Jafari, J. (1989) Sociocultural dimensions of tourism: an English language literature review. In: Bystranowski, J. (ed.) *Tourism as a Factor of Change: A Sociocultural Study*. Vienna Centre, Vienna, pp. 17–60.

Jenkins, C.L. (1999) Tourism academics and tourism practitioners: bridging the great divide. In: Pearce, D. and Butler, R. (eds) *Contemporary Issues in Tourism Development*. Routledge, London, pp. 52–64.

Jenkins, C.L. (2015) Tourism policy and planning for developing countries: some critical issues. *Tourism Recreation Research* 40(2), 144–156. DOI: 10.1080/02508281.2015.1045363.

Lam, T. and Xiao, H. (2000) Challenges and constraints of hospitality and tourism education in China. *International Journal of Contemporary Hospitality Management* 12(5), 291–295. DOI: 10.1108/09596110010339643.

Lim, C. and Pan, G.W. (2003) An econometric analysis of inbound tourism for China. *International Congress on Modeling and Simulation* (Vol. 3). Available at: https://pdfs.semanticscholar.org/0b59/6f71c3a6e5ac02eb29d22628a91ec3a9434d.pdf (accessed 10 March 2020).

Lin, J. (2016) Tourism and service industry training on the Internet: MOOC status. *Tourism Tribune* 31(5), 10–14.

Miller, T. (2012) MOOCs break down barriers to knowledge. *The Australian,* 12 December.

Murphy, J., Kalbaska, N., Williams, A., Ryan, P., Cantoni, L. *et al.* (2014) Massive open online courses: strategies and research areas. *Journal of Hospitality & Tourism Education* 26(1), 39–43. DOI: 10.1080/10963758.2014.880618.

Shah, D. (2019) By the numbers: MOOCs in 2018. Website of Class-Central. Available at: https://www.class-central.com/report/mooc-stats-2018/ (accessed 1 January 2019).

Trading Economics (2020) China GDP Annual Growth Rate. Available at: https://tradingeconomics.com/china/gdp-growth-annual (accessed 15 January 2020).

Travel China Guide (2019) 2018 China Tourism Facts & Figures. Available at: https://www.travelchinaguide.com/tourism/2018statistics/ (accessed 15 January 2020).

Tribe, J. (2002) The philosophic practitioner. *Annals of Tourism Research* 29(2), 338–357. DOI: 10.1016/S0160-7383(01)00038-X.

University of Hawaii (2018) School of travel industry management. Website of University of Hawaii. Available at: http://www.catalog.hawaii.edu/schoolscolleges/tim/TravelIndustryManagement.html (accessed 14 May 2020).

Wu, C. and Gao, F. (2019) Tourism contributed 11.04% of China's GDP last year. Travel China. Available at: http://travel.china.com.cn/txt/2019-02/14/content_74464252.htm (accessed 1 March 2020).

Zhang, W. and Fan, X. (2006) Tourism higher education in China: past and present, opportunities and challenges. *Journal of Teaching in Travel & Tourism* 5(1-2), 117–135.

17 Roads Not Adopted: Tourism Research in Neo-liberal(?) Times

Roy C. Wood*
University of Cumbria, Ambleside, UK

Introduction

A recurring theme in both academic and public commentaries on tourism pertains to the extent to which various forms of (in particular 'international') tourism development are expressive of, variously, colonial, post-colonial and imperialist ideologies (e.g. Palmer, 1994; Burns, 2008; Sinclair-Maragh and Gursoy, 2015). Sometimes explicitly, sometimes implicitly, we are quite properly encouraged to deplore such constructions. A more recent strand of critique in tourism, in many other academic disciplines and in public life more generally, focuses on the negative consequences of 'neo-liberalism'. Although the possible parallels between neo-liberalism, colonialism/post-colonialism and imperialism are not extensively explicated in the tourism literature, the first of these is invariably cast as a (principally 'western') hegemonic force *vis-à-vis* various forms of international economic development. That is, it possesses some of the characteristics of colonialism and imperialism.

This chapter explores conceptions of neo-liberalism in the context of the development of tourism research. Although its intellectual origins are somewhat earlier, neo-liberalism as an economic philosophy is mostly seen as growing in global dominance from the late 1970s and early 1980s. However, the somewhat, perhaps unusual, starting point for this discussion is the extensive and ongoing debate about neo-liberalism's general influence on higher education in particular. This is justified in terms of the corresponding growth of tourism in higher education since the 1970s (see also Chapter 14). Putting it another way, in the English-speaking world (and, some would argue, beyond), the apparent growth of neo-liberalism in higher education is coincidental with the rise of tourism as a subject in that milieu. Accordingly, we might not unreasonably expect the development of tourism as a relatively new area of enquiry to more strongly reflect the supposed tropes of the neo-liberal project than is the case with more established subjects. Following from this, the chapter will seek to explore the extent to which the neo-liberal project has influenced tourism research, finally reflecting on the implications of such an analysis.

Neo-liberalism and Higher Education

Most of the maladies supposedly afflicting contemporary higher education are increasingly attributed to the, invariably perceived as malign, influence of 'neo-liberal' ideology. A difficulty with this view, however, is that neo-liberalism is not monolithic, but embraces varied and

*roy.wood@cumbria.ac.uk

 Tourism in Development: Reflective Essays (eds P.U.C. Dieke *et al.*)

DOI: 10.1079/9781789242812.0017

sometimes competing emphases (Saunders, 2010; Gane, 2014; Simmons, 2017). A great many commentaries recognize this fact before largely ignoring it. In the study of higher education, the disposition to neo-liberal ideology is mainly antagonistic and critical (e.g. Saunders, 2010 in the case of the USA; Radice, 2013; Robinson *et al.*, 2017; Brandist, 2017 in the UK; and Geppert and Hollinshead, 2017 on the UK and Germany) and only occasionally in sympathy and defence (e.g. Miller, 2014).

Radice (2013, p. 408) defines neo-liberalism as comprising changes in capitalism, specifically 'privatisation, deregulation, financialisation and globalisation'. In the UK public or state sector, neo-liberalism manifests itself in a 'new public management':

> ...a particular combination of Stalinist hierarchical control and the so-called free market, in which the values, structures and processes of private sector management are imposed upon the public sector; key elements include a shift from professional to executive power, a focus on 'performance' as measured by quantitative targets, and the widespread use of financial incentives.

The contemporary university is thus rendered 'a simulated private enterprise, with effective possession by a self-selecting academic-business elite' (Radice, 2013, p. 415). In the US context, Saunders (2010) argues that neo-liberal hegemony has led to universities facing decreased state funding; greater fundraising from private sources; casualization of academic labour; increasingly hierarchical rather than shared governance models; and a growing emphasis on research with potential commercial exploitation value. Teaching and research are consistently subordinated to corporate objectives. In the UK, Brandist (2017), p. 584) notes that research is directed directly and indirectly towards a justificatory regime predicated on its impact on the 'economy and social cohesion' with faculty channelled into research:

> ...likely to yield publications in the most prestigious journals... [.] The resulting conservatism extends beyond encouraging academics to restrict themselves to established patterns of research behaviour, impinging on the ideological content of what is produced. In this Kuhnian world of 'normal science', the established paradigm remains unchallenged. (Brandist, 2017, p. 586)

More than occasionally, this vein of commentary smacks of a somewhat simplistic determinism. It ignores what acts of resistance there may be to some of the more egregious instruments of neo-liberalism in higher education, even where such resistance takes the unfortunate form – as it often does – of the academic class acquiring and retaining such privileges as they can. This is demonstrated by Sykes (1988; see also Sykes, 2016) in a by no means impartial but still challenging commentary on how the US academic community maintains what others might regard as extraordinary occupational entitlements. Deterministic perspectives also have difficulty with formulating adequate explanations of the complicity of academic personnel in the rise of managerialism (Wood, 2017). In the US context, Saunders (2010) writes: 'What is new to the neo-liberal university is the scope and extent of...profit-driven, corporate ends, as well as how many students, faculty, administrators, and policy makers explicitly support and embrace these capitalistic goals and priorities'. For writers like Radice (2013, p. 415), 'The question remains as to why academic staff have acquiesced so readily in these changes'. His response seems to rest on a vague formulation of Marxist false consciousness with faculty accepting neo-liberal ideas because they subscribe to notions of bourgeois individualism, including self-determination, rather than seeing themselves as 'social beings'.

A further objection to the supposed influence of neo-liberalism in higher education is that commentators confuse that phenomenon with political conservatism, the view of Miller (2014, p. 144) who, citing Marginson (2006), notes that:

> While critics of managerialism blame neo-liberal policies for the rise of the phenomenon, it is not obvious why such a link should exist. Neo-liberalism is defined by an overriding belief in freedom and its protection although, as Marginson (2006) has noted, many critics 'perhaps confuse the neo-liberal project with the political conservatism often joined to it'.

To some, this might understandably appear as a somewhat *too* nuanced perspective. Yet the debate remains to be had. There are numerous instances of where the term 'neo-liberalism' simply figures as a vulgar shorthand for often crude politically conservative ideas and policies, with the effect of reducing a serious and presumably dangerous ideology to a form of 'virtue signalling'. Inserting a knowing reference in a piece of ostensible scholarship as to the influence of neo-liberal policies on this or that aspect of higher education is both the nod and the wink to the proverbial blind horse. In this last case, consider, for example, the assertions of Fletcher *et al.* (2017); see also Silk *et al.*, 2017) on the field of 'leisure studies', an area of enquiry that sometimes lays claim to be the mother field of tourism: 'the identity and profile of Leisure Studies in UK Higher Education (HE) curricula have become less apparent over the past decade [.] This trend is not peculiar to Leisure Studies; the social sciences as a whole are threatened by a neo-liberal economic discourse which increasingly informs HE strategic management.' Manifestations of this discourse include, according to the authors, 'Cost-efficiencies, utilitarian learning and a turn to the wisdom of 'big data'', but little evidence is adduced to support the implied universality of these influences.

In addition to these questions of conceptual fluidity are those pertaining to both antecedent and likely corrective states. There is a palpable silence on both matters, but many writers implicitly look to a nostalgic past, and perhaps future, where universities were/are largely democratic, self-governing communities of scholars dedicated to the virtues of higher learning and untainted by commercialism – conjuring images, in the words of the American satirist Tom Lehrer (n.d., circa 1959), of 'Ivy-covered professors in ivy-covered halls'. More conventionally, Saunders (2010) comments that:

> Many...who discuss the impact of neo-liberalism on higher education juxtapose the neo-liberal university...with a previous university that allegedly focused on civic engagement, democratic education, learning for its own sake. To some extent, this contrast is accurate... [.] However, the claim that universities were ever such democratic institutions with altruistic aims is questionable.

A final point here is that while there can certainly be broad agreement as to the rise of managerial control ideologies in higher education, it remains a work in progress to demonstrate how 'the' neo-liberal agenda contributes greater explanatory power to understanding of the current state of higher education than more conventional, if not wholly mutually exclusive, approaches. The 'decline of donnish dominion' (Halsey, 1995) is, after all, hardly a new phenomenon. These might include Marxist analyses of the labour process and proletarianization (cf. Brandist, 2017) and consideration of the general advance of Weberian 'rationality', its interpretation in terms of 'McDonaldization' (Ritzer, 1993; Hayes and Wynyard, 2002; Hayes, 2017; see also Ritzer, 2013), or in its relationship to neo-liberalism (Gane, 2014).

Characterizing Tourism Research

Tourism's genealogy as a subject corresponds almost exactly with the changes to higher education allegedly wrought by neo-liberalism. Some 35 years ago in the UK, tourism in higher education was mainly studied as a subordinate feature of programmes in hotel – or as the preferred label is now, 'hospitality' – management. In many public institutions of higher education, hospitality has gone from boom to bust even as it thrives in private colleges and universities across the world (for further elaboration see Baum, 2012; Wood, 2015a, b). Tourism, in contrast, has apparently consolidated its position within higher education in almost a single generation. Coles *et al.* (2006, p. 293) noted that:

> ...the study of tourism sits curiously within the contrived academic division of labour. On the one hand, tourism is identified as one of the major features of everyday life in the era of globalisation...yet on the other it struggles to obtain greater institutional legitimacy in higher education...

Over a decade since this was written, questions as to tourism's academic legitimacy appear to have been largely overcome. Tourism education incorporates business and management elements designed to be relevant to the future employment of graduates, leading some at least to believe that neo-liberalism is a strong influence on the curriculum (Ayikoru *et al.*, 2009). At the same time, other subjects that have no such obvious focus are frequently present, often as specializations. This environment is mirrored in tourism research where three broad traditions can be discerned. The initially dominant educational *and* research paradigm was always oriented towards policy questions, incorporating management and business issues broadly defined, long before contemporary commentators started to detect insidious neo-liberal influence on higher education. This research has subsequently been complemented by a substantial growth in sociological enquiry which, superficially at least, is explicitly resistant to all forms of neo-liberal ideology. Where neo-liberalism has had a manifest impact is in the third kind of tourism research activity – 'research of duty' – which probably constitutes, quantitatively, the largest output as it does in most other subject areas, including hospitality (Wood, 2017). Each will now be considered briefly in turn.

Policy, business and management approaches to tourism research

The obvious contemporary extension of earlier policy approaches in tourism research is, as elsewhere, the preoccupation with sustainability, 'sustainable tourism' being the egregious near neologism employed to describe work in this field. Tourism, like most human activities, is not sustainable in any intrinsic sense although the effects of tourism, as with most human activities, can possibly be ameliorated. Given the probability that human activity (or inactivity) is contributing to a range of potentially catastrophic and terminal environmental events, one might have expected a more fundamentalist socio-economic approach to sustainability than is in fact apparent. However, the majority of research in the field automatically submits to the dominant ideological worldview of the beneficial role of tourism (Higgins-Desbiolles, 2006). Sustainability research in tourism most usually focuses on notions that tourism industries can be 'nudged' towards sustainable behaviours. There is, thus, a systemic failure to interrogate the likely extent to how far, within a global capitalist system, such change can go without changes to the system itself, representing a significant retreat from economic and social realities.

It is also within the policy field that we find the very few radical critiques of tourism theory and practice, including those explicitly addressing the purported impacts of neo-liberalism on tourism development. This is absolutely explicit in the work of Higgins-Desbiolles (2006, p. 1205) who excoriates tourism academics and practitioners for their failure to see beyond the evils of a marketized neo-liberal tourism to a social (and indeed socialist) tourism that fosters 'contact between peoples who increasingly need to understand each other and co-operate harmoniously in a world where space, resources and options are shrinking quickly'. Pavlović and Knežević (2017, p. 65) argue that one effect of the neo-liberal order is that 'Tourism has strongly emerged in areas where it once did not intensively appear or did not appear at all', citing health and the biosphere as developments in tourism 'often in opposition to the interests of the development of these areas'. In the context of tourism for development, Lacey and Ilcan (2015, pp. 41-43) studied the role of Namibian conservancies as an instance of 'colonial governmentality' arguing that global aid regimes frame 'development goals in terms of establishing the conditions for successful participation in global markets'. Duffy (2015, pp. 529-530) argues that 'nature-based tourism allows neoliberalism to turn the very crises it has created into new sources of accumulation'. This is achieved through nature being reconfigured by marketization and commodification to create economic value. Reyes (2011) offers a searing critique of the relationships between poverty, sustainable tourism initiatives and neo-liberal tourism policies in Nicaragua, whilst Buades (2010) examines developments in the contemporary hotel industry through the lens of neoliberalism, implying that the development of the sector has been facilitated by a combination of the laundering of private capital and misuse of public monies. Real or supposed moral shortcomings of tourism are the focus of Dimitriou's (2017) work, and include poor tourism planning; ethical challenges in international tourism management arising from culturally varied attitudes towards forms of corruption; and the problems of constructed authenticity. However, Dimitriou's proposed end-correctives to these issues are ill-defined and lack an economic context. This is not true of Castañeda (2012, p. 47), who argues that from the 1960s to the 1980s, academic and policy attitudes towards tourism were polarized: tourism was either seen as a force for moral good or for moral evil, with the former view predominating. Those articulating the latter view saw tourism as a form of modern imperialism 'in terms of "pseudo-events", cultural inauthenticity, crass consumerism, and social status mongering'. Subsequent research, Castaneda argues, has demonstrated both positive and negative effects of tourism, a position which sits somewhat uneasily with his closely argued critique of the UNWTO Global Code of Ethics which, he suggests, mistakes rights for ethics and promotes an individualistic, neo-liberal view of rights to tourism which reinforce capitalist orthodoxies.

Many of the commentaries cited immediately above are found in journals outside the field of tourism or in practitioner or activist publications. Having noted the observations of Brandist (2017) earlier, we should not be surprised by this; a consequence of contemporary management styles in higher education is that those who set themselves against dominant paradigms often find themselves confined to the margins of those paradigms (Wood, 2017). Beyond this, most of the studies mentioned, as was the case with the literature on higher education and leisure studies noted earlier, fail to deal with neo-liberalism as anything other than a cipher, an observation also largely true of the few more intense treatments of that ideology in tourism (e.g. Mosedale, 2016a, Mosedale, 2016b; Simmons, 2017). A limited pattern thus emerges and the invocation of supposed and uniquely neo-liberal influences on tourism policy and practice become almost impossible to differentiate from more conventional critiques of capitalism – whether general or specific to tourism (e.g. Nowicka, 2007).

Tourism and the sociological

Sociologically oriented tourism research has sought to distance itself from the managerialism that initially imbued the subject, often embracing post-modern tropes that both express a wider ambivalence towards any notion of the socially real and doubt the very idea that critical theorizing has the capacity to inform social change (Callinicos, 1990; for a review of the current status of sociological theorizing in tourism more positive than that rendered here, see Cohen and Cohen, 2012). Sometimes proposed as the antithesis of, or corrective to, the policy and management research paradigm, a flavour of the overall tone of sociological approaches is contained in Franklin and Crang's (2001, p. 5) editorial to the first edition of the journal *Tourist Studies*:

> The first trouble with tourism studies...is that its research object, 'tourism' has grown very dramatically and quickly and that the tourism research community is relatively new. [.] Part of this trouble is that tourist studies has simply tried to track and record this staggering expansion... [.] One reason for this is that tourist studies has been dominated by policy led and industry sponsored work so the analysis tends to internalize industry led priorities and perspectives...

The authors go on to argue that this policy-led/industry-sponsored work has been created 'by people whose disciplinary origins do not include the tools necessary to analyse and theorize the complex cultural and social processes that have unfolded. How many schools of tourism hire the specialist skills of social and cultural theory?'

Coles *et al.* (2006, p. 297) note the tone of condescension in these remarks but refrain from commenting on its self-regard. With the benefit of a now longer view on Franklin and Crang's (2001) observations, it is difficult to argue that current tourism scholarship is any more *systematically* theorized, sociologically, than in the past. Instead, there have been a number of loose theoretical fashions as perfectly recounted by Cohen and Cohen (2012), dominant among which is John Urry's (1992) concept of the 'tourist gaze'. Urry's original work flattered many tourism scholars and the subject of tourism itself, while those who regarded 'the tourist gaze' as little more than a marginally interesting contribution to sociological analysis of tourism were (politely) viewed as eccentric (Wood, 2017). It is perhaps worth noting in passing the parallel influence in hospitality studies of George Ritzer's (1993) concept of 'McDonaldization', the influence of which has only recently begun to diminish with discovery and analysis of Derrida's extremely slight commentaries on hospitality – see Derrida and Dufourmantelle (2000).

A more recent and emergent strand of sociological research leverages identity politics as a supposed corrective to social injustice, often claiming a novel critical approach to tourism, as articulated by Vorobjovas-Pinta and Robards (2017, pp. 369-370):

> Over the last decade, scholars within the field of tourism have become increasingly discontented with positivist enquiry towards creating tourism knowledge; arguably, the new era of critical tourism studies began. This...resulted in a more reflexive, ethical, responsible and nuanced form of research. It can be advanced that critical research has adopted a stance privileging voices of those who are marginalised based upon their gender, social status, race, ethnicity, sexual orientation and so on, in an active effort to counteract societal discrimination.

There are, however, a number of different claims as to the nature of critical tourism studies. For example, Lacey and Ilcan (2015, p. 44) suggest, somewhat grandly, that the field:

> ...sets itself apart from that broad range of work in tourism studies investigating structural and institutional determinants and outcomes of neo-colonial forms of tourism for development, and instead seeks to explore the hybrid and contingent strategies of negotiation, cultural production and agency that characterize contemporary tourism projects...[.]

Despite these remarks, 'positivist enquiry' in tourism shows signs of continuing to thrive with some success, and to what extent the 'new' approaches signalled in the quoted passages are genuinely critical remains to be seen. A point of comparison is the development of 'critical management studies' (e.g. Tadajewski *et al.*, 2011) but here, unlike (for the most part) in tourism, substantial and systematic linkages to theory across the social sciences is in evidence.

Research of duty

'Research of duty' is a label employed here to describe published output increasingly encouraged as an appropriate response to the web of metrical fantasies designed to assess such things as research quality and impact, and is an extension of the so-called 'publish or perish' culture (Wood, 2017). Increasingly, research quality is measured by the crude surrogate of journal rankings, not least in business and management to which tourism is often aligned for assessment purposes. These rankings are, for the most part, created by self-appointed and largely unaccountable elites and have an atavistic character. The culture encouraged is not one of the celebration of collegiality, criticality or diversity, but of competition, conformity and the punishment of deviation (see Willmott, 2011; Brandist, 2017).

The success of the new public management (Radice, 2013) has lain in its ability to insist on the righteousness of such evaluation through employing the language of accountability, consumer satisfaction and value for money. The practical consequences of the approach are well known in the business and management field and are seen also in tourism research. They include the encouragement of aspirations to an unattainable scientism reflected in scores of published research papers on often marginal topics, based on small samples and exhibiting an over-reliance on quantitative methods. Compared to that reliant on qualitative methods or the development of theory, such research can be produced relatively easily within the time-sensitive cycle of assessment. With many academic journals in the field increasing the frequency of their publication, it is also not too difficult to place such output. Most controversially, a substantial amount of this output is, quite literally, worthless in any intrinsic sense, for example in terms of contributing to the systematic theoretical and empirical development of a subject or discipline. Moreover, the focus on journal publications is at the expense of other, often more substantial, published outputs.

The supreme irony that results from these processes is that, despite the best efforts of researchers, their work will nearly always fall short in terms of impact or the expectations of 'stakeholders' external to the academic world (Thomas and Ormerod, 2017). As the 'Schumpeter' columnist of *The Economist* (2014, cited in Wood, 2017) once acidly noted: 'Professors...will perish unless they publish in the right journals. And they have too little incentive to produce usable research. Oceans of papers with little genuine insight are published in obscure periodicals that no manager would ever dream of reading'.

Instead, business (and others) turn to consultants who are frequently, if mistakenly, regarded as offering the possibility of more relevant and credible knowledge. If research of duty is the one true offspring of neo-liberal hegemony in higher education, then it is a weak child indeed, at least in the extent to which it is able to satisfy that dimension to the ideology which focuses on the commercial applicability of knowledge (cf. Saunders, 2010). While the production of such research might be said to satisfy the bean-counting culture of the new managerialism, it can rarely, if ever, contribute significantly to the development of intellectual profundity.

Neo-liberalism, Tourism Research and 'Real World' Tourism

A number of points flow from the preceding discussion. First, the growth of tourism research does not seem, at first sight, save perhaps in the case of research of duty (which in any case has quite possibly come to define the majority of academic output), to have been excessively or uniquely influenced by the neo-liberal culture that supposedly commands higher education. Tourism's initially dominant research paradigm was always rooted in the business-policy-management triangle and, thus, could do little else than reflect, with occasional critique, the dominant influence of 'capitalist' ideologies. It continues to do so. Indeed, given the claims made for neo-liberal ideology's influence on higher education, the emergence of an extensive vein of sociologically informed research might be considered unusual if it were not for the fact, secondly, that this output is theoretically piecemeal and not in sociology's more radical traditions. Indeed, both policy and sociological research into tourism evidence a certain conservatism, specifically characterized by a fundamental retreat from critical engagement with the socially (and economically) 'real'.

This leads us, thirdly, to the observation that tourism research remains essentially a scholarship of practice failing to engage extensively with the theoretical foundations or the grand theoretical narratives of its source disciplines (Wood, 2017). We have virtually no extensive analyses of tourism phenomena couched in, for example, Marxist, Weberian, Durkheimian, Freudian, Keynesian or monetarist terms. In a review of the political economy of tourism development, Bianchi (2018) felt confident to devote around one of 15 pages to Marxist perspectives. Coles *et al.* (2006, pp. 295-297) illustrate well the aversion to grand theoretical narratives in tourism when they pose, but deliberately do not answer, the question as to 'whether theory has to be the ultimate goal of all knowledge producers', outlining an argument in favour of 'post-disciplinary perspectives' which, we are told, allow 'more flexible and creative approaches to investigating and defining objects through an insistence on overcoming the intellectual inhibitions associated with disciplinary parochialism'. While such an approach may appear refreshing it is, in point of fact, a piece of intellectual misdirection. What we require in tourism – and elsewhere – is a pre-, not post-disciplinary approach to research that engages with the (global) history of ideas.

Weaknesses in academic tourism's analysis of neo-liberalism does not, of course, preclude the role and influence of neo-liberalism on the 'real world' phenomenon of tourism. Even here, however, the research literature still indulges claims for the impacts of neo-liberalism on tourism that take the form, and articulate the content, of those already encountered earlier in this discussion with regard to higher education. Thus, Mosedale (2016b, p. 169) makes a number of observations about the fluidity of ideas concerning neo-liberalism (though not, interestingly, on the similar fluidity of 'original' neo-liberal thought) which leads him to assert that:

> neo-liberalism has no fixed definition and is not a neutral term. Instead, it is discursively (re) constructed in the policy arena, the media and through political, economic and social practices. Language is thus instrumental in creating, controlling and disseminating certain narratives and discursive positions and in (re)producing the structural inequalities of neo-liberal projects.

Sociologically, this is highly sensible but fluidity is not the same as amorphousness and, as with many other discussions of the subject, such permissive observations effectively allow a range of phenomena to be described in terms of, or be attributed to, neo-liberalism without the benefit of meaningfully circumscribing that philosophy. It is an example of a well-established if unfortunate trend in some branches of academic social science towards 'humptydumptying' (Warburton, 2000) – a term alluding to the character of Humpty Dumpty in Lewis Carroll's *Through the Looking Glass*, a person who insisted that when he used a word it meant exactly what *he* chose it to mean. In short, we face here a similar problem to that identified earlier – if neo-liberalism has had such a corrosive effect on us all, then very precise theoretical and empirical demonstration of the fact is surely preferable to vague accounts couched in terms of an interpretive fluidity that degrades the phenomenon to little more than a leitmotif.

Concluding Remarks

The evolution of tourism as a subject within higher education coincides with a period during which the forces of neo-liberalism have allegedly visited various degrees of destruction upon the groves of academe. Yet tourism as an academic research pursuit has thrived in all its frequently messy diversity. In contrast, tourism's 'sister' area of hospitality has firmly pinned its disciplinary colours to the managerial mast, producing large quantities of 'quick win' research of highly variable quality and often doubtful utility (the exception to this is a small but growing group of scholars pursuing a 'hospitality studies' approach, e.g. Lashley, 2017; Rowson and Lashley, 2019).

While in contemporary societal terms it is the custom to laud the value of diversity, such diversity can be problematic for aspirant academic disciplines, not least when the boundaries of a subject remain ill-defined; where there exists a plethora of topics competing for analysis; where approaches to the study of these topics are wide-ranging; and where, underlying all of the foregoing, there is a tendency to valorize various forms of pragmatism over intellectual coherence and connectivity. Despite its manifest achievements, tourism research exhibits many of these weaknesses to varying degrees. It is, as noted earlier, for the most part absent of the connectivity to the wider history of foundational ideas in the social sciences that characterize mature disciplines. Moreover, establishing such connections does not seem to be an intellectual priority. Indeed, there appears to be a general satisfaction with tourism's atheoretical character and status as a 'scholarship of practice', characterized by a failure to resolve core questions of disciplinary scope and meaning (Leiper, 2008; Wood, 2017) and as McKercher and Prideaux (2014, p. 26; see also Mihalic and Aramberri, 2015) observe, a reliance on conceptual myths that:

> ...promote ideological beliefs that have little basis in fact, entrench flawed methodologies that produce 'good' but ultimately wrong findings... [.] Many have become elevated to the stature of received doctrines. The promulgation of some of these myths has also had a detrimental effect on industry, with many myths about special interest tourism resulting in the expensive pursuit of phantom markets.

In a simplistic way, the concerns of the authors quoted can be illustrated by revisiting the earlier example of sustainable tourism. While the published literature on this major specialization in tourism does contain critical material (see Bianchi, 2018), much of the extant oeuvre appears to be engaged in selling the idea of sustainability rather than systematically engaging with the notion that, in a Marxist sense, the various forms of capitalism can never be intrinsically sustainable. Further, there is an often implied, sometimes explicit and always spurious moral imperative underlying the wider sustainability agenda to the effect that, in the interests of a sustainable future, the developing world should forgo the economic and social benefits that the developed world has so long exploited and enjoyed.

Further to the observations of McKercher and Prideaux (2014), we may add that when disciplinary myths and/or doctrines prevail, they arguably expose a subject more readily to the exigencies of fashion and there is a greater possibility that bandwagons are created and ridden with little regard for their theoretical and conceptual roadworthiness. Such appears to be the case with neo-liberalism applied outside of its disciplines of economics and political economy. As Bianchi (2018) somewhat indirectly reminds us in the case of tourism, there is little about contemporary forms of capitalist economic exploitation in tourism or any other sector *not* foreshadowed in the works of 'classical' political economists – and certainly in the work of Marx. There is, thus, little need to invoke 'neo-liberalism' as a special case for understanding such forms, especially when argument relies on selective interpretation of a limited number of key ideas from a large and diverse pool then translated into a monolithic critical structure of almost parodic character. Academic tourism and academic tourism research have both thrived during a purported 'neo-liberal age' and show every sign of continuing to do so – provided, that is, their practitioners exorcise the subject's myths, address the need for theoretical development and resist the temptation to become dedicated followers of fashion.

References

Ayikoru, M., Tribe, J. and Airey, D. (2009) Reading tourism education: Neoliberalism unveiled. *Annals of Tourism Research* 36(2), 191–221.

Baum, T. (2012) Picking o'er the bones: A postmortem on hospitality management education (DoB 1893). *Proceedings of the CAUTHE Conference 2012: The New Golden Age of Tourism and Hospitality*, Melbourne, Australia, pp. 53–56.

Bianchi, R. (2018) The political economy of tourism development: a critical review. *Annals of Tourism Research* 70, 88–102. DOI: 10.1016/j.annals.2017.08.005.

Brandist, C. (2017) The *perestroika* of academic labour: The neoliberal transformation of higher education and the resurrection of the 'command economy'. *Ephemera: Theory & Politics in Organization* 17(3), 583–608.

Buades, J. (2010) The role of tourism corporations: the neoliberal attack on local and global democracy. Available at: https://www.tourism-watch.de/en/content/role-tourism-corporations-neoliberal-attack-local-and-global-democracy (accessed 26 April 2018).

Burns, P.M. (2008) Tourism, political discourse, and post-colonialism. *Tourism and Hospitality Planning & Development* 5(1), 61–71. DOI: 10.1080/14790530801936502.

Callinicos, A. (1990) *Against Postmodernism: A Marxist Critique*. Polity Press, Cambridge, UK.

Castañeda, Q. (2012) The neoliberal imperative of tourism: rights and legitimization in the UNWTO global code of ethics for tourism. *Practicing Anthropology* 34(3), 47–51. DOI: 10.17730/praa.34.3.w0251w655647750j.

Cohen, E. and Cohen, S.A. (2012) Current sociological theories and issues in tourism. *Annals of Tourism Research* 39(4), 2177–2202. DOI: 10.1016/j.annals.2012.07.009.

Coles, T., Hall, C.M. and Duval, D.T. (2006) Tourism and post-disciplinary enquiry. *Current Issues in Tourism* 9(4-5), 293–319. DOI: 10.2167/cit327.0.

Derrida, J. and Dufourmantelle, A. (2000) *Of Hospitality*. Stanford University Press, Palo Alto, California.

Dimitriou, C.K. (2017) The quest for a practical approach to morality and the tourism industry. *Journal of Hospitality and Tourism Management* 31, 45–51. DOI: 10.1016/j.jhtm.2016.08.001.

Duffy, R. (2015) Nature-based tourism and neoliberalism: concealing contradictions. *Tourism Geographies* 17(4), 529–543. DOI: 10.1080/14616688.2015.1053972.

Fletcher, T., Carnicelli, S., Lawrence, S. and Snape, R. (2017) Reclaiming the 'L' word: leisure studies and UK higher education in neoliberal times. *Leisure Studies* 36(2), 293–304. DOI: 10.1080/02614367.2016.1261182.

Franklin, A. and Crang, M. (2001) The trouble with tourism and travel theory? *Tourist Studies* 1(1), 5–22. DOI: 10.1177/146879760100100101.

Gane, N. (2014) Sociology and Neoliberalism: a missing history. *Sociology* 48(6), 1092–1106. DOI: 10.1177/0038038513512728.

Geppert, M. and Hollinshead, G. (2017) Signs of dystopia and demoralization in global academia: reflections on the precarious and destructive effects of the colonization of the Lebenswelt. *Critical Perspectives on International Business* 13(2), 136–150.

Halsey, A.H. (1995) *Decline of Donnish Dominion: The British Academic Professions in the Twentieth Century*. Clarendon Press, Oxford, UK.

Hayes, D. and Wynyard, R. (eds) (2002) *The McDonaldization of Higher Education*. Praeger, New York.

Hayes, D. (ed.) (2017) *Beyond McDonaldization: Visions of Higher Education*. Routledge, London.

Higgins-Desbiolles, F. (2006) More than an "industry": The forgotten power of tourism as a social force. *Tourism Management* 27(6), 1192–1208. DOI: 10.1016/j.tourman.2005.05.020.

Lacey, A. and Ilcan, S. (2015) Tourism for development and the new global aid regime. *Global Social Policy: An Interdisciplinary Journal of Public Policy and Social Development* 15(1), 40–60. DOI: 10.1177/1468018114523633.

Lashley, C. (ed.) (2017) *The Routledge Handbook of Hospitality Studies*. Routledge, London.

Lehrer, T. (n.d., circa 1959) Bright College Days. Available at: http://lyricsfreak.com/t/tom+lehrer/bright+college+days_20138381.html (accessed 5 April 2018).

Leiper, N. (2008) Why 'the tourism industry' is misleading as a generic expression: The case for the plural variation, 'tourism industries'. *Tourism Management* 29(2), 237–251. DOI: 10.1016/j.tourman.2007.03.015.

Marginson, S. (2006) *Hayekian neo-liberalism and academic freedom: keynote address to Philosophy of Education Society of Australia Women's College, University of Sydney*. Centre for the Study of Higher Education, University of Melbourne, Melbourne, Australia.

McKercher, B. and Prideaux, B. (2014) Academic myths of tourism. *Annals of Tourism Research* 46, 16–28. DOI: 10.1016/j.annals.2014.02.003.

Mihalic, T. and Aramberri, J. (2015) Myths of top tourism countries, tourism contribution and competitiveness. *Tourism Review* 70(4), 276–288. DOI: 10.1108/TR-08-2014-0048.

Miller, B. (2014) Free to manage? A neo-liberal defence of academic freedom in British higher education. *Journal of Higher Education Policy and Management* 36(2), 143–154. DOI: 10.1080/1360080X.2013.861055.

Mosedale, J. (2016a) Neoliberalism and the political economy of tourism: projects, discourses and practices. In: Mosedale, J. (ed.) *Neoliberalism and the Political Economy of Tourism*. Routledge, Abingdon, UK, pp. 1–20.

Mosedale, J. (2016b) Tourism and neoliberalism: states, the economy and society. In: Mosedale, J. (ed.) *Neoliberalism and the Political Economy of Tourism*. Routledge, Abingdon, UK, pp. 157–161.

Nowicka, P. (2007) *The No-Nonsense Guide to Tourism*. New Internationalist Publications, Oxford, UK.

Palmer, C. (1994) Tourism and colonialism: the experience of the Bahamas. *Annals of Tourism Research* 21(4), 792–812.

Pavlović, D. and Knežević, M. (2017) Is contemporary tourism only a neoliberal manipulation? *Acta Economica Et Turistica* 3(1), 59–66. DOI: 10.1515/aet-2017-0007.

Radice, H. (2013) How we got here: UK higher education under neoliberalism'. *ACME: An International E-Journal for Critical Geographies* 12(3), 407–418.

Reyes, J. (2011) The roles of sustainable tourism in neoliberal policies and poverty reduction strategies: do they adequately address quality of life? *The Applied Anthropologist* 31(1), 23–29.

Ritzer, G. (1993) *The McDonaldization of Society*. Pine Forge Press, Thousand Oaks, California.

Ritzer, G. (2013) MOOCs and the McDonaldization of education. Available at: https://georgeritzer.wordpress.com/2013/01/12/moocs-and-the-mcdonaldization-of-education/ (accessed 24 February 2018).

Robinson, S., Ratle, O. and Bristow, A. (2017) Labour pains: Starting a career within the neoliberal university. *Ephemera: Theory & Politics in Organization* 17(3), 481–508.

Rowson, B. and Lashley, C. (eds) (2019) *Experiencing Hospitality*. Nova Science Publishers, New York.

Saunders, D.B. (2010) Neoliberal ideology and public higher education in the United States. *Journal for Critical Education Policy Studies* 8(1), 42–77.

Silk, M., Caudwell, J. and Gibson, H. (2017) Views on leisure studies: pasts, presents & future possibilities? *Leisure Studies* 36(2), 153–162. DOI: 10.1080/02614367.2017.1290130.

Simmons, D.G. (2017) Neoliberal politics and the fate of tourism. In: Butler, R. and Suntikul, W. (eds) *Tourism and Political Change*. Goodfellow Publishers, Oxford, UK, pp. 9–24.

Sinclair-Maragh, G. and Gursoy, D. (2015) Imperialism and tourism: the case of developing island countries. *Annals of Tourism Research* 50, 143–158. DOI: 10.1016/j.annals.2014.12.001.

Sykes, C.J. (1988) *Profscam: Professors and the Demise of Higher Education*. St. Martin's Press, New York.

Sykes, C.J. (2016) *Fail U.: The False Promise of Higher Education*. St. Martin's Press, New York.

Tadajewski, M., Maclaran, P., Parsons, E. and Parker, M. (2011) *Key Concepts in Critical Management Studies*. Sage, London.

Thomas, R. and Ormerod, N. (2017) The (almost) imperceptible impact of tourism research on policy and practice. *Tourism Management* 62, 379–389. DOI: 10.1016/j.tourman.2017.02.009.

Urry, J. (1992) *The Tourist Gaze*. Sage, London.

Vorobjovas-Pinta, O. and Robards, B. (2017) The shared OASIS: an insider ethnographic account of a gay resort. *Tourist Studies* 4(4), 146879761668756–387. DOI: 10.1177/1468797616687561.

Warburton, N. (2000) *Thinking from A to Z*, 2nd edn. Routledge, London.

Willmott, H. (2011) Journal list fetishism and the perversion of scholarship: reactivity and the ABS list. *Organization* 18(4), 429–442. DOI: 10.1177/1350508411403532.

Wood, R.C. (2015a) 'Folk' understandings of quality in UK higher hospitality education. *Quality Assurance in Education* 23(4), 326–338. DOI: 10.1108/QAE-04-2015-0015.

Wood, R.C. (2015b) *Hospitality Management: A Brief Introduction*. Sage, London.

Wood, R.C. (2017) The unspoken question: a response to Thomas and Ormerod. *Tourism Management* 62, 390–393. DOI: 10.1016/j.tourman.2017.02.008.

18 Human Mobility and Tourism Development: A Complex Knot of Enfolded Mobilities

Allan M. Williams*
University of Surrey, Guildford, UK

Introduction

The relationships between development and tourism flows have long been researched and have constituted one of the founding pillars of much of the earlier research in tourism. Quite simply, tourism flows convert into expenditure which, via various multiplier effects, result in output, employment and income changes in the destination economy. Over the years, this deceptively simple model has been further elaborated; at the same time, there has also been increasing awareness that other forms of human mobility are entwined with these tourism-development relationships, transforming the nature of this nexus. While there are many strands to the relationship between tourism in particular and other forms of mobility in general, this chapter focuses specifically on just one of these, namely migration. The diversity of tourism–migration relationships has been increasingly recognized in recent years (Williams and Hall, 2000) and as spanning not only production-led (labour and entrepreneurial) but also consumption-led migration which is sharply differentiated across the life cycle. This reframing of our analytical lens has also been paralleled by a shift from a focus on GDP and employment outcomes to attempts to understand how the migration–tourism nexus shapes productivity, innovation and competitiveness, as well as the broader concepts of well-being or quality of life.

This chapter argues for the need to understand tourism and migration flows as being enfolded together through space and time (Williams, 2009). That is, individual mobilities are enfolded over time, with earlier migration experiences influencing later patterns of tourism and vice versa. For example, at the individual level, today's tourists are often tomorrow's labour or lifestyle migrants, while today's student, or labour or lifestyle migrants are tomorrow's tourists. All enfolded individual mobilities spatially transfer and create resources and demands which, in turn, contribute to challenging, remaking or reinforcing uneven regional development. But their key feature is that they also generate additional, cumulative mobilities, notably at the meso or network level. This is because mobilities spatially reconfigure networks of friends and relatives, potentially generating diverse multi-directional human flows which range from short sojourns to various forms of shorter- or longer-term migration. This process of cumulative mobility generates further shifts in resources and demand, contributing to dynamic economic changes, but also providing the

*allan.williams@surrey.ac.uk

 Tourism in Development: Reflective Essays (eds P.U.C. Dieke *et al.*)
DOI: 10.1079/9781789242812.0018

new baseline for further rounds of inter-related mobilities (for example, Rodriguez *et al.*, 2001).

Understanding the enfolded natures of these mobilities provides insights not only into the broader range of economic impacts of tourism, but also into the relationship between structure and agency (Giddens, 1984). In turn, that takes us to the notion of the path-dependent versus the path-creation (Martin and Sunley, 2006) nature of migration-related tourism development trajectories. The question to be addressed, then, is how and to what extent is the future outline of tourism development already mapped out by the histories of enfolded individual mobilities and networks? This is a question with profound implications for the scope and realistic ambitions of tourism development policies.

Tourism, Migration and Development

Williams and Hall (2002) probably still provide the most comprehensive review of the relationship between migration and tourism. They identify a variety of links between tourism and migration which Etzo *et al.* (2014) helpfully condense into three main causal mechanisms: (i) the tourism-led migration hypothesis, whereby tourism leads to migration; (ii) the inverse causal relationship, whereby migration generates tourism via visits to friends and relatives; and (iii), bi-directional causality as the two types of mobility dynamically interact and mutually reinforce each other.

Not surprisingly, the second of these mechanisms has received particular attention from tourism researchers, primarily because it links directly to the tradition of analysing the segmentation of tourism demand and the resulting economic outcomes. Most of the earlier studies within this strand of research modelled the relationships between aggregate migration and tourism data, differentiating between the impacts of migration on both inbound and outbound tourism. The initial focus was on the most obvious and most direct impacts through the generation of visiting friends and relatives (VFR) tourism, both to the migration destination and in the form of return visits. For example, Seetaram and Dwyer (2009) analysed the impacts of migration on inbound tourism demand in the migrant destination, while Seetaram (2012) analysed the reverse flow; that is, how return visits by the migrants generated tourism in the countries of migrant origin. Later studies within this body of research broadened the analytical focus from VFR tourism in order to consider how migration generated other forms of tourism, notably business and vacation tourism (Prescott *et al.*, 2005; Etzo *et al.*, 2014). Interestingly, but unsurprisingly, these different demand segments exhibited differing levels of resilience to external shocks, with VFR tourism demonstrating the highest level of stability, reinforcing its importance as a stable core element of tourism demand and a pillar of resistance in the face of most types of demand shock.

While data limitations make it impossible to estimate precisely the total impact of migration on tourism demand, the UNWTO (2010) has estimated that between 15% and 25% of total tourism flows are migration-related. However, any such estimates must be treated with considerable caution, especially due to the increasing diversification of migration flows (King, 2012) and the blurring of tourism and migration flows along a continuum (Williams and Hall, 2002). Temporary migration and circular migration on the one hand, and long-term and residential tourism on the other, constitute powerful obstacles to any attempt to dichotomize tourism and migration, and to model their impacts.

There is a separate and contrasting strand of research on how tourism impacts on migration, led more by sociological than by economic perspectives. Here, the primary links centre on consumption and, especially, on how tourism informs and shapes what has become known as lifestyle migration (Benson and O'Reilly, 2009). This bridges the worlds of work (consumption-led migration by the economically active) and post-work (retirement migration). At the heart of these flows can be found a complex set of motivations which include fleeing from critical life events (divorce, death, bankruptcy and so on), searching for authenticity and more meaningful lives, reshaping identities and the search for different places to reside for more banal reasons. The latter include the allure, typically, of warmer climates that allow more out-of-doors lifestyles, or the perceived attractions of rural as opposed to urban lifestyles to which are attached different notions of community (King *et al.*, 2000).

Tourism comes into play in lifestyle migration in the search for idealized, sometimes romanticized, places to enact these motivations in the form of new, place-specific practices. In reality, lifestyle migration flows are informed by contrasting and sometimes conflicting shifting emphases on the cultural, environmental and cost-of-living characteristics of the destinations. The searches for these idealized places are rarely comprehensive – in the face of information overload (Lee and Lee, 2004) – and almost never start from a blank piece of paper or totally open search agenda. Rather they are informed by a mixture of codified and tacit knowledge (for example, novels, guidebooks to 'expatriate' living versus lifetime experiences, and the person-to person exchange of knowledge). Previous tourism experiences – which implicitly are learning experiences to varying extents – often figure large in defining the resulting 'search spaces' (King *et al.*, 2000; Williams *et al.*, 2004a). In this way, tourism generates economic consequences across time and space as tourism destinations subsequently become individual migration destinations and the impacts of tourist expenditures are replaced by more substantial and sustained individual expenditures. The impact on particular places can be transformational and, for example, has been recognized as a factor that can transform the nature of the tourism life cycle model (Foster and Murphy, 1991), with tourism-related migration extending and reshaping the initial tourism product life cycle (Butler, 1980). Yet surprisingly, few attempts have been made to quantify the economic outcomes of these tourism-led migration relationships, in contrast to the migration-led tourism causal mechanism.

The third set of causal mechanisms focuses on the long-term cumulative mutual interrelationships between tourism and migration. These have been mapped out in an idealized form in Williams and Hall (2002): tourism experiences feed into lifestyle migration, and this may generate labour migration to provide the services demanded by the migrants. But there is also the generation of VFR tourism by the initial lifestyle migrants and by the labour migrants. In turn, these may lead to a new round of migration by the visiting friends and relatives, which in due course may spur further VFR and migration movements. The cumulative effects, in terms of mobility and economic outcomes, can be very substantial, and also demonstrate how initial tourism flows can reverberate through time, with transformational economic consequences.

Sticky Flows, Scapes and Path Dependency

Tourism and migration flows are not mapped out on a blank space but, instead, are part of a continuous process of creating structures which shape subsequent flows. A bridge can be constructed here to the notions of scapes and flows, on which John Urry's (2012) work is particularly relevant. He understands scapes as being constituted of networks of technologies, organizations and actors which represent a set of interconnected nodes through which flows are channelled. Winter (2007, p. 29) comments that 'Far from being mutually exclusive, scapes and flows continually intersect, with the former providing a partial structure – a global order – within which fluidity can emerge and thrive'. Urry emphasizes the multi-dimensionality of these flows involving human mobility, knowledge, images and finance, but here we focus on the first of these, and specifically on the tourism–migration nexus. As Williams *et al.* (2004a, b) argue, while on the one hand it is political–economic conditions which shape the broad outlines of flows, they are given final form by the combination of social networks and human agency.

In the case of human mobility, the scapes are structured by the mundane but fundamental materialities of tourism; that is, by transport infrastructures, hotels and restaurants. These scapes are perhaps symbolized by the icons of different transport paradigms, such as roadside hostelries and inns, railway hotels, motels and diners, and airport hotels and food courts. But these are more than physical spaces and infrastructures, or even markers for flows, because, collectively, they also generate agglomeration economies which enable and channel flows (Shaw and Williams, 2004). Less visibly but no less important, scapes also become largely synonymous with the search spaces of mobile individuals who are seeking destinations. They are spaces that users have knowledge of and where, therefore, the risks and uncertainties of mobility are reduced (Williams and Baláž, 2014). They are not immutable and, indeed, they channel rather than confine flows, but sunk costs and the difficulties of unlearning lend them considerable durability. But, having been formed by flows, scapes are also ultimately undermined and transformed by changes in flows.

It is only a short bridge from scapes and flows to the evolutionary economics concepts of path dependency and path creation, which also brings us closer to economic outcomes. Hodgson (1996) provides a simple definition of path dependence, writing that '... the future development of an economic system is affected by the path it has traced out in the past'. This has sometime been misconceived as signifying that outcomes are preordained or inevitable (North, 1990), but it does imply that some outcomes are more likely than others. In a later formulation, Nielsen *et al.* (1995) proposed the alternative concept of 'path-dependent path-creation', which resonates with the notion of how scapes and flows are mutually informing. While these parallel but not identical concepts are helpful in framing the relationship between tourism and migration, in practice it is highly challenging to disentangle the outcomes of this nexus. As Kesselring (2008, p. 81) contends, 'in most cases it is impossible to distinguish between the autonomous moves of individuals and the structural impacts of societal and professional constraints within mobility decisions'. Impossible to distinguish, perhaps, but that does not deny their prevalence.

These ideas can be further developed by drawing on the notion of enfolded mobilities (Williams, 2009), especially the reasons why prior mobilities influence subsequent mobilities. At the heart of these links lie knowledge and risk. Travel is a source of learning and may result in the acquisition of two types of tacit knowledge. The first of these is knowledge about mobility; some of the competences needed to manage mobility (self-confidence, learning, communication and so on) are transferable. In learning to be a migrant, one also learns some essential competences for being a tourist and vice versa. Of course, there are also different competences involved and migration in particular requires a range of additional skills to adapt and integrate. In practice, these competences differ according to place, whether for example the place of temporary or permanent residence is in a migrant or ex-patriate enclave, or whether it involves immersion in the larger local society (King *et al.*, 2000). The second type of tacit knowledge is that of specific places (of its culture, customs, regulations, resources and spatiality) and it is this type of tacit knowledge which informs the search spaces discussed earlier.

Taken together, these two forms of tacit knowledge can significantly change the willingness to take risks in relation to mobility. In essence, the willingness to take risks in making decisions about mobility is the outcome of a combination of risk tolerance/aversion and competence to manage risks, and this applies as much to tourism as to migration (Williams and Baláž, 2013, 2014). The competence to manage risks is of course perceived, and may be infused with over-optimism and over-confidence or, indeed, by over-pessimism and under-confidence, but it clearly does influence mobility. However, tacit knowledge is not only obtained via personal travel. There are also what may be termed *knowledge spillover effects* – that is, indirect spillovers from the mobility experiences of others (mostly friends and relatives, but also occasional or casual acquaintances). To some extent, this is a matter of who is trusted as a source of reliable and useful knowledge, but not all person-to-person knowledge is sourced from long-established friendships and familial relationships. There are also times when casual acquaintances become sources of such knowledge, particularly when they are recognized as experts. This is illustrated by the experiences of backpackers intersecting at different nodes (often backpacker hostels) and exchanging up-to-date, experience-based tacit knowledge of particular places (Tsaur *et al.*, 2010). Of course, there are also multiple sources of knowledge to be drawn on, especially in the internet age, when social media platforms provide opportunities to convert their personal tacit knowledge into codified knowledge via blogs, Facebook pages and other platforms. Inevitably, decision making in respect of mobility is made on the basis of the blending of different types of knowledge but, in most cases, the objective is always to shift the boundary between (total) knowledge and risk.

While the relationship between knowledge and risk lies at the heart of the enfoldment of different types of mobility and how this impacts on tourism and migration, enfoldment also assumes other forms (Williams, 2009). This includes 'contingent mobilities' such as, for example, the generation of labour migration to fill the jobs created by the demand generated by an earlier phase of mobility, whether for migration or tourism. Such mobility may take many forms, but one of the most significant is the generation of labour migration to fill jobs in tourism destinations (Williams and Hall, 2000) while reference must also be made to Sassen's (1991) dual- labour migration thesis, whereby skilled migration to global cities generates demand for unskilled migration to provide essential services, especially in hospitality. There are also collective mobility links whereby dependants move with the lead mobile persons, and this applies as much to tourism as to migration. These represent forms of gendered social multipliers, as the earlier migration experiences of individuals generate consequences for family members.

Finally, as previously noted, VFR is also an important part of the 'social glue' that binds mobilities together. Typically, this takes the form of migration leading to subsequent VFR travel which can flow in both directions between the origin and destination. This can significantly add to the personal tacit knowledge of the individuals involved, both of how to manage mobility generally and of particular places, leading to further rounds of migration. The importance and persistence of VFR lies in the way the practices of mobility are deeply imbued with both meaning and instrumental purposes. At one level, VFR is about delivering different forms of care and support, ranging from the practicalities of childminding to the emotional (Janta *et al.*, 2015). But it also plays a critical role in reaffirming relationships, as well as in shaping identities (Humbracht, 2015). These visits, particularly the return visits to places of origin, can also contribute to the notion that migrations are temporary aberrations from the 'normal pattern of residence'. This, in turn, contributes to both return and the myth of return.

In summary, then, a combination of structural factors and enfolded mobility serve to shape the trajectory of tourism in a way which is best described as being path-dependent path-creating. Structure and agency interact in ways which are partly predictable but also dynamic, in flux and variable. This has significant consequences not only for understanding the flows of tourism but also, as discussed in the next section, their economic consequences.

Human Mobility and Economic Consequences

The inter-relationships between tourism and migration have a number of major impacts on tourism development. The direct impacts in terms of the volumes of tourism are relatively straight forward: changes in demand result in production shifts, with income and employment consequences (Schubert *et al.*, 2011). In the longer term, this leads to investment and innovation as firms respond to competitive pressures, shifts in demand and technology. However, the impacts of enfolded mobilities extend beyond this, as they also result in the transfer of knowledge, skills and capital, with implications for innovation, productivity and competitiveness.

The direct economic consequences of tourism are seen in the effects of tourist expenditure. These were traditionally measured via multiplier estimations which also captured indirect effects, but increasingly they have come to be estimated through the application of general equilibrium models which incorporate more dynamic effects (Blake, 2009). There are two keys to understanding these impacts. The first is the structure of the local economy and, in particular, the extent to which the economic benefits are captured endogenously within the destination or whether – through supply chains or ownership effects – they are partly 'lost' in the form of, say, imports or profit remittances. There may also be migrant remittances and their level will partly have been determined by the migration–tourism nexus outlined earlier in this chapter.

The second important set of determinants is the nature of the tourism; it is a question not only of numbers but also of duration of stay and per capita spending. This is exemplified by the analysis of the economic effects of VFR tourism. As noted earlier, this can flow in both directions and can generate out-movements as well as in-movements of tourists, with associated redistributions of expenditure. However, most of the research to date has focused primarily on estimating the economic consequences of the inflow of VFR rather than the outflows. Nevertheless, it is now acknowledged that much of the earlier research in this area tended to underestimate the economic impacts of VFR, partly because of the assumption that staying with friends or relatives significantly reduced tourism expenditure. While this is true in terms of hospitality services, at the same time some VFR involves staying and eating in commercial hospitality establishments, while visitors may also indulge in various forms of tourism practices which generate spending at the destination. Moreover, an earlier focus on tourist numbers tended to underestimate the impact of the longer average length of stay of VFR versus leisure tourists (Dwyer *et al.*, 2014). In addition, VFR tourism (because of how it embeds care and other social obligations) is likely to be less volatile than leisure tourism, which is more likely to be susceptible to transformations in price exchange rates, fashion and other factors.

Of course, the economic consequences of the tourism–migration nexus extend beyond the impacts of VFR tourism expenditure shifts. Where tourism generates migration, then there are also implications in terms of the transfer of knowledge and skills. Conceptually, this can be understood in terms of a rejection of Castells's (1989) notion of 'the space of flows' which focuses on the flows, particularly in the context of digitally driven globalization, and fails sufficiently to acknowledge what geographers refer to as 'the inherent "stickiness" of the world' (Latham, 2002). This leads to an alternative conceptualization advanced by Massey (1994) in which places are seen to constitute a combination of both local and spatially stretched relationships. In contrast to Castells's (1989) approach, this means that social relationships are locked into and have specific meanings in particular places (Allen *et al.*, 1998). Place matters and, by extension, there is a need to focus on how flows of tourism-related migration lock knowledge, skills and capital in particular places, with implications for economic development.

This is perhaps best understood in terms of the transfer of knowledge and how this feeds into innovation. Much of the extant generic literature on this topic focuses on foreign direct investment, research and development, and patents, reflecting the dominant manufacturing-led development paradigm. However, as Hudson (1999, p. 61) contends, know-how 'is only partially commodifiable. It therefore can only be purchased, if at all, via the labour market as embodied knowledge and not in the form of patents, turn-key plant or other forms of "hard technology"'. Migrants are one important means to effect the transfer of knowledge and skills which can feed into the process of innovation in three main ways (Williams and Shaw, 2011).

First, migration adds to the supply of labour and, irrespective of skill differences between migrants and the indigenous labour force, this can serve to lower the costs of labour. On its own, that would seem to lead to the conclusion that immigration would reduce innovation, because the reduction in labour costs would be a disincentive to investment in capital-intensive technologies. Second, however, the argument changes if migrants have higher general levels of skill than non-migrant workers, or if they are employed in posts for which they are over-educated (Chiswick and Miller, 2009). In that case, they can enhance the volume of knowledge available to firms. In other words, this represents a quantitative shift in the availability of knowledge. In contrast, the third argument is that international migration is a source of new or different knowledge. In other words, it leads to a qualitative change in the knowledge base of a place. From this perspective, migration constitutes a potentially significant mechanism for the realization of knowledge spillovers. The resulting greater diversity in the knowledge base contributes to creativity (Brown and Duguid, 1991) and innovation.

Paraskevopoulou *et al.* (2012) provide one of the very few analyses which have sought to identify the relationship between migration and innovation in tourism. Their study of small London hotels found that, at first sight, there was little difference in the innovativeness of managers who were born in, rather than outside, the UK. However, when the UK-born were disaggregated into those who had and who had not had international work experience, then a clearer pattern emerged. Those who had worked abroad, whether UK- or non-UK born, were likely to be more consistently innovative than those who had only worked in the UK. While these findings need to be further explored in other settings – not least because London represents a highly internationalized labour market and tourism destination – they do provide initial support for the argument that international mobility does contribute to greater innovation capacity.

Yet, while there does appear to be a positive story to tell about the role of migration in tourism innovation, there is also a different story to tell, especially when taking into account the segmented nature of labour markets, that is, the existence of a dual labour market (Piore, 1979; Joppe, 2012). As the Global Commission on International Migration (GCIM, 2005) states:

> In some parts of the world, certain sectors of the economy, including agriculture, construction, hotel and catering services, as well as domestic and sex work, have come to rely to a significant extent on migrants with irregular status, who are prepared to work in difficult, dangerous and dirty jobs with little security and low wages.

Matthews and Ruhs (2007) provide an insightful study of how this plays out in the restaurant sector in the UK, with there being a sharp division of labour between front-of-house and back-of-house operations. The backstage operations, such as dishwashing, were far more likely to be populated by irregular migrants, with little scope for using the full extent of their human capital or to contribute to innovation.

The contribution of migrants to productivity in tourism is the outcome of their multiple roles; they are sometimes replacement labour taking jobs that indigenous workers shun, and they are also bearers of higher levels or different types of knowledge that can enhance the human capital of a firm. In either case, although via different mechanisms, this means that they can potentially increase the productivity of the individual tourism firm and the tourism sector. The evidence on this is somewhat patchy and uneven but both studies of individual hotels (Yaduma *et al.*, 2015) and of the sector as a whole (Rolfe *et al.*, 2013) have found evidence to support the argument that the employment of international migrant workers has positive effects on tourism productivity.

Conclusions

This chapter has advanced two main arguments: first, that tourism and migration are strongly interrelated, and that this is due to both their shared structural determinants (for example, transport infrastructures) in the form of scapes, and to the enfolded nature of mobility. Both tourism and migration are means to accumulate knowledge, whether the competence to manage mobility or knowledge of particular places that subsequently shape their search spaces. This is underpinned by the resultant shifting of boundaries between uncertainty and knowledge. Hence, previous migration experiences may channel future tourism, or tourism experiences may channel future migration. In both cases, these inter-relationships are further intensified by complex knots of visiting friends and relations practices and by the dynamic generation of further rounds of tourism and migration. For example, when the migrant attracts sojourns from friends and relatives, the latter's newly acquired knowledge of the destination subsequently feeds into their migration decision. This introduces an element of path dependency into tourism in so far as today's migration flows contribute to shaping future tourism flows and vice versa. But while some pathways of development become more likely than others as a consequence, there is still a role for human agency to interact with structures in a path-dependent path-creation framework.

Second, the chapter has argued that there are two main economic development outcomes of the migration–tourism nexus. The first is seen in the roles of individuals as consumers, that is, as tourists. Tourism expenditures convert into income and jobs so that migration flows have long-term implications for future tourism expenditure patterns. VFR patterns can also contribute to the reallocation of expenditure patterns. VFR return visits to the country of origin are likely to result in net reductions in expenditure in the migration destination while inward VFR visits to the migrants contribute additional expenditure effects in the migrant destination. In other words, through their consumption (expenditures), they contribute to aggregate demand impacts on production. And secondly, migration plays an important role in the territorial redistribution of resources; migrants are bearers of knowledge, skills, competences and capital. As such, they contribute to the process of uneven regional development (Williams *et al.*, 2004a). Hence, tourism-led migration becomes a component of the larger picture of the relationship between human capital and development. This takes different forms; migrant workers may be filling gaps in labour markets or they may be transforming these via the transfer of different types of skills or knowledge. In different ways, all of these can contribute to innovation and productivity and to the relationship between them.

References

Allen, J., Cochrane, A. and Massey, D. (1998) *Rethinking the Region*. Routledge, London.

Benson, M. and O'Reilly, K. (2009) Migration and the search for a better way of life: a critical exploration of lifestyle migration. *The Sociological Review* 57(4), 608–625. DOI: 10.1111/j.1467-954X.2009.01864.x.

Blake, A. (2009) The dynamics of tourism's economic impact. *Tourism Economics* 15(3), 615–628. DOI: 10.5367/000000009789036576.

Brown, J.S. and Duguid, P. (1991) Organizational learning and communities-of-practice: toward a unified view of working, learning, and innovation. *Organization Science* 2(1), 40–57. DOI: 10.1287/orsc.2.1.40.

Butler, R.W. (1980) The concept of a tourist area cycle of evolution: implications for management of resources. *The Canadian Geographer/Le Géographe canadien* 24(1), 5–12. DOI: 10.1111/j.1541-0064.1980.tb00970.x.

Castells, M. (1989) *The Informational City: Information Technology, Economic Restructuring, and the Urban–Regional Process*. Blackwell, Oxford, UK.

Chiswick, B.R. and Miller, P.W. (2009) The international transferability of immigrants' human capital. *Economics of Education Review* 28(2), 162–169. DOI: 10.1016/j.econedurev.2008.07.002.

Dwyer, L., Seetaram, N., Forsyth, P. and King, B. (2014) Is the migration-tourism relationship only about VFR? *Annals of Tourism Research* 46, 130–143. DOI: 10.1016/j.annals.2014.03.009.

Etzo, I., Massidda, C. and Piras, R. (2014) Migration and outbound tourism: evidence from Italy. *Annals of Tourism Research* 48, 235–249. DOI: 10.1016/j.annals.2014.07.002.

Foster, D.M. and Murphy, P. (1991) Resort cycle revisited: the retirement connection. *Annals of Tourism Research* 18(4), 553–567.

GCIM (2005) *Migration in an Interconnected World: New Directions for Action*. Global Commission on International Migration, Geneva, Switzerland.

Giddens, A. (1984) *The Constitution of Society: Outline of the Theory of Structuration*. University of California Press, Berkeley and Los Angeles, California.

Hodgson, G.M. (1996) *Economics and Evolution: Bringing Life Back into Economics*. University of Michigan Press, Ann Arbor, Michigan.

Hudson, R. (1999) The learning economy, the learning firm, and the learning region: a sympathetic critique of the limits to learning. *European Urban and Regional Studies* 6(1), 59–72.

Humbracht, M. (2015) Reimagining transnational relations: the embodied politics of visiting friends and relatives mobilities. *Population, Space and Place* 21(7), 640–653. DOI: 10.1002/psp.1918.

Janta, H., Cohen, S.A. and Williams, A.M. (2015) Rethinking visiting friends and relatives mobilities. *Population, Space and Place* 21(7), 585–598. DOI: 10.1002/psp.1914.

Joppe, M. (2012) Migrant workers: challenges and opportunities in addressing tourism labour shortages. *Tourism Management* 33(3), 662–671. DOI: 10.1016/j.tourman.2011.07.009.

Kesselring, S. (2008) The mobile risk society: mobility and ambivalence in the second modernity. In: Canzler, W., Kaufmann, V. and Kesselring, S.S. (eds) *Tracing Mobilities*. Ashgate, Aldershot, UK, pp. 77–102.

King, R. (2012) Geography and migration studies: retrospect and prospect. *Population, Space and Place* 18(2), 134–153. DOI: 10.1002/psp.685.

King, R., Warnes, T. and Williams, A. (2000) *Sunset Lives: British Retirement Migration to the Mediterranean*. Berg, Oxford, UK.

Latham, A. (2002) Retheorizing the scale of globalization: topologies, actor-networks, and cosmopolitanism. In: Herod, A. and Wright, M. (eds) *Geographies of Power: Placing Scale*. Blackwell, Oxford, UK.

Lee, B.-K. and Lee, W.-N. (2004) The effect of information overload on consumer choice quality in an on-line environment. *Psychology and Marketing* 21(3), 159–183. DOI: 10.1002/mar.20000.

Martin, R. and Sunley, P. (2006) Path dependence and regional economic evolution. *Journal of Economic Geography* 6(4), 395–437. DOI: 10.1093/jeg/lbl012.

Massey, D. (1994) *Space, Place and Gender*. Polity Press, Cambridge, UK.

Matthews, G. and Ruhs, M. (2007) *Are You Being Served? Employer Demand for Migrant Labour in the UK's Hospitality Sector*. University of Oxford, Centre on Migration, Policy and Society, Oxford, UK.

Nielsen, K., Jessop, B. and Hausner, J. (1995) Institutional change in post-socialism. In: Hausner, J., Jessop, B. and Nielsen, K.K. (eds) *Strategic Choice and Path Dependency in Post Socialism: Institutional Dynamics in the Transformation Process*. Edward Elgar, Aldershot, UK, pp. 3–44.

North, D.C. (1990) *Institutions, Institutional Change and Economic Performance*. Cambridge University Press, Cambridge, UK.

Paraskevopoulou, A., Markova, E., Williams, A. and Shaw, G. (2012) Migration and innovation at the bottom end: understanding the role of migrant managers in small hotels in the global city. *Mobilities* 7(3), 389–414. DOI: 10.1080/17450101.2012.662359.

Piore, M.J. (1979) *Birds of Passage: Migrant Labor and Industrial Societies*. Cambridge University Press, Cambridge, UK.

Prescott, D., Wilton, D., Dadayli, C. and Dickson, A. (2005) Travel to Canada: the role of Canada's immigrant populations. *Applied Economics* 37(6), 651–663. DOI: 10.1080/0003684042000329045.

Rodriguez, V., Salva-Tomas, P. and Williams, A.M. (2001) Northern Europeans and the Mediterranean: a new California or a new Florida. In: King, R., de Mas, P. and Beck, J.M. (eds) *Geography, Environment and Development in the Mediterranean*. Sussex Academic Press, Falmer, UK, pp. 176–195.

Rolfe, H., Rienzo, C., Lalani, M. and Portes, J. (2013) *Migration and Productivity: Employers' Practices, Public Attitudes and Statistical Evidence*. National Institute of Economic and Social Research, London.

Sassen, S. (1991) *The Global City*. Princeton University Press, New York, London, Tokyo, Princeton, New Jersey.

Schubert, S.F., Brida, J.G. and Risso, W.A. (2011) The impacts of international tourism demand on economic growth of small economies dependent on tourism. *Tourism Management* 32(2), 377–385. DOI: 10.1016/j.tourman.2010.03.007.

Seetaram, N. (2012) Estimating demand elasticities for Australia's international outbound tourism. *Tourism Economics* 18(5), 999–1017. DOI: 10.5367/te.2012.0161.

Seetaram, N. and Dwyer, L. (2009) Immigration and tourism demand in Australia: a panel data analysis. *Anatolia* 20(1), 212–222. DOI: 10.1080/13032917.2009.10518905.

Shaw, G. and Williams, A.M. (2004) *Tourism and Tourism Spaces*. Sage Publications, London.

Tsaur, S.-H., Yen, C.-H. and Chen, C.-L. (2010) Independent tourist knowledge and skills. *Annals of Tourism Research* 37(4), 1035–1054. DOI: 10.1016/j.annals.2010.04.001.

UNWTO (2010) *International Recommendations for Tourism Statistics 2008*. United Nations, Statistical Division Series M, N. 83, New York.

Urry, J. (2012) *Sociology Beyond Societies: Mobilities for the Twenty-first Century*. Routledge, London.

Williams, A.M. (2009) International migration, uneven regional development and polarization. *European Urban and Regional Studies* 16(3), 309–322. DOI: 10.1177/0969776409104695.

Williams, A.M. and Baláž, V. (2013) Tourism, risk tolerance and competences: travel organization and tourism hazards. *Tourism Management* 35, 209–221. DOI: 10.1016/j.tourman.2012.07.006.

Williams, A.M. and Baláž, V. (2014) Mobility, risk tolerance and competence to manage risks. *Journal of Risk Research* 17(8), 1061–1088. DOI: 10.1080/13669877.2013.841729.

Williams, A.M. and Hall, C.M. (2000) Tourism and migration: new relationships between production and consumption. *Tourism Geographies* 2(1), 5–27. DOI: 10.1080/146166800363420.

Williams, A.M. and Hall, C.M. (2002) Tourism, migration, circulation and mobility. In: Hall, C.M. and Williams, A.M. (eds) *Tourism and Migration*. Springer, Dordrecht, The Netherlands, pp. 1–52.

Williams, A.M. and Shaw, G. (2011) Internationalization and innovation in tourism. *Annals of Tourism Research* 38(1), 27–51. DOI: 10.1016/j.annals.2010.09.006.

Williams, A.M., Baláž, V. and Wallace, C. (2004a) International labour mobility and uneven regional development in Europe: human capital, knowledge and entrepreneurship. *European Urban and Regional Studies* 11(1), 27–46.

Williams, A.M., King, R. and Warnes, A. (2004b) British second homes in southern Europe: shifting nodes in the scapes and flows of migration and tourism. In: Hall, C.M. and Müller, D. (eds) *Tourism, Mobility and Second Homes: Between Elite Landscape and Common Ground*. Channel View Publications, Bristol, UK, pp. 97–112.

Winter, T. (2007) Rethinking tourism in Asia. *Annals of Tourism Research* 34(1), 27–44. DOI: 10.1016/j.annals.2006.06.004.

Yaduma, N., Williams, A., Lockwood, A. and Park, S. (2015) Performance, labour flexibility and migrant workers in hotels: an establishment and departmental level analysis. *International Journal of Hospitality Management* 50, 94–104. DOI: 10.1016/j.ijhm.2015.07.003.

19 Reflections on Trends and Issues in Global Tourism

Alastair M. Morrison*
University of Greenwich, Old Royal Naval College, London, UK

Prelude

What does the future promise for international tourism? Will tourism businesses survive amid the current plethora of global trends, issues and challenges that they face? This chapter addresses these questions by reflecting on the trends and issues that are buffeting international tourism in the early 2020s.

Sitting down to write this chapter in the first three months of 2020, the author had much more material at hand as a result of the COVID-19 (coronavirus) pandemic and its clutches on many parts of the world. The devastating impacts on global tourism were already apparent (UNWTO, 2020a; World Economic Forum, 2020) and solidified the notion that a more systematic way of dealing with trends, issues and challenges was badly needed. This chapter begins by supplying clear and concise definitions of challenges, trends and issues for global tourism, as well as for opportunities. It identifies 100 specific trends and issues and presents a classification of three groups of these (market, supply and external factors). This is followed by a brief discussion of six types of trends and issues. The chapter wraps up with reflections on challenges for international tourism using the 10 As and ADVICE models developed by the author. A brief summary concludes the chapter highlighting its major messages.

A Challenging Search

Pinpointing trends and issues in international tourism is challenging in itself as there are no books or comprehensive documentation sources for these data. The proverbial 'searching for a needle in a haystack' is a good metaphor for anyone trying to gather this information in order to look ahead. Also, the changes impacting tourism are so dynamic and often unpredictable that searching for trends and issues can be an endless, and is certainly an ongoing task (Fig. 19.1).

Making this challenge even more daunting are the imprecise and often conflicting definitions and meanings attached to trends, issues and challenges, as well as the lack of robust knowledge of the inter-relationships among the three terms. Also, the huge chasm that exists in tourism between academia and industry tends to create even greater uncertainty about meanings. Let's start, then, with providing clear definitions of the key concepts in this chapter.

*a.morrison@greenwich.ac.uk

 Tourism in Development: Reflective Essays (eds P.U.C. Dieke *et al.*)
DOI: 10.1079/9781789242812.0019

Fig. 19.1. Many ladders but no walls – the challenge in identifying tourism trends, issues and challenges.

(From Shutterstock, Inc.)

Definition of Terms

Definition of a challenge for international tourism

Oxford Learner's Dictionaries (2020) define a challenge as 'a new or difficult task that tests someone's ability and skill'. This can be interpreted for the current context as being *new or difficult tasks facing international tourism that test the sector's capabilities, skills and resources*. In this chapter, it is assumed that these difficult tasks are caused by trends and issues – that is, trends & issues → challenges.

Challenges are often cited with or juxtaposed against opportunities in the titles of tourism academic articles, books and book chapters, monographs and essays. Part of the well-known SWOT (strengths, weaknesses, opportunities, threats) acronym and strategic planning technique, opportunities are usually thought of as *positive outcomes presented by external factors that if appropriately acted upon can produce beneficial outcomes or performance for a tourism business or destination*. Opportunities are often positioned as the opposite of threats. Budeanu *et al.* (2016), for example, reviewed the challenges and opportunities for sustainable tourism; Sharpley (2001) investigated the challenges and opportunities for tourism in Cyprus.

Definition of a trend in international tourism

A trend is a new development or a change in international tourism. The trend line can be increasing or decreasing; the change can be positive, negative or neutral with respect to international tourism. Trends are either measurable (quantitative) or non-measurable (qualitative) and can also be a combination of the two. Trends are short-, medium- or long-term. The intensity of trends may vary by region of the world or even differ by country. For example, the growth of community-based tourism (CBT) is a trend in South-east Asia; however, it is not as prevalent in the USA. Marijuana tourism is a trend in North America; however, the trend is not as strong in more traditional Asian societies.

The author has identified 50 trends in international tourism, and these are shown in Table 19.1. This listing is not exhaustive but, rather, indicative of some of the trends within international tourism.

It is noteworthy that several of the trends in Table 19.1 are types or forms of tourism. Others relate to specific groups of travellers and the ways that people traverse the globe. Some are being 'pushed' by people's internal needs and motivations; others are 'pulled' by newer supply offerings.

Definition of an issue in international tourism

For international tourism, an issue *is an important topic or problem that people are discussing or talking or writing about*. The topic or problem remains unresolved and people have differing opinions on how to address it. Issues often have proponents (advocates) and opponents (critics). Issues are short-, medium- or long-term.

Table 19.2 is a list of 50 selected issues. Again, this is not a complete listing and more items could be added. However, it highlights the diversity of problems facing international tourism.

Many of these issues are resource related, including problems with and constraints of the environment and human resources. Some of the issues have been around for decades without

Table 19.1. 50 trends in international tourism.

• Adventure tourism	• LBGT (lesbian, gay, bisexual, transgender) travel
• Artificial intelligence (AI)	• (Usage of) LCCs (low-cost air carriers)
• Augmented and virtual reality (AR-VR)	• Medical and wellness tourism
• Bleisure	• Millennials and Gen Z
• Breakationing	• Multi-destination travel
• Bucket-list travel	• Multi-generational travel
• Casinos and gaming	• New accommodation types
• China and India outbound market growth	• Online travel booking
• Co-creation	• Quest for authentic experiences
• Creative industries and tourism	• Sharing economy
• Crowdsourcing and crowdfunding	• Slum or ghetto tourism
• Cruising growth	• Smart tourism
• Domestic tourism	• Smartphone and app use
• Driverless vehicles	• Social media use
• Ecotourism	• Solo tourism
• Event festivalization	• Space tourism
• Flashpacking	• Special-interest tourism
• Food/gastronomic/culinary tourism	• Staycationing
• Genealogy and roots tourism	• Super long-haul flights
• Geotourism	• User-generated content (UGC)
• Glamping	• Vegetarianism
• High-speed rail travel	• Voluntourism
• Independent travel	• Wildlife tourism
• Influencers	• Wine tourism
• Last-minute travel	• Worldwide tourism growth

(From the author)

any resolution (e.g. sex tourism), while others, such as slum tourism, are relatively new.

All of the key terms for this chapter have been clarified and examples provided. Table 19.3 is a summary of the definitions of these key terms.

Sources of Trends and Issues in International Tourism

From where do trends and issues originate? It is easy to come up with a 'laundry list' of issues and trends – as has just been done; however, is it possible to accomplish this in a more systematic and comprehensive fashion? As a starting point, it is important to note that there are three main sources of trends and issues in international tourism:

- trends and issues on the market (demand) side;
- trends and issues on the supply (industry sector and destination) side; and
- other trends and issues (political, economic, social-cultural, technological, environmental, and legal – or PESTEL (Yüksel, 2012)).

Figure 19.2 provides a simple diagram showing six cells (1.1 – 3.2) in which trends and issues affecting international tourism can be located. This is an oversimplification of reality, however, as trends and issues can have multiple

Table 19.2. 50 issues with international tourism.

- Accessible tourism
- Ageing infrastructure
- All-inclusive resorts
- Animal rights
- Career prospects for tourism students
- Climate change
- Commodification
- Competition
- Crimes and scams
- Crises
- Cross-border travel
- Crowding and overcrowding
- Dependency (over-dependency)
- Diseases
- Drinking/pub crawl/stag tourism
- Forced shopping
- Funding for tourism (inadequate)
- Gender inequality
- Gentrification
- Globalization
- Global warming
- Governance
- High labour turnover
- Labour supply
- Lack of qualified staff
- Low wages and salaries
- Loss of authenticity/homogenization
- Marijuana/cannabis/pot tourism
- Migrant workers
- Orphanage tourism
- Overpricing or price gouging
- Overtourism
- Political indifference
- Pollution
- Problem gambling
- Refugees
- Safety and security
- Seasonality
- Sex tourism
- Sharing economy
- Slum or ghetto tourism
- Sustainable tourism
- Terrorism
- Tour guide quality
- Touristification
- Traffic
- Trivialization (lack of respect) for tourism
- Unattractiveness of tourism employment
- Urbanization
- Water shortages/inequalities

(From the author)

Table 19.3. Summary of definitions.

Trend A trend is a new development or a change in international tourism.	**Issue** An issue is an important topic or problem that people are discussing or talking or writing about.
Challenge A challenge is a new or difficult task facing international tourism that tests the sector's capabilities, skills and resources.	**Opportunity** An opportunity is a positive outcome presented by external factors that if appropriately acted on can produce beneficial outcomes or performance for a tourism business or destination.

Fig. 19.2. Sources of trends and issues.

Source: (From the author)

causes or antecedents. For example, it is often argued that there is a connection between the rise of sharing economy providers and overtourism, with examples being cited for Lisbon, Prague and Reykjavik (Sheivachman, 2016; Kiss the Bottle, 2019; Tait, 2019, 2020).

Are there 'trissues'?

Can a trend also be an issue, and can an issue also be a trend? For international tourism, the answer is definitely in the positive. Two examples might help clarify this. The worldwide expansion of so-called sharing economy providers, such as Airbnb and Uber, represents both a trend and an issue for international tourism. On the one hand, these services are increasingly popular with consumers; on the other hand, however, they are causing visitor management problems for many destinations. Likewise, the continuous growth in cruising and cruise markets is a second example of a 'trissue' (trend + issue). Cruising is enjoying a long-term growth trend, which is a positive for several home ports and ports of call. On the other side of the coin, cruise lines are blamed for polluting the oceans and areas adjacent to where they dock.

The bicephalous nature of some phenomena within global tourism highlights both their complexity and also the difficulties in their treatment. These trissues often lead to divisiveness in societies or place the tourism sector in an adversarial position to local communities. Two examples here are the inclusive tour and casino resorts that have been developed in several destinations.

Why should we be concerned about trends and issues?

Simply put, trends and issues lead to challenges for the tourism sector, upsetting the status quo and normal operating conditions. Tourism is a dynamic phenomenon and change is constant and often rapid. Additionally, tourism is an open system and is highly susceptible to impacts from other parts of society and the economy (Morrison *et al.*, 2018). Hence, there is a a need to be aware of and respond to trends and issues, not least because tourism is an experience good and people's desires for experiences change.

Another reason we need to keep on top of trends and issues is to inform policy making and legislation and regulations. Indeed, this is one of the reasons why we conduct research in order to track and assess the potential impacts of change. Destination management organizations (DMOs) and tourism businesses must assess trends and issues for planning, human resource management, product development decisions, marketing, branding and promotion campaigns, and visitor management initiatives. But who should track trends and issues and why? This is a great question and one that needs more attention within the tourism sector. Here is a list of those who need to do the tracking and why they are required to do so:

- Businesses: trends lead to changes in business volumes and types of customers.
- DMOs: trends and issues might require modifying strategies or plans.
- Governments: issues may bring need for new policies, laws and regulations.
- Universities: help to keep other parties updated on tourism trends and issues.

Discussion of Trends and Issues

Market-side trends and issues

Market-side trends (1.1 in Fig. 19.2) are new developments and changes in the markets for international tourism. Tourism markets are not static; rather they are undergoing constant change. These trends are reflected in the volumes and flows of tourists between origins and destinations. For example, international tourism arrivals globally are increasing, and this is likely to remain a long-term trend (UNWTO, 2020b). Domestic tourism is rapidly growing as well in many countries including China, India and Indonesia. The outbound market from mainland China is booming and continuing to expand numerically and geographically. The cruise market is continuously becoming larger worldwide.

Trends also occur in the changing attitudes and behaviours of tourists. For example, it is thought that more tourists are looking for authentic and memorable experiences within destinations. In addition, it is believed that some tourists are developing a greater concern for environmental conservation and their own impacts on local people when they travel, although research has demonstrated this is not as widespread as might be imagined (Miller *et al.*, 2010)

Market-side issues (1.2) are topics or problems related to the markets for tourism that are being discussed. For example, some types and levels of tourism raise issues and create challenges for destinations and the sector. They can damage the local environment and historic-heritage resources. The trends are resented by local residents or run contrary to local beliefs and practices. Some encourage undesirable behaviour and lead to criminal activity. They may compete with locals for resources. Some examples of these market-side issues for international tourism are:

- responsible and ethical travel (how to convince tourists to travel more responsibly);
- demand for 'socially undesirable' forms of tourism (drinking tourism, marijuana tourism, sex tourism, drinking/drunk tourism, stag/hen tourism, gambling, etc.);
- tourism that depletes scarce resources through, for example, hunting and fishing, reef diving, and forest clearance.

Overtourism (how to handle over-capacity situations in destinations) is a term coined by Skift and Rafat Ali in 2016 (Ali, 2018). The UNWTO (2018) defined this issue as 'the impact of tourism on a destination, or parts thereof, that excessively influences perceived quality of life of citizens and/or quality of visitors' experiences in a negative way'. Overtourism usually occurs when there is too much demand when compared to supply or when too many tourists occupy too limited space. Rising prices of products and services, as well as for real estate, may adversely affect residents.

Supply-side trends and issues

Supply-side trends (2.1) are new developments or changes in the tourism sector and within destinations. The tourism supply and destinations are expanding quickly in many parts of the world (World Bank, 2019). Several countries need tourism to expand trade, create income and employment, and earn foreign exchange. Tourism companies are expanding globally and becoming more sophisticated.

Some of the specific supply-side trends are an increase in the number of casino resort destinations (e.g. South Korea and Singapore), the expansion of cruise line capacity, healthier food offerings, hotels adding more local and neighbourhood design features, growth of low-cost carrier (LCC) airlines, the rise of sharing economy providers, increasing spa and wellness tourism offerings, and new technologies being used in many different types of tourism attractions (e.g. VR/AR, Smart). In particular, one much-debated supply trend is the rise of sharing economy providers. The sharing economy provides 'new marketplaces that allow services to be provided on a peer-to-peer or shared usage basis' (OECD, 2016). While the sharing economy has been a boon to many places for increasing tourism, it has also caused resentment among local people and caused more overcrowding.

Supply-side issues (2.2) are topics or problems related to the tourism sector and within destinations that are being debated. For example, and as mentioned earlier, there are some forms of tourism that people do not like or that are undesirable, and there are types of tourism that damage the environment. Moreover, the ultra-competitive nature of tourism may lead to business failures, which cause economic problems and may inconvenience consumers, as exemplified by the demise of Thomas Cook in 2019 (Collinson, 2019). The global spread of popular hospitality and travel brands is one of the features of globalization (Dwyer, 2015; Linetti, 2019). These global chains and well-known brand names are increasing the competitive pressures on smaller, local businesses.

External factor trends and issues

External factors include political, economic, social-cultural, technological, environmental and legal aspects (known as the PESTEL framework). These external factors cause trends and issues that influence customers (market); they also affect the tourism sector (supply). The external factor trends with market-side influences (3.2) include the increasing use of the internet and social media; people becoming more environmentally conscious ('green' consumers); greater reliance on apps on smartphones, tablets and laptops; growing time pressures on people; political instability and civil unrest on the rise; natural disasters becoming more common; and the shifting influence of generational cohorts. Specific examples of external factor trends on the market or demand side are the online booking of travel, selfies and photo-sharing, last-minute travel, bleisure, breakations and staycations, green vacations/holidays, responsible travel and multi-generational travel.

The external factor trends on the supply-side (3.3) include the increasing influence of online travel agencies, the growing market power of review sites such as Tripadvisor, increasing importance accorded to digital and social media marketing, and greater emphasis on 'green' and sustainable practices. Specific external factor trends on the supply-side are app development, artificial intelligence (AI) use, augmented and virtual reality, ecotourism, carbon-offset programmes. influencer marketing and last-minute travel offers.

External factor issues on the market-side include animal rights concerns, concerns over the exploitation of people, data privacy issues and concerns, ethical concerns related to climate change and environmental footprints, financial constraints due to the economy, food safety worries (e.g. food neophobia) and health concerns. Other examples of external factor issues are air travel carbon footprint, interruptions of tourism due to civil unrest and wars (Hong and Ye, 2019), ocean pollution through cruise ships, political disputes between countries affecting travel flows, responsible travel codes and programmes, terrorism aimed at tourists and tourism facilities, and voluntourism, including orphanage tourism.

The external factor issues on the supply-side raise crucial questions for international tourism, comprising:

- how to prepare for unexpected crises;
- how to reduce negative environmental impacts of tourism businesses;
- how to ethically source supplies;
- how best to practise corporate social responsibility (CSR);
- which technologies to apply and how best to use these; and
- how to be a good and ethical employer.

Specific examples of external factor issues on the supply side are climate change, equal opportunity employment, fair trading, food and water security and supply, gender equality, global warming, poverty reduction, recycling, and safety and security. The Sustainable Development Goals (SDGs) of the United Nations (2020) address several of these issues.

Climate change is one of the most critical among these issues and some tourism destinations are being adversely affected by climate change, for example the Maldives. Tourism itself is a major contributor to global warming and climate change, most notably as a result of fossil fuel-based transport, such as flying. And climate is often a magnet for tourists. So climate change is a huge issue and dilemma for tourism.

Challenges for International Tourism

Had this chapter been written at the end of 2019, the outlook would have been much rosier, although still guarded as there were multiple issues facing tourism globally. However, fast-forward to March 2020 and tourism was in chaos worldwide. The long-term, upward-pointing trend line in the UNWTO's international tourist arrivals seemed to be heading for an even sharper decline than that caused by 9/11 and the global financial crisis of 2008–09. Nevertheless, the point here is that rather than just randomly picking out the challenges for international tourism, which are complex and numerous, it is better to have a systematic procedure for doing so. Therefore, the author decided to make use of two models outlined in one of his previous books, the 10 As decagon and the ADVICE model. These provide useful frameworks within which to discuss some of the international tourism challenges and opportunities.

The challenges for destinations – the 10 As decagon

The 10 As are criteria to measure the success of tourism destinations (Morrison, 2019). The 10 As decagon is a double polygon and is shown in Fig. 19.3.

The following is an explanation of each of the 10 As along with the identification of some associated challenges for international tourism:

Awareness (AWA): This attribute is related to tourists' level of knowledge about the destination and is influenced by the amount and nature of the information they receive. For the purposes of this chapter, the concept of awareness is broadened to capture some of the earlier identified trends and issues. First, the tourism sector faces the task of increasing awareness of its importance in society, the economy and the environment. This is an 'evergreen' or perennial problem for tourism as the 'industry of fun' struggles to secure appropriate recognition. While the COVID-19 calamity caused havoc in international tourism, it also shone a brighter light on the scope and significance of travel, and how closed borders and shuttered doors make our lives much less enjoyable.

A second awareness challenge for international tourism is of the need to accept visitor management as a significant responsibility for destinations. In this respect, the six actors or parties within the ADVICE model described below must work together in visitor management. This may not only be the antidote for overtourism; it also will make travel safer, more memorable and beneficial to all involved. Third, greater awareness and discussion is needed on some of the trends and issues within international tourism. For example, animal rights and especially the exploitation of animals to entertain tourists is worthy of a global-level debate in tourism (see also Chapter 21 in this volume). This issue gained greater public attention with the release of the *Blackfish* documentary in 2013 (Hassan, 2018); however, since then it seems to have faded into the background. A fourth task facing the sector is to increase the recognition of business tourism and all of its sub-components. While

Fig. 19.3. The 10 As decagon and challenges.

bodies such as the US Travel Association (USTA) with its Meetings Mean Business Coalition (USTA, 2020) have done much to raise the profile of business travel, it remains an 'off-the-radar' part of international tourism for many people. The 'Beyond tourism impacts of meetings' is another commendable effort to broaden the appreciation of business tourism and is run by the International Congress and Convention Association (ICCA, 2018).

Attractiveness (ATT): The number and geographic scope of appeal of the destination's attractions comprise this attribute. More broadly, it seems radical to say that the attractiveness of international tourism itself will require attention in the latter half of 2020 and beyond. Destinations and travel marketers will have to build up the confidence and remove barriers to tourism caused by the concepts of self-quarantining and social distancing. Diversifying the attractiveness and markets of many destinations is a second important challenge for the future. The 'too many eggs in one basket' syndrome was a predominant feature of many destinations prior to the COVID-19 pandemic. Places like Thailand, Macau, Hong Kong, France, Australia and New Zealand paid a very high price for being over-dependent on the outbound market from mainland China.

Availability (AVA): This attribute is determined by the ease with which bookings and reservations can be made for the destination, and the number of booking and reservation channels available. The internet, social media and online travel agencies (OTAs) have made international tourism experiences vastly more available in the past 25 years. The sharing economy providers have accentuated this availability. The first challenge here may be the task of making certain destinations less (and not more) available in the future. This is surely needed for fragile spots such as Antarctica (McClanahan, 2020), Mount Everest (The Conversation, 2019), Maya Bay in Thailand (Wipatayotin, 2019) and Venice (BBC, 2020).

Access (ACC): The convenience of getting to and from the destination, as well as moving around within the destination, constitutes this attribute. Again, the popular thinking in tourism is that increasing access to destinations is a positive. However, restricting tourism and closing down certain destinations has become more popular in recent years; for example, climbing Ayers Rock – Uluru in Australia (Parks

Australia, 2020); the Faroe Islands (Marcus, 2019); Boracay in the Philippines (Haynes, 2018); and Maya Bay in Thailand (Wipatayotin, 2019). Other destinations, including Bhutan, try to limit access by levying fees; in Bhutan's case, all adult tourists must pay a package price of US$250 per day in peak months and US$200 per day in lower season (Lonely Planet, 2020).

A second issue that requires more attention by the tourism sector and destinations is providing access for all, including people with physical and visual challenges. Often referred to as 'accessibility', this means paying more attention to barrier-free travel.

Appearance (APP): This attribute measures the impressions that the destination makes on tourists, both when they first arrive and then throughout their stays in the destination. One of the challenges for many destinations is to improve the first impressions at entry points. This is not just an impending task for places in developing countries, as sub-standard entrances have also been disparaged at airports such as La Guardia in New York (Santora, 2014).

Activities (ACV): The extent of the array of activities and experiences available to tourists within the destination is the determinant of this attribute. It is certainly desirable for destinations to offer visitors a portfolio of activities in which to participate. The first challenge here is that tourists are increasingly searching for experiences rather than just participating in activities. Second, there are activities that are considered socially undesirable that need to be reconsidered by some destinations.

Assurance (ASU): This attribute relates to the safety and security of the destination for tourists. Safety has quickly become a major concern for many travellers due to an assortment of terrorist actions, accidents, food safety and disease issues. The events of 2019–2020 and before clearly suggest that there must be greater awareness and programming within international tourism to assure higher levels of safety and security for tourists. Side by side, this must be more crisis management planning in the tourism sector and destinations.

Appreciation (APR): The feeling of the levels of welcome and hospitality contribute to this attribute. The roles of communities in tourism have been further highlighted since the coining of the term overtourism in 2016. Unfortunately, recent events suggest there is now greater antagonism toward tourists rather than appreciation. How to make residents friendlier and more welcoming to tourists has been a dilemma for many years and the typical solution has been through education and community awareness efforts. While these are useful, they are often transient initiatives and do not go far enough in addressing the real issues. Greater engagement and involvement of residents in tourism seems to offer greater potential, as does the continual monitoring of resident sentiment.

Action (ACT): The availability of a long-term tourism plan and a marketing plan for tourism are some of the required actions. From the author's experience as a consultant, there are many destinations that do not have tourism plans despite the advisability of having them. Also, there does not seem to be an upward trend in preparing such plans.

Accountability (ACU): This attribute is about the evaluation of performance by the DMO. Accountability is an important aspect of the governance issue that was previously identified. The task here is for more tourism destinations and organisations to demonstrate greater accountability and with that to also be more transparent.

The challenged – the ADVICE model

ADVICE is an acronym for authority, destination, visitor, industry, community and the environment (Morrison, 2019). These are the actors that will have to deal with the future challenges of international tourism. A brief description of the six follows after the ADVICE model diagram in Fig. 19.4.

Authority (A): Visitors interact with various government agencies beginning from the trip planning stage (e.g. visa issuance authorities) until they exit destinations (e.g. tax and immigration agencies). Government agencies have a predominant role in international tourism in many countries of the world. While within destinations, visitors may interact with resource protection bodies of government (e.g. in national parks) and with law enforcement agencies. With their policy, planning, regulation and research responsibilities, government authorities will

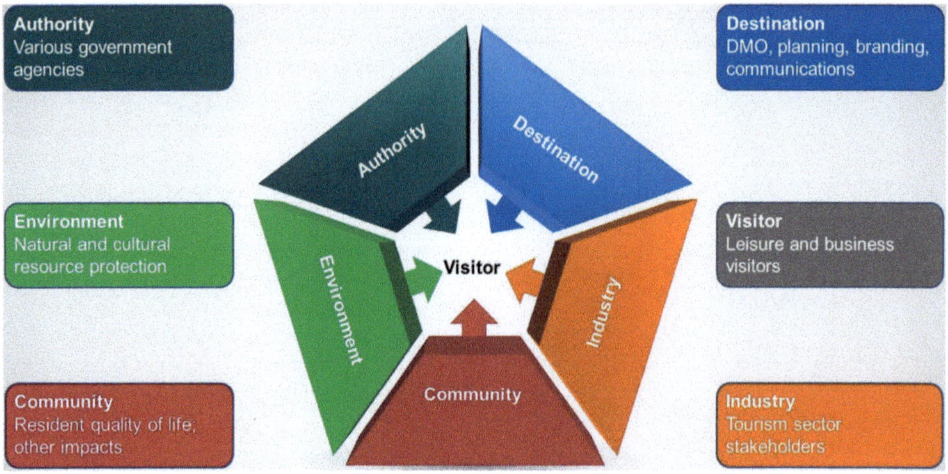

Fig. 19.4. The ADVICE model.

have to play a leading role in addressing future challenges.

Destination (D): DMOs have several roles that involve direct and indirect engagement with visitors. Visitors are part of the focus for tourism destination planning and research, marketing planning, destination branding, integrated marketing communications, product development, and information provision. DMOs are concerned about creating satisfied visitors, as well as assuring that they are safe and secure within their destinations and receive good service and memorable experiences. As with government authorities, DMOs have to be leaders in dealing with future challenges.

Visitor (V): Of course, visitors are the core focus for visitor management and the ADVICE model. However, visitors also interact with other visitors within the destination, as well as sharing information on social media and in other ways (word of mouth and eWOM). Visitors in the future must take greater responsibility for their actions and impacts on destinations.

Industry (I): Tourism sector stakeholders play an important role in serving visitors and in creating experiences for them. In fact, when in destinations visitors may have the greatest contact with hotels, restaurants, attractions, tour operators, and transport providers. The tourism sector must be among the leaders in facing up to the challenges ahead for international tourism.

Community (C). Visitors have an impact on community residents' well-being and quality of life, directly and indirectly. Overcrowding may affect resident access to places and traffic congestion can restrict their everyday movements. Property values may be pushed upward and become too expensive for locals. Traditional neighbourhoods disappear with gentrification. However, residents also benefit in different ways from the positive economic impacts of tourism, including having better infrastructure and public amenities. The 'them and us' dichotomy that exists in tourism destinations must be dealt with using much greater resident engagement and involvement in tourism.

Environment (E). The natural and cultural-heritage environments of destinations are normally the core of the visitor experience. However, these resources are often fragile and can be irreparably damaged or transformed through visitor use. The natural environments of destinations often pose physical dangers for visitors based upon the activities in which they participate and from crises and natural disasters. The so-called 'silent partner' in the ADVICE model, this may in fact represent the most important of the six actors.

Table 19.4 provides a summary of the challenges for each of the 10 As along with suggestions as to whom within the ADVICE model should lead the efforts to address the future challenges for international tourism.

Table 19.4. The 10A Challenges and ADVICE Leaders.

10 as	Challenges	ADVICE leaders
Awareness	Increase awareness of the importance of tourism in society, the economy, and the environment Accept visitor management as a significant responsibility for destinations Create greater awareness and discussion on significant trends and issues within international tourism Increase the recognition of business tourism and its sub-components	Authorities, Destination, Industry ADVICE ADVICE Destination, Industry
Attractiveness	Enhance the attractiveness of international tourism in the latter half of 2020 and beyond Diversify the attractiveness and markets of destinations	Authorities, Destination, industry Authorities, Destination, Industry
Availability	Make certain destinations less available in the future	Authorities, Destination, Environment
Access	Restrict tourism and close down certain destinations Provide access for all – barrier-free travel	Authorities, Destination, Environment
Appearance	Improve first impressions at entry points	Authorities, Destination, Industry, Community, Environment
Activities	Provide more meaningful and memorable destination experiences Reconsider socially undesirable forms of tourism	Industry, Destination, Community, Environment Authorities, Destination, Industry, Community, Environment
Assurance	Improve safety awareness and programming within destinations Enhance crisis management planning	Authorities, Destination, Visitor, Industry, Environment Authorities, Destination, Industry, Community, Environment
Appreciation	Involve and engage local residents in tourism to a larger extent Continually monitor resident sentiment	ADVICE
Action	Prepare more tourism plans and strategies for destinations	Authorities, Destination, Industry, Community, Environment
Accountability	Demonstrate greater accountability and transparency	Authorities, Destination, Industry

Postlude

The storm clouds hanging above international tourism are forbidding and it will take some time for them to completely clear. According to Bill Hybels, the American author and church figure, storms draw something out of us that calm seas don't. Thus, it is hoped that international tourism will emerge even stronger after 2020. However, there were plenty of problems within international tourism at the end of 2019; the COVID-19 pandemic only exacerbated them. How can we avoid a reoccurrence of this situation in the future?

This is a tough question to grapple with; however, we could start with a new 'early warning system' for international tourism based on a comprehensive tracking of trends and issues. As mentioned at the beginning of the chapter, it is impossible at present to locate such a source. In particular, the events of 2020 solidified the notion that closer alliances are needed within the tourism sector. The absence of proactive planning within tourism for global crises was exposed and the tourism sector sputtered to find solutions in the face of adversity. Media sources had a heyday in chronicling the downturns in tourism and the frailties of tourism enterprises. Indeed, at the time of writing, the prevailing sentiment about international tourism was tainted with much doom and gloom. Dire predictions were being made about the prospects for international tourism and even for particular sub-sectors, particularly airlines (CAPA, 2020). While this crisis was unprecedented in world tourism, there was still some sense of *déjà vu* with it and a knowledge of the resilience of the phenomenon. This was buoyed with the news that tourism in China, decimated by the virus outbreak from January to March 2020, was on the road to recovery. Even so, this interlude will leave an indelible mark on tourism and with those most involved in the sector. Hopefully, it will encourage our community to be more vigilant about trends and issues, and their potential effects.

In addition, one of the major themes in this chapter is that greater attention to visitor management is required. While many experts argue that overtourism is nothing new and that it was being overhyped by the media, its increased public profile since 2016 has underlined the need for visitor management. There are also a few components of international tourism that are under-researched as well as under-appreciated. Business tourism is cited as one of these 'stepchild' aspects of tourism. Two others are domestic tourism and visiting friends and relatives (VFR) travel. If there is one beneficial outcome of the challenge of a crisis, it is that it can trigger a greater focus on these 'taken for granted' tourists. Why does that have to be the case, one might ask? On reflection, are these not a huge proportion of international and domestic tourism? In future, these three markets need greater research, respect and recognition by all involved.

Acknowledgement

The author has had a lifelong interest in reflecting on international tourism that was ignited as a student within the Scottish Hotel School, University of Strathclyde in Glasgow, Scotland. This was also an influential period in cultivating a passion for research, analysis and writing. A healthy scepticism for tourism was developed along with a strong sense of loyalty and enthusiasm for the sector. Prof Kit Jenkins was and is one of the author's most enduring mentors and this chapter is dedicated to his sage advice over many years

References

Ali, R. (2018) The genesis of overtourism: Why we came up with the term and what's happened since. Skift. Available at: https://skift.com/2018/08/14/the-genesis-of-overtourism-why-we-came-up-with-the-term-and-whats-happened-since/ (accessed 17 March 2020).

BBC (2020) Coronavirus: Venice canals clearer after lockdown. 18 March. Available at: https://www.bbc.com/news/av/world-europe-51943104/coronavirus-venice-canals-clearer-after-lockdown

Budeanu, A., Miller, G., Moscardo, G. and Ooi, C.-S. (2016) Sustainable tourism, progress, challenges and opportunities: an introduction. *Journal of Cleaner Production* 111(B), 285–294. DOI: 10.1016/j.jclepro.2015.10.027.

CAPA (2020) COVID-19. By the end of May, most world airlines will be bankrupt. Centre for Aviation, 17 March. Available at: https://centreforaviation.com/analysis/reports/covid-19-by-the-end-of-may-most-world-airlines-will-be-bankrupt-517512?fbclid=IwAR3JWiBHvcUHdpJgIxUOV7Oa4j2z5A492ioIl5JRRwShrQ5bIBaYGT1DOAA

Collinson, P. (2019) Why did Thomas Cook collapse after 178 years in business? The Guardian, 23 September. Available at: https://www.theguardian.com/business/2019/sep/23/thomas-cook-as-the-world-turned-the-sun-ceased-to-shine-on-venerable-tour-operator

Dwyer, L. (2015) Globalization of tourism: drivers and outcomes. *Tourism Recreation Research* 40(3), 326–339. DOI: 10.1080/02508281.2015.1075723.

Hassan, A. (2018) The 'Blackfish effect' is over for SeaWorld—for now. Quartz, 6 November. Available at: https://qz.com/1451546/the-blackfish-effect-is-over-for-seaworld-for-now/

Haynes, S. (2018) Boracay islanders feel the pinch after the Philippines shuts down a top tourist destination. Time, 15 May. Available at: https://time.com/5262455/boracay-island-philippines-tourists-closure-duterte/

Hong, J. and Ye, Q. (2019) Chinese tour groups to Hong Kong plunge 86% for Golden Week. Bloomberg, 26 September. Available at: https://www.bloomberg.com/news/articles/2019-09-25/protests-see-chinese-tourists-ditch-hong-kong-for-golden-week

ICCA (2018) Incredible impacts: the beyond tourism impact of meetings. International Congress and Convention Association. Available at: https://www.iccaworld.org/knowledge/article.cfm?artid=540

Kiss the Bottle (2019) Airbnb: why Overtourism is a problem in Lisbon. *Kiss the Bottle*, 17 March. Available at: http://kissthebottle.org/overtourism-airbnb-lisbon/ (accessed 23 March 2020).

Linetti, L. (2019) The effect of globalization on the tourism industry. Thrive Global, 20 August. Available at: https://thriveglobal.com/stories/the-effect-of-globalization-on-the-tourism-industry/

Lonely Planet (2020) Bhutan/Asia: money and costs. Available at: https://www.lonelyplanet.com/bhutan/money-costs

Marcus, L. (2019) Why the Faroe Islands will close to tourists in 2020. *CNN Travel,* 12 November. Available at: https://www.cnn.com/travel/article/faroe-islands-closed-tourists-2020/index.html

McClanahan, P. (2020) Tourism in Antarctica: Edging Toward the (Risky) Mainstream. *The New York Times,* 26 February. Available at: https://www.nytimes.com/2020/02/26/travel/antarctica-tourism-environment-safety.html

Miller, G., Rathouse, K., Scarles, C., Holmes, K. and Tribe, J. (2010) Public understanding of sustainable tourism. *Annals of Tourism Research* 37(3), 627–645. DOI: 10.1016/j.annals.2009.12.002.

Morrison, A.M. (2019) *Marketing and Managing Tourism Destinations*, 2nd edn. Routledge, Abingdon, UK.

Morrison, A.M., Lehto, X.Y. and Day, G.J. (2018) *The Tourism System*, 8th edn. Kendall Hunt Publishing, Dubuque, Iowa.

OECD (2016) *Re-thinking policies for the tourism sharing economy*. Organisation for Economic Co-operation and Development, Paris.

Oxford Learner's Dictionaries (2020) Challenge (definition). Available at: https://www.oxfordlearnersdictionaries.com/definition/english/challenge_1?q=challenge (accessed 23 March 2020).

Parks Australia (2020) Uluru climb closure. Available at: https://parksaustralia.gov.au/uluru/discover/culture/uluru-climb/

Santora, M. (2014) Some see Biden's 'Third World' description of La Guardia as too kind. *The New York Times*, 7 February. Available at: https://www.nytimes.com/2014/02/08/nyregion/some-see-third-world-as-too-kind-for-la-guardia.html

Sharpley, R. (2001) Tourism in Cyprus: challenges and opportunities. *Tourism Geographies* 3(1), 64–86. DOI: 10.1080/14616680010008711.

Sheivachman, A. (2016) Iceland tourism and the mixed blessings of Airbnb. Skift, 19 August. Available at: https://skift.com/2016/08/19/iceland-tourism-and-the-mixed-blessings-of-airbnb/ (accessed 23 March 2020).

Tait, R. (2019) The fall of Prague. 'Drunk tourist are acting like they've conquered our city'. *Observer Special Report*, 25 August. Available at: https://www.theguardian.com/world/2019/aug/25/prague-drunk-tourists-conquer-our-city

Tait, R. (2020) 'I'm a stranger in my own city': Prague takes on Airbnb to dam the floods of tourism. The Observer, 1 February. Available at: https://www.theguardian.com/environment/2020/feb/01/overwhelmed-prague-tries-to-limit-airbnb-to-curb-tourism

UNWTO (2020a) COVID-19: putting people first. UN World Tourism Organization, 17 March. Available at: https://www.unwto.org/tourism-covid-19-coronavirus (accessed 18 March 2020).

UNWTO (2020b) *World Tourism Barometer No. 18,* January 2020. UN World Tourism Organization. Available at: https://www.unwto.org/world-tourism-barometer-n18-january-2020 (accessed 23 March 2020).

USTA (2020) Meetings mean business coalition. U.S. Travel Association. Available at: https://www.ustravel.org/programs/meetings-mean-business-coalition

The Conversation (2019) Death on Everest: the boom in climbing tourism is dangerous and unsustainable.. Available at: http://theconversation.com/death-on-everest-the-boom-in-climbing-tourism-is-dangerous-and-unsustainable-114033 (accessed 30 July 2020).

United Nations (2020) About the sustainable development goals. Available at: https://www.un.org/sustainabledevelopment/sustainable-development-goals/

UNWTO (2018) 'Overtourism'? Understanding and managing urban tourism growth beyond expectations. UN World Tourism Organization. Available at: https://www.e-unwto.org/doi/pdf/10.18111/9789284420070 (accessed 23 March 2020).

Wipatayotin, A. (2019) Maya Bay to remain closed until mid-2021. Bangkok Post, 9 May. Available at: https://www.bangkokpost.com/thailand/general/1674364/maya-bay-to-remain-closed-till-mid-2021

World Bank (2019) *World Bank Open Data*. World Bank, Washington, DC. Available at: https://data.worldbank.org/

World Economic Forum (2020) This is how coronavirus could affect the travel and tourism industry. Available at: https://www.weforum.org/agenda/2020/03/world-travel-coronavirus-covid19-jobs-pandemic-tourism-aviation/ (accessed 18 March 2020).

Yüksel, I. (2012) Developing a multi-criteria decision making model for PESTEL analysis. *International Journal of Business and Management* 7(24), 24–52. DOI: 10.5539/ijbm.v7n24p52.

20 Reflections on the Intersection between Gender and Race in Tourism

Donna Chambers*
University of Sunderland, Sunderland, UK

> That man over there says women need to be helped into carriages, and lifted over ditches, and to have the best place everywhere. Nobody ever helps me into carriages, and lifted [me] over ditches, or over mud-puddles, or gives me any best place. And aint I a woman? Look at me! Look at my arm! I have ploughed, and planted, and gathered into barns, and no man could head me! And aint I a woman? I could work as much as and eat as much as a man – when I could get it! and bear the lash as well! And aint I a woman? I have borne thirteen children, and seen most all sold off into slavery, and when I cried with my mother's grief, none but Jesus heard me! And aint I a woman? (Sojourner Truth, 1851, cited in Carby, 1996, p. 112)

Introduction

I begin this essay on the intersectionality between gender and race within the context of tourism with the often-cited words of Sojourner Truth (an African-American abolitionist and women's rights activist) who, at the 1851 Women's Convention in Akon, Ohio, highlighted how racism has resulted in the very different constructions and lived experiences of black women in contrast to white women. It is clear that even during the era of slavery, it was recognized that black women face multiple oppressions and that, since then, a series of terms have been coined to name these various oppressions, such as 'double oppression', 'interlocking oppression' and 'triple jeopardy' (the latter referring to gender, race and class oppression). However more than a century since Sojourner Truth's powerful statement, it is the term *intersectionality* that today best reflects the interlocking oppressions faced by black women. This term was developed by African-American civil rights activist and scholar Kimberlé Crenshaw in 1989, who used it to refer to the way in which racial and sexual subordination are inextricably linked. Through the prism of intersectionality, Crenshaw (1989) has been able to argue that it is the failure to recognize the reinforcing relationship between racism and sexism that has resulted in the erasure of black women's experiences from both the discourses of feminism and the discourses of anti-racism. In a similar vein, black British scholar Hazel Carby (1996, p. 112) has suggested that 'racist theory and practice is frequently gender specific', thus making it difficult to separate these two oppressive forces. Carby went even further to contend that white feminists wrote their 'herstory' and made it the story of *all* women, thus ignoring the lives of black women. In this way, Carby suggests, white women were acting within the confines of racist relationships and reproducing 'his-story'.

It is against this background that in this essay I bring to the fore my own subject position as a black woman academic in tourism, originally from the Global South, because it is this

*donna.chambers@sunderland.ac.uk

 Tourism in Development: Reflective Essays (eds P.U.C. Dieke *et al.*)
DOI: 10.1079/9781789242812.0020

positioning that has influenced my decision to seek to unpack the complexity of the socially constructed and power-political concept of gender, which, according to West and Zimmerman (1987, p. 147) 'produces, reproduces and legitimates choices'. I seek to problematize gender through an exploration of the often ignored or bracketed intersection between gender and race in tourism gender research. Race is, of course, like gender, a social construct which wields great political power and, like Anne McClintock (1995), I agree that gender and race are not isolated from each other; as a black woman in tourism, I do not experience gender and race as distinct and separate parts of my own lived experiences. I suggest, in similar vein to McClintock (1995), that gender and race are articulated together and have been integral to sustaining power relationships in tourism, particularly between the western world and the largely developing countries of the Global South.

Gender intersectionality has not been often studied in tourism, although there is increasing recognition that this is a necessary and valid way of exploring issues of power and/or disadvantage (see for example Files-Thompson, 2013; Cole, 2017; Hamid-Turksoy *et al.*, 2014; Mooney, 2018). In this essay, I want to present a more nuanced approach to tourism gender research which focuses on the gender–race intersection while being cognisant of the fact that gender has many different intersections with a host of other social categorizations such as sexuality, class, nationality and (dis)ability, among others. However, as a black woman in tourism, I have become exhausted with discussions primarily by white tourism scholars of the 'Othering' of black women's bodies and their oppression and/or empowerment through tourism. I have become exhausted with, as Henry (2019) has deemed it, 'the unspeakable whiteness of volunteer tourism' and the seemingly endless march of 'white saviours' across the countries of the Global South. I have become exhausted with sitting through endless conference presentations where predominantly white tourism scholars present anthropological and/or sociological accounts of their research in 'indigenous' communities in the Global South with a lack of reflexivity on their own motivations for embarking on such research and the effects of their research on these communities (see Tuhiwai Smith, 1999). Therefore, in this essay, which is concerned with the intersection between gender and race, I wish to take a different perspective by focusing on white female privilege rather than the traditional black female disadvantage. I do this through the prisms of black feminist theory and critical race theory and, specifically regarding the latter, the study of 'whiteness' which was developed in academia from the works of noted African-American scholar and civil rights activist W.E.B. Du Bois. Indeed, Henry (2019, p. 326), in the context of volunteer tourism, argues that it would be beneficial for tourism academics to 'engage with critical race theory in order to understand the power of white privilege'. I suggest in this essay that the study of whiteness is one way in which gendered power relationships in tourism can be understood. I support this argument with an analysis of two films which explore white female sex tourism to countries in the Global South.

On Whiteness

The concept of whiteness is relevant to our contemporary understanding of the location of power in tourism when discussing the intersection between race and gender. I contend that in tourism there has been little, if any, problematization of the way in which whiteness shapes ways of knowing (epistemology) and ways of being (ontology). Whiteness provides a lens through which we can explore exploitation and subjugation in tourism without the traditional focus on tourism's effects on non-white 'others'. My suggestion here is that in tourism studies, we have concentrated almost exclusively on racial *disadvantage* and, consequently, we have failed to adequately theorize the power that comes from racial *privilege*. Furthermore, we have failed to explore in tourism how whiteness can be understood through the lens of intersectionality to unpack white women's privilege. Indeed, there seems to be a discursive silence around the notion of white women's privilege in tourism studies.

W.E.B Du Bois is said to have initiated the study of whiteness. He was particularly interested in issues of class and why it was that in the USA of the 19th and early 20th centuries there was no political coalition between blacks and poor whites given their commonalities in terms of social disadvantage. DuBois argued that this was because of what he termed the 'public and psychological wage' possessed by poor whites which earned them valuable social status simply from being 'not black'. That is, whiteness was able to provide 'compensation' for poor whites who were otherwise themselves exploited within the capitalist system. In his seminal essay, 'The Souls of White Folks', first published in 1920 in the book *Darkwater* (DuBois, 1920), DuBois engages in a hypothetical exchange with a white person whom he labels a 'sweeter soul of the dominant world'. In this hypothetical conversation, this 'sweeter soul' says to DuBois:

> My poor, un-white thing! Weep not nor rage. I know, too well, that the curse of God lies heavy on you. Why? That is not for me to say, but be brave! Do your work in your lowly sphere, praying the good Lord that into heaven above, where all is love, you may 1 day, be born – white!

To this, he replies:

> I do not laugh. I am quite straight faced as I ask soberly 'But what on earth is whiteness that one should so desire it?' Then always, somehow, someway, silently but clearly, I am given to understand that whiteness is the ownership of the earth, forever and ever Amen. (DuBois, 1920)

DuBois thus suggests that at the heart of white identity is a passionate belief in one's right to everything and anything or, as it were, the title to the universe. It was this sense of entitlement that underpinned colonial expansion and the exploitation of non-whites politically, socially, economically, sexually and, I would add, also in terms of research and knowledge.

Since DuBois' unpacking of 'whiteness' there have been several understandings and applications of the concept. In particular, Brander Rasmussen *et al.* (2001) in their collection titled *The Making and Unmaking of Whiteness*, identify six different though interconnected strands in studies of whiteness:

1. whiteness as invisible and unmarked;
2. whiteness as an 'empty' identity established only through appropriation from non-white others;
3. whiteness as structural privilege;
4. whiteness as violence and terror;
5. whiteness as the institutionalization of European colonialism; and
6. critical whiteness studies as an antiracist practice.

Alternatively, Ruth Frankenberg (1993), in her seminal text on the social construction of whiteness, in which she problematizes the role of white women in the feminist movement, used the term 'whiteness' to describe a set of three interlinked dimensions which includes whiteness as a location of structural advantage and as a set of cultural practices that are usually unmarked and unnamed (Frankenberg, 1993, p. 1). She does, however, add a self-reflexive element by arguing that whiteness is also a standpoint or place from which white people look at themselves, others and society. Frankenberg suggested further that whiteness is not an empty signifier but is instead a 'daily experience of racial structuring' (Frankenberg, 1993, p. 1).

More specifically, Richard Dyer (1997, p. 4), in his treatise *The Matter of Whiteness*, argued that 'It has become common for those marginalised by culture to acknowledge the situation from which they speak, but those who occupy positions of cultural hegemony blithely carry on as if what they say is neutral and unsituated – human not raced'. He adds:

> In the realm of categories, black is always marked as a colour (as the term coloured egregiously acknowledges) and is always polarising. Whereas white is not anything really, not an identity, not a particularising quality, because it is everything – white is no colour because it is all colours. (Dyer, 1997)

I suggest that in contemporary quotidian tourism practices and theorizations of power in tourism, whiteness has become normalized and is rendered invisible. For example, Henry (2019, p. 326) suggests that 'the (limited) data on the 'race' of volunteers suggest a majority of international volunteers are white'. Indeed, whenever we speak about race in tourism, it is always seen as applicable only to those who are not white – white people are in this sense not racialized and therefore remain un-researched. In other words, there is, in tourism studies, largely silence around the power of whiteness as a racialized categorization. According to Dyer

(1997), 'in our time it is only extreme right and racist discourse, that has an acknowledged and clear concept of a white race...now however to talk about race is to talk about all races except the white'. The invisibility of whiteness has meant that in tourism studies we have not yet sufficiently theorized the location of power.

Whiteness, like other racial categorizations, is, of course, constructed and, according to Lipsitz (2006, p. vii), a 'scientific and cultural fiction...that has no valid foundation in biology or anthropology'. Indeed, whiteness, as well as other racial identities, has no essence and, according to Nakayama and Krizek (1995, p. 293), 'there are only historically contingent constructions of [their] social locations'. However, while recognizing the fabricated nature of all racial categorizations, this is not to deny that they do, none the less, have a material reality and real social consequences for the way in which wealth, prestige and opportunity is distributed. Indeed, Lipsitz (1995, p. 371) stated that 'race is a cultural construct, but one with sinister structural causes and consequences'.

Krista Ratcliffe (2005) indicates that while gender classifications have a long history, tropes of race (including of course whiteness) have a much shorter but more complex history. The racial category 'white' has come to occupy a position of privilege and advantage even if all white people do not 'share identical social and economic privileges' (Ratcliffe, 2005). She states further that, like gender, race 'is always a question not only of difference but of unearned privilege and power – and the lack thereof' (Ratcliffe, 2005).

I now borrow from black feminist literature to suggest that hegemonic whiteness is often portrayed almost exclusively as hetero-patriarchal and, as such, has paid insufficient attention to women. In tourism research in particular, there is a dearth of studies which investigate the racial meanings attached to being a white woman, such as in the context of tourism to the post-colonial countries of the Global South – to countries in Africa, in the Caribbean and in Latin America. This is despite the existence of several studies which, for example, examine western female solo travellers as subjects and others which have investigated the way in which western women are objectified and sexualized by the (black) male gaze through the tourism experience. The issue of race does emerge in a very few of these studies (for example, see Wilson and Little, 2008; Brown and Osman, 2017) but it is not conceptualized in terms of studies of whiteness. The question which preoccupies me here is: How do white women 'perform gender' in liminal post-colonial tourism spaces?

Vron Ware (1992) has suggested that in feminism, the representation of white women has been in terms of gender, class and sexuality while eliding the dynamics of race. Ware (1992, p. 38) indicates that a potent symbol of colonial repression was the 'threat of real or imagined violence toward white women' which became one of the most dangerous forms of insubordination. A colony was secure to the extent that it was able to keep its white women protected from the fear of sexual assault. Indeed, she argues that 'white women provided a symbol of the most valuable property known to white man and it was to be protected from the ever-encroaching and disrespectful black man at all costs' (Ware, 1992). More broadly, Ruth Frankenberg suggests that race shapes the lives of white women in similar fashion to the way men's lives are shaped by gender, and sexuality shapes the lives of heterosexuals.

However, in our contemporary world, the financial, emotional, sexual, and legal advances occasioned by feminism for largely white, western women has meant that they are now capable of performing whiteness, a status which has provided them with resources, power and opportunity even if only in the context of the liminal spaces of post-colonial tourism which are necessarily fleeting.

We can unpack white female privilege in the tourism context through a critical analysis of female sex tourism. In this activity, primarily middle-aged white women from western countries travel to countries in the Global South in order to engage in sexual relations with normally young black men, often perhaps derogatively termed 'beach boys', 'rent-a-rastas' or 'cowboys'. In one of the first tourism articles on the subject of female sex tourism, Pruitt and LaFont (1995, p. 427) suggest that 'these women are able to explore more dominant roles in the tourism relationship due to their economic and social status which invests them with an independence that translates into power and control over these men'. However, in their study,

Pruitt and LaFont failed to articulate a theory of race which might provide a useful explanatory framework for the source of the women's economic and social power, especially in the context of these post-colonial tourist destinations. I suggest that such power and control is fundamentally underpinned by conceptualizations of whiteness. Franz Fanon (2004, p. 31), in the *Wretched of the Earth,* speaks to the colonial situation by noting that 'in the colonies, the economic superstructure is also a substructure. The cause is the consequence. You are rich because you are white, you are white because you are rich'. And as Jacqueline Sanchez Taylor and Sánchez Taylor (2001, p. 761) has suggested, 'focusing on gendered power can easily slip into accepting gender essentialist models of sexuality within which the sexual behaviour of women is interpreted and judged differently from that of men simply because they are women'. Such an approach tends to privilege the power of gender, occluding the power of race (and following on from Fanon's statement above, also the associated characteristic of economic power which is rather loosely encompassed by class), and specifically of whiteness which I argue here is fundamental to female sex tourism to the countries of the post-colonial world.

It is possible to understand the intersection between gender and race through an examination of female sex tourism as it is represented in film. Gross (2018, p. 508) suggests that 'engaging with film, media, and other forms of cultural production is a key feminist approach to understand how representation and politics operate as interlocking structures of power'. Several films have been released since the new millennium about female sex tourism to the post-colonial world. Here, I focus on two films, primarily owing to the complex colonial histories of the destination countries involved and their continued problematic post-colonial presents – *Heading South,* set in the Caribbean island of Haiti and released in the UK in 2006, and *Paradise: Love,* set in the sub-Saharan east-african country of Kenya, which was first screened at the Cannes Film Festival in 2012 and put on general release in 2013.

Before proceeding with the analysis of both films, however, it is important to acknowledge that two previous publications have examined both films – the first is by Shirley Tate (2011), who undertook a critical exploration of *Heading South* through the lens of Achille Mbembe's necropolitics, which refers to the 'subjugation of life to the power of death' (Tate, 2011). As such, Tate was interested in the way in which the people of post-colonial societies (such as the central black male character in the film) used death as a form of colonial resistance and liberation and as their prime means of exerting agency in their fight against their own alterity and the power of the former white colonizers (embodied in this case in the white female sex tourists). The second, by Zoë Gross (2018), examines *Paradise: Love* as a representation of the intersecting power structures of race, ethnicity, gender, sexuality and class that 'shape and transform transnational interracial intimacies within economies and geographies of eroticized exchange' (Gross, 2018, p. 508). Gross draws on the literature on sex tourism and feminist geographical frameworks as key underpinnings for her work. My analysis of the films in this essay furthers the conversations contained in these two previous publications, but my exploration is different in three respects. First, I provide an analysis of both films showing the synergies between them but also highlighting where they differ. Second, I draw much more explicitly on critical race theory and, specifically, studies of whiteness along with black feminist literature. Third, my personal situatedness as a black woman from the Global South provides a reflexive lens through which my interpretations are expressly filtered. That said, I have drawn on insights from both publications where they are relevant to illuminating aspects of the discussion.

Exploring Whiteness through Film

Heading South, directed by Frenchman Laurent Cantet, is based on three short stories written by Haitian American author Danny Laferriere, and is set in Haiti in the late 1970s when the country was ruled by dictator Jean Claude Duvalier (nicknamed Baby Doc). Haiti obtained its independence from France in 1804 and was the first slave colony to stage a successful revolution against its white colonizers to establish an independent state in the Caribbean ruled by former African slaves. The country was invaded and occupied by the United States in 1915 under the guise of restoring the country to democracy after the assassination of the then president – this occupation lasted for 19 years until 1935. In fact, Haiti has a long history of political and economic turmoil, while a series of natural disasters and crises have devastated the country and affected its development. These have contributed to it being for some time one of the poorest countries in the world. The continued political instability in Haiti in the late 1970s provides the setting for *Heading South*, which tells the story of three North American women (two from the USA and one from Montreal in Canada) who travel to the country purposely to engage in sex with the local men – the main protagonist, Ellen, is played by renowned English actor Charlotte Rampling.

Paradise: Love is the first in a trilogy of films by Austrian Director Ulrich Siedl and recounts the story of a German, middle-aged, divorced woman (Teresa), who is played by Austrian actor Margarethe Tiesel. Teresa, who is the main protagonist of the film, travels to Kenya for the first time for a holiday but willingly initiates a series of sexual encounters with several local men, ostensibly to find romance to fill the emptiness in her life occasioned by her divorce and a challenging teenage daughter. Kenya's colonial history can be said to have started with the European scramble for Africa in the late 19th century and with its declaration as a British colony in 1920. Just over two decades later, overt resistance to British colonial rule emerged in the form of the Mau Mau movement which staged a series of violent liberation struggles until independence was declared in 1963. Hence, the colonial history of Kenya bears some similarity to that of Haiti in so far as both engaged in violent struggles to unseat their colonial masters and to gain their independence. Unlike Haiti, however, Kenya is one of the more economically developed countries in sub-Saharan Africa although there is also severe poverty and unemployment. Nevertheless, like Haiti, Kenya has also suffered from major political and economic turmoil which has also affected the country's development.

For both Kenya and Haiti, agriculture is the mainstay of their economies, although in the case of Kenya, tourism also plays a significant role with over 1.3 million tourist arrivals in 2016 (Kenya Tourist Board, 2017). In Haiti, while tourism has existed for many decades, tourism as an industry is still in its infancy although it is the expressed desire of the Haitian government to make tourism the second largest economic activity on the island (Dodd, 2017). In *Paradise: Love*, unlike in *Heading South*, there is no indication of the politico-historical period within which the film was set. Instead, here the political environment in the country is more of a subtext, although we are provided with an insight into the poverty which pervades the lives of the Kenyans.

The exoticization of the black male body: eating the other

In both films, black male bodies are exoticized and objectified by the white women from the outset. In *Heading South*, although the women are from North America where there is a significant minority of black men, they suggest that they are not interested in these men. The following is a conversation among the three women in the film, Brenda, Sue and Ellen. Sue (the Canadian) is asked why is it that given that there are so many black men in Montreal she had never dated any of them:

> **Sue:** At home black guys don't interest me. They are very very different.
> **Brenda:** It's true. I don't know why. Is it because they are closer to nature? Is it the sun? Anyway, they're more gracious.
> **Ellen:** The big difference sweetheart is that here you see them stripped to the waist. ... think of these cute guys, they are a dime a dozen, take your pick...if you are too shy to pay them, just give them gifts...

In *Paradise: Love*, not long after the main protagonist Teresa arrives in her holiday resort in Kenya, she meets a white female sex tourist at the hotel bar. The conversation between the two women is as follows:

> **Woman:** Do you like it?
> **Teresa:** Incredible. It smells so different. It's so...you feel different. Don't you feel different?
> **Woman:** I always feel that way here. I'm dripping. You have to smell their skin. Its unforgettable.
> **Teresa:** Whose skin?

Woman: The Negroes...it smells of coconut. I could lick and bite them forever. It's amazing, you'll see.
Teresa: They are nice. I met two but the problem is, they look alike. I thought they were the same person. They look the same.
Woman: It's like that at first.
Teresa: Really?
Woman: You can tell by their size...if they are big or small...mine's really big... everywhere!

In these conversations in both films, the post-colonial black male body is exoticized and differentiated from those of black men at home. They are seen as more primitive (the notion of closer to nature), less dangerous (they are more gracious) and, in the liminal space of the tourist resort, the white women feel free to treat these men like cheap commodities, to be bought either with cash or with gifts. Indeed, one black man is easily substitutable for another as, after all, not only are they plentiful but they all look the same until, of course, one is able to distinguish them by the size of their penises. According to Gross (2018, p. 515), in reference to *Paradise: Love*, this exchange references bell hooks's (1992) theory of 'eating the Other', and 'illustrate the ways in which racial and sexual scripts become mutually reinforced between female sex tourists and local men. Paradise: Love confirms: sexuality is racialized and race is sexualized.'

The objectification of the female body by men is a mechanism of power and the female sex tourists in both films engage in the same practice as a demonstration of their power over the black male body. It is, thus, in the liminal space of the tourist beaches that traditional patriarchal roles are inverted.

One of the most disturbing scenes in *Paradise: Love* occurs when Teresa is provided with a birthday present of a young black male stripper by three other female sex tourists whom she has met at the resort. The stripper, who looks like just a boy, does a striptease for the women in Teresa's hotel room while the four women ogle him. One of the women says, 'That's Africa, that's wild for you'... After a while another woman says 'Let's put the beast into action. Whoever gets him hard is the winner'. The young boy starts to gyrate on top of one of the women who says, 'Like a predator. Hear him? Like a gorilla. No like a tiger.'

According to Margarethe Tiesel, the Austrian actor who plays Teresa in *Paradise: Love* in a press interview in which she was asked about the storyline of the film, 'It's about female loneliness that takes hold when you reach a certain age and no longer look like someone from an advert. The exploited begin to exploit in a place where they have power.'

According to Jacqueline Sanchez Taylor:

> where at home they might feel stigmatised for having relations with black men, younger men, 'womanisers' or for having many sexual partners, in holiday resorts...they are permitted to consume the black male, the younger boy, the playboy or as many men as they desire while maintaining their reputation back home. So for once they can experience feeling more powerful than a man – and particularly a black man, a person who they may stereotypically think of as 'dangerous' at home. (Cited by Wheat, 1999.)

Certainly, the popular colonial representation of black men as dangerous and as sexual predators constantly desiring and seeking to consume white female bodies is not played out in the context of the post-colonial tourism resort where the female sex tourists see the local black men as unthreatening, as docile. Indeed, in these post-colonial tourism resorts we see a subversion of representations – here it is the white female tourists who, comfortable in the structural privilege their whiteness affords them in these resort spaces, are at last free to consume the black male body without fear of consequences.

Whiteness as violence

In the tourist resorts, whiteness is also associated with violence. In *Heading South*, a series of autobiographical vignettes are used as a way of telling the stories of the three women. In the first such vignette, Brenda recounts her first visit to Haiti three years earlier with her then husband during which she engaged in a sexual encounter with one of the black boys on the island, whom she and her husband had taken 'pity on' as he appeared to be starving. During the course of this holiday, they basically took the boy under their wing. It was Brenda who seduced the boy on the beach and, in her autobiographical vignette, she describes their first sexual coupling:

> his body fascinated me. Long, lithe, muscular, his skin glistening. I couldn't take my eyes off him...I slipped two fingers into his bathing suit...I threw myself on him. I literally threw myself on him...It was so violent I couldn't help but scream, I think I never stopped screaming. It was my first orgasm.

At the time, the young boy was 15 and Brenda 45. Yet in recounting this story, Brenda makes no mention of any personal regret at the exploitation of a 15-year-old boy which in her own American society would be deemed as paedophilia, a criminal offence. In Haiti she suffers no penalty. Brenda is emotional in the telling of the story, not because of any remorse or any feelings of guilt, but because of the intensity of the memory, a repeat of which she has travelled to Haiti three years later to experience but which she is afraid will be out of her grasp as the young man, Legba, now 18, has been 'purchased' by 55-year-old Ellen, on what seems to be an exclusive contract.

In *Paradise: Love*, Teresa finds out that her 'beach boy', Munga, has been dishonest about his marital status, as the woman he claimed was his sister was actually his wife and they had a son. Teresa confronts him on the beach where he is apparently enjoying bathing in the sea with his wife and young son. She grabs him by the hair and abuses him both physically and verbally. Munga does not retaliate, he does not fight back, but instead starts whimpering and tries to protect himself from Teresa's rage. Interestingly, none of the Kenyans in the sea, including Munga's wife, intervenes while this is playing out – they look on dispassionately as if they have seen it all before. After Teresa finishes venting her anger, she says to Munga, who is still whimpering, 'now get lost'. It is, therefore, Teresa who decides when and how the relationship ends.

The spatialization of whiteness

In both films, there is a clear spatial demarcation within the tourist resorts. The female sex tourists are free to traverse through all the spaces of the resort but the 'beach boys' who service their sexual needs can only occupy the beach; they are not permitted to venture beyond the confines of the beach. Ironically, to do so would be to transgress established boundaries. In fact, the black male sexualized body is only allowed into the non-beach spaces of the resort with the permission of the white female sex tourist. In other words, it is white female privilege which acts as a passport to entry to non-beach resort spaces. In *Paradise: Love*, Teresa's sexual encounters with the black men occur predominantly in the impoverished local spaces outside of the resort. In these local spaces, as a white woman she is not in danger but is protected.

However, when she invites one of the male hotel employees called Yosef to her room for sex, the man is clearly uncomfortable in the space and is unable to perform sexually, even though Teresa is trying her best to seduce him. When she questions what is wrong, he says, 'it is hard for me'. Teresa asks him to leave as he is unable to satisfy her sexual fantasies and desires but not without warning him not to take any money from her purse.

In *Heading South*, Brenda invites her beach boy, Legba (whom she is now sharing with Ellen), to have dinner with her and Sue in the hotel restaurant. The head waiter, Albert, refuses to serve him and indicates that he is not allowed in the restaurant according to hotel policy:

> **Sue:** He eats with us every day on the beach.
> **Albert:** On the beach it is no problem. You do what you want. But this is the restaurant.

It is only after Brenda insists that Albert relents and allows Legba to eat at the table. What is interesting here, though, is that Brenda then says to Sue: 'It's incredible how racist they can be'. Clearly, Brenda, by projecting racism onto the black waiter, is unaware of, or failing to acknowledge, her own racism.

The spatial privilege enjoyed by the white female sex tourists in both films is also evident in their unfettered ability to not only move across space but also across time within resort spaces and also between these resorts and their home countries. This freedom is not enjoyed by the beach boys and their liberation can only arguably occur through death, as happened to Legba in *Heading South* (Tate, 2011). That is, Legba and his Haitian female friend are murdered as they were apparently embroiled (wittingly or unwittingly) in the volatile politics of Haiti in the late 1970s. Ellen, who, it will be recalled, was sharing Legba with Brenda, had passionately implored him to escape the violence of Haiti by returning home with her, but Legba had refused her 'protection' and was subsequently gunned down in an apparent organized killing. The murderer dumps the bodies on the same beach resort where the female sex tourists have been staying. According to Tate (2011), the death of Legba might be deemed as symbolizing resistance to white female privilege and the wider sex tourism industry in which it is embroiled. Indeed, Tate (2011, p. 47) argues that:

> Legba's necropolitical agency emerges in his choice of death rather than Ellen's rescue through which emerges his position as a subject who resists both state terror and white female 'race' and class sovereignty. His refusal to engage with the capitulation of the male love object and the 'happily ever after' of White Western romance speaks back, from a position of alterity...[his] is a spirit that says he would rather die a free man in Haiti than be an unfree man back in the United States, domesticated by white love.

However, this resistance by the subaltern to white female privilege and power might also be interpreted as pervading the wider Haitian society. That is, Legba's murder and the dumping of his body on the beach of the same resort where the sex tourism is enacted perhaps also symbolizes the defiance of a fiercely independent nation to the new form of colonialism that white female sex tourism increasingly represented.

Ellen briefly blames herself for Legba's death and is perplexed at the casual behaviour of the policeman who comes to investigate the murders and who does not ask her any probing questions. She questions whether her own life and that of the other tourists is in danger. In response, the policeman utters three powerful words which attest to the privileged nature of the tourists' existence and the value placed on their lives as opposed to the lives of the locals who dare to engage in this activity: 'Tourists never die...'

The ambivalence of whiteness

In both films, we are not made to feel empathy for the black men but rather for the white female sex tourists. All the women are middle-aged, some are overweight, divorced, lonely or unable to find a suitable partner at home. In *Heading South*, Ellen indicates that: 'If you're not as dumb as a fashion model, the only guys that are interested are natural born losers or husbands wives are cheating on'.

We are made to feel these women's lack of fulfilment in their lives at home and to imagine with them that what they are looking for in their touristic journey is love or romance, which legitimates their behaviour and perhaps distinguishes them from white male sex tourists who are often portrayed as being interested only in sex or the physical. Indeed, the coining of the term 'romance tourism' to refer to female sex tourism is a manifestation of this distinction as it negates the exploitative nature of the relationship between the white women and the beach boys and denies the existence of their white privilege. In *Heading South*, Ellen appears to be conflicted about whether what she is seeking is actually love or sex. She says, 'I'm crazy about love. Sex and love. I'm not really sure anymore.' In *Paradise: Love*, Teresa indicates that she wants to find someone who: 'Really looks at you as a person. Not just the surface. Deep into your pupils, into your soul. Seeing what's inside, past your crow's feet and fat bum. Into your eyes, right into your eyes'.

Jacqueline Sanchez Taylor makes a potent argument about female sex tourists from the West and the power that they exercise:

> Although men involved in tourist related sexual exchanges and their female clients are often equally concerned to ignore or downplay the material inequalities that divide them, and to construct their relationships as if they were based simply and solely on mutual desire and/or romance and/or friendship, their motives for constructing a fiction are different, and have different consequences. Black [Caribbean] men who provide sexual services as an economic activity are performing their 'Otherness' in ways that are desired by tourists...this means reproducing racist stereotypes about the 'Other' and having to subjectively experience sexualized racisms as empowering. The tourists' economic survival does not depend on them performing any particular version of 'femininity', and tourist resorts [in the Caribbean] actually provide them with a stage on which to perform their femininity in any way they choose. (Sánchez Taylor, 2006, p. 56)

Final Reflections

Through female sex tourism to the post-colonial world, the traditional gendered power relations between men and women are reversed through the intervening factor of race – the women, racialized as white have privilege and power while the men, racialized as black, do not. While it might be that the situation is not so clear-cut and binary – that is, white women can sometimes be vulnerable and exploited by 'foreign, exotic others' and black men are not always dependent and 'passive objects of women's desires' (Hamid-Turksoy *et al.*, 2014) – an intersectional approach, it is argued, allows for the 'fluidity of power exertion, acknowledging that not all oppressions function oppressively in all situations at all times' (Files-Thompson, 2013). I would contend that in the context of white female sex tourism to countries of the Global South, historical power-political colonial relationships, which are both sexualized and racialized, still have a very potent force in the dynamics that play out between white women and black men in the sexual encounters inherent in tourism as a spatial and temporal phenomenon. Yet, when race is spoken about in tourism, it is often focused on race disadvantage and not on race privilege. This is further problematized when one adds a gender dimension. Through sex tourism, then, white power is reproduced and re-inscribed in the Caribbean and other parts of the Global South not only by western men but also by western women.

In this essay, I have thus provided an interpretation of white female sex tourism to the countries of the Global South such as those in the Caribbean, Latin America and in Africa, and I have suggested that we need to examine the sources of power in this tourism phenomenon. I argue that the sources of power are inextricably intertwined with racist notions of white privilege which are exercised by both men and women. An intersectional approach which recognizes the interplay between gender and race, and which acknowledges the potency of 'whiteness', enables a more critical unpacking of the power inherent in these sexual encounters within liminal post-colonial tourism spaces.

It is important to state, none the less, that this discussion on whiteness is not intended to create a simplistic dichotomy between black and white; it is not intended to homogenize whiteness. Rather, like Gayatri Spivak (2006), I am calling for a problematization of positionality. As a black woman from the Global South who has become accustomed to tourism narratives that focus on the racialized 'Other' largely espoused by western scholars, I am encouraging white people to (re)consider myopic understandings of racial power within tourism by refracting their epistemological lens onto themselves. Using the lens of whiteness theory, white people and, in this case, white women, are implored to reflect on their own racial identities, the privileges that accompany them, and how these serve to legitimate racial inequalities in tourism.

However, Sara Ahmed (2007) questions whether by focusing on whiteness we are at the same time enabling its continued power. She questions whether:

> speaking about whiteness allows it to become an 'essential something'? If whiteness gains currency by being unnoticed, then what does it mean to notice whiteness? What does making the invisible marks of privilege more visible actually do? Could whiteness studies produce an attachment to whiteness by holding it in place as an object? Such questions are addressed by scholars not in order to suspend the project of whiteness studies, but to consider what it means for a project of critique to be complicit with its object. We could say that any project that aims to dismantle or challenge the categories that are made invisible through privilege is bound to participate in the object of its critique. (Ahmed, 2007, pp. 149–150.)

This is indeed a very valid point, but I agree with Richard Dyer (1997, p. 10) that 'the point of looking at whiteness is to dislodge it from its centrality and authority, not to reinstate it (and much less, to make a show of reinstating it, when, like male power, it doesn't actually need reinstating)'.

While the main thrust of this essay has been on the intersection between gender and race with a focus on white female sex tourism, implicit in the discussion of whiteness is also a class dimension where class is seen in an economic context. However, it is clear from the films that not all the white female sex tourists are either middle class or wealthy; for instance, Teresa from *Paradise: Love* is portrayed early on in the film as a working-class divorced mother. What is important here, however, is the fact that they are white and, according to Franz Fanon, whiteness implies richness from the perspective of the colonized. That said, there are also several other intersections that are evident in both films that could be added to the gender–race nexus to better illustrate the complexity of the intersectionality at play in both films which has not been discussed in this essay, such as the characteristics such as age, nationality and sexual orientation. But this is for a future project. In addition, while recognizing that the lived experiences of black women are different from those of white women, as Sojourner Truth reflected at the beginning of this essay, I have not explored the position of black women within the context of female sex tourism, neither in the sense of black female sex tourists (albeit they are only a minority) to the Global South nor in terms of the way in which black women in the films have been represented. Indeed, in both films we only see black women in traditional roles – as mothers, as wives/girlfriends, as lowly paid workers in the hotels – and it would be interesting in a future project to apply critical race theory and black feminism in order to understand more fully the extent of their agency within this context.

Finally, in this essay I hope I have demonstrated how whiteness is pervasive in relationships in tourism to the developing world, including in the context of female sex tourism. However, it has thus far remained largely unnamed and therefore its wider implications for tourism development, and indeed for tourism research, have been ignored. It is however important for western tourists to developing countries to reflect deeply on their positionality of racial privilege, as this has consequences for how tourism is practised and performed. In the two films explored in this essay I have noted how performances of female sex tourism to the developing world are underpinned and enabled by white privilege. I suggest that exposing the power of whiteness (including its gendered dimension) allows for an excavation of a central source of exploitation that persists in all forms of tourism, including in research. Such an exercise is a necessary endeavour for the sustainable and ethical development of tourism in the Global South.

Note

[1] This chapter is part of a project funded by the Spanish Ministry of Science and Innovation (HAR2017-83005-R: Tourism and Performativity of Local Identity: Nation and Region from a Postcolonial and Gender Perspective).

References

Ahmed, S. (2007) A phenomenology of whiteness. *Feminist Theory* 8(2), 149–168. DOI: 10.1177/1464700107078139.

Brander Rasmussen, B., Klinenberg, E., Nexica, I.J. and Wray, M. (eds) (2001) *The Making and Unmaking of Whiteness*. Duke University Press, Durham, North Carolina.

Brown, L. and Osman, H. (2017) The female tourist experience in Egypt as an Islamic destination. *Annals of Tourism Research* 63, 12–22. DOI: 10.1016/j.annals.2016.12.005.

Carby, H. (1996) White woman listen! Black feminism and the boundaries of sisterhood. In: Baker, H.A., Diawara, M. and Lindeborg, R. (eds) *Black British Cultural Studies: A Reader*. University of Chicago Press, Chicago and London.

Cole, S. (2017) Water worries: an intersectional feminist political ecology of tourism and water in Labuan Bajo, Indonesia. *Annals of Tourism Research* 67, 14–24. DOI: 10.1016/j.annals.2017.07.018.

Crenshaw, K. (1989) Demarginalizing the intersection of race and sex: a black feminist critique of antidiscrimination doctrine, feminist theory and antiracist politics. *The University of Chicago Legal Forum* 1(8), 139–167.

Dodd, C. (2017) Haiti tourism: how this beautiful Caribbean country aims to attract more travellers. *The Independent*, Monday 8th May.

DuBois, W.E.B. (1920) The souls of white folks. In: *Darkwater: Voices from the Veil*. Dover Publications, New York, pp. 17–31.
Dyer, R. (ed.) (1997) The matter of whiteness. In: *White: Essays on Race and Culture*. Routledge, London, pp. 1–40.
Fanon, F. (2004) *The Wretched of the Earth*. Grove Press, New York.
Files-Thompson, F. (2013) An intersectional analysis of sexuality in the tourist space. In: Kozak, M. and Kozak, N. (eds) *Aspects of Tourist Behaviour*. Cambridge Scholars Publishing, Newcastle-upon-Tyne, UK.
Frankenberg, R. (1993) *The Social Construction of Whiteness: White Women, Race Matters*. Routledge, London.
Gross, Z. (2018) Unpacking the landscape of female sex tourism in Kenya: a film analysis of *Paradise: Love*. *Gender, Place & Culture* 25(4), 507–524. DOI: 10.1080/0966369X.2018.1442816.
Hamid-Turksoy, L., van-Zoonan, L. and Kuipers, G. (2014) I dumped my husband for a Turkish toyboy: romance tourism and intersectionality in British tabloid newspapers. *Feminist Media Studies* 14(5), 806–821.
Henry, J. (2019) The unspeakable whiteness of volunteer tourism. *Annals of Tourism Research* 76, 326–327. DOI: 10.1016/j.annals.2018.09.003.
hooks, b. (1992) Eating the other: desire and resistance. In: *Black Looks: Race and Representation*. Routledge, New York.
Kenya Tourist Board (2017) Major achievements and initiatives of the tourist sector. Available at: http://ktb.go.ke/wp-content/uploads/2017/07/Achievements-since-2013.pdf (accessed 22 September 2019).
Lipsitz, G. (1995) The possessive investment in whiteness: Racialized social democracy and the 'white' problem in American studies. *American Quarterly* 47(3), 369–387. DOI: 10.2307/2713291.
Lipsitz, G. (2006) *The Possessive Investment in Whiteness*. Temple University Press, Philadelphia, Pennsylvania.
McClintock, A. (1995) *Imperial Leather: Race, Gender, and Sexuality in the Colonial Contest*. Routledge, New York.
Mooney, S. (2018) Illuminating intersectionality for tourism researchers. *Annals of Tourism Research* 72, 175–176. DOI: 10.1016/j.annals.2018.03.003.
Nakayama, T.K. and Krizek, R.L. (1995) Whiteness: a strategic rhetoric. *Quarterly Journal of Speech* 81(3), 291–309. DOI: 10.1080/00335639509384117.
Pruitt, D. and LaFont, S. (1995) For love and money: romance tourism in Jamaica. *Annals of Tourism Research* 22(2), 422–440.
Ratcliffe, K. (2005) *Rhetorical Listening: Identification, Gender, Whiteness*. Southern Illinois University Press, Carbondale, Illinois.
Sánchez Taylor, J. (2001) Dollars are a Girl's Best Friend? Female Tourists' Sexual Behaviour in the Caribbean. *Sociology* 35(3), 749–764. DOI: 10.1177/S0038038501000384.
Sánchez Taylor, J. (2006) Female sex tourism: a contradiction in terms? *Feminist review* 83(1), 42–59. DOI: 10.1057/palgrave.fr.9400280.
Spivak, G. (2006) Can the subaltern speak? In: Ashcroft, B., Griffiths, G. and Tiffin, H. (eds) *The Postcolonial Studies Reader*. Routledge, London.
Tate, S. (2011) *Heading South*: Love/sex, necropolitics, and decolonial romance. *Small Axe: A Caribbean Journal of Criticism* 15(2), 43–58. DOI: 10.1215/07990537-1334230.
Tuhiwai Smith, L. (1999) *Decolonizing Methodologies: Research and Indigenous Peoples*. Zed Books, London.
Ware, V. (1992) *Beyond the Pale: White Women, Racism and History*. Verso, London.
West, C. and Zimmerman, D.O.N.H. (1987) Doing gender. *Gender & Society* 1(2), 125–151. DOI: 10.1177/0891243287001002002.
Wheat, S. (1999) Travel: women are sex tourists too. *The Independent,* Saturday 6th February.
Wilson, E. and Little, D.E. (2008) The solo female travel experience: exploring the geography of women's fear. *Current Issues in Tourism* 11(2), 167–186. DOI: 10.2167/cit342.0.

21 Rethinking the Animal in Tourism Studies

Erik Cohen*
Hebrew University of Jerusalem, Israel

Introduction

The representations of tourist–animal relations in tourism studies are one-sided; they prioritize the tourists but neglect the animals. This imbalance not only fails to give an appropriate account of the animal side of the relationship, but also makes the representation of the human side incomplete. The concentration on the human 'point of view' and disregard for 'the animal point of view' (Descola, 2014) is a deficiency that imposes substantial limitations on the insights and conclusions emerging from research into the dynamics of tourist–animal relations.

This weakness is not unique to the field of tourism but, rather, reflects a prevailing anthropocentric perspective in the modern social sciences (Badmington, 2003; Descola, 2014; Wilkie, 2015) which had, until recently, marginalized animals and paid little attention to their social significance. That deficiency provoked the anthropologist Descola's (2014) call to 'repopulat[e] the social sciences with non-human beings'. The recent recognition that animals are part of human society brought about a turn towards an interspecies sociology (Wilkie, 2015) and abetted the emergence of Human-Animal Studies (HAS) (Marvin and McHugh, 2014) as a new interdisciplinary field (Wilkie, 2015). However, these developments have not yet found much resonance in tourism studies.

In this chapter I will first outline a trend of change in the image of animals in several disciplines, which indicates the emergence of a paradigmatic shift in the treatment of animals and their relations to humans. I shall then discuss the apparent deficiencies in the study of tourist–animal relations, and end up with some suggestions to rethink the basic assumptions and theoretical approaches to animals in tourism studies.

The Emerging Paradigmatic Shift in Human–Animal Studies

To get a full grasp of the significance of this turn, it is necessary to pay attention to recent parallel developments in several disciplines which indicate a progressive paradigm shift in the philosophical and scientific perspectives on animals and their relations to humans.

Biology

Humans and non-human animals have been increasingly shown to share many genetic, physiological and behavioural similarities. One

*erik.cohen@mail.huji.ac.il

 Tourism in Development: Reflective Essays (eds P.U.C. Dieke *et al.*)
DOI: 10.1079/9781789242812.0021

commentator has even remarked that 'science is in the process of whittling those [remaining] differences down to nothing' (Ruiz, 2017). Research has increasingly demonstrated that at least some animal species possess such human properties as consciousness (Edelman and Seth, 2009) and a sense of 'self', or self-consciousness (Bekoff and Sherman, 2004; Irvine, 2007) which were, up until recently, ascribed only to humans.

Philosophy

Modern humanist philosophy, rooted in the Enlightenment, has neglected animals (Iveson, 2010). Post-humanism, a post-modern current in philosophy (Ferrando, 2013), advocates a radical break with Enlightenment philosophy (Jackson, 2013; Brinkmann, 2017), negates the Cartesian human–animal dualism (Badmington, 2003, pp. 16–17; Wilkie, 2015) and criticizes humanism for its anthropocentrism (Badmington, 2003, p. 19; Descola, 2014; Wilkie, 2015), speciesism (i.e. favouring the interests of one's own species against those of other species) (Singer, 1975) or prioritizing humans over animal species (Ferrando, 2013) (Marchesini, 2016), and human exceptionalism (the belief that humans are fundamentally different from all other animals) (Iveson, 2010, p. 46; DeMello, 2012, p. 32-33).

Post-humanism called for a break with the 'imperialist ontology and metaphysical essentialism' of the Enlightenment (Jackson, 2013, p. 670) and for the abolition of the Enlightenment's picture of the animal (Wolfe, 2009, p. 564) as fundamentally different from humans, which was based on René Descartes' view of animals as 'complex living machines that do not think, but instead respond automatically to external stimuli' (Miller, 2013). This view still resonates in contemporary thinking about animals, such as, for example, in behaviourism (Staats, 2003; Staddon, 2014). In contrast, post-humanism destabilizes the 'putatively stable, hierarchical, ontological boundaries dividing human, plant, animal and stone' (Jackson, 2013, p. 679). It proposes a very different view of the animal than that of the Enlightenment; specifically, it does not see the animal as an 'oppositional other' (Derrida, 2008), but recognizes an affinity between humans and 'non-human animals' (e.g. Badmington, 2003; Wolfe, 2009). It favours the perception of animals as persons capable of thinking and feeling, and volitionally responding (instead of just deterministically reacting) to humans, which has gained ascendance in several scientific fields in recent decades.

Psychology

In a recent review of human–animal relations in psychology, Amiot and Bastian (2015, p. 7) acknowledge that the discipline of psychology 'has been remarkably slow to investigate the topic of human–animal relations' and 'somewhat anthropocentric in its orientation'; consequently, they review the 'psychological processes...involved in the interaction between humans and animals' primarily 'as a domain of *human* behaviour' (Amiot and Bastian, 2015, emphasis added). The authors recognize that the 'human–animal boundary is (at least partly) a social construction' (Amiot and Bastian, 2015, p. 31); but although this implies some similarity or continuity, rather than a total divide between human and animal behaviour patterns, they tell us little about the behaviour of animals towards humans. For example, they discuss biophilia as 'an innately emotional affiliation that humans have toward other forms of life' (Amiot and Bastian, 2015, p. 8), but do not pose the question whether any animal species are innately attracted (repulsed or inimical) to humans. Amiot and Bastian (2015) thus opened important questions regarding human relations with animals, but seem to be unconcerned, or professionally unprepared, to deal in depth with the animal side of the relationship in the same terms in which they dealt with the human side. This issue has been taken up by a stream in contemporary animal ethology.

Ethology

Cognitive ethology has made considerable progress in the last 20–30 years in the study of the capacities of animals (Wolfe, 2009, p. 567). In a recent innovative paper, Lestel *et al.* (2014) presented a new paradigmatic approach to animal life. Rejecting the dominant 'realist–Cartesian' paradigm, which, these authors claim, 'has atrophied our zoological imagination in reducing animal life to behaviours...and behaviour to causal mechanisms', these authors criticize the experimental, behaviouristic approach in animal ethology which studied the animal in laboratorial isolation, as an organism deterministically adapted to a set of objective conditions. In contrast, they claim that ethology should seek to understand the animals' 'point of view' (Lestel *et al.*, 2014, p. 130) as they interact with humans in real-life situations. The approach assumes that animals possess agency and intentionality (Pearson, 2017), as well as the ability to interpret situations, to make choices and to engage in meaning-making activities by which they construct their worlds. Phenomenological ethology thus posits that both humans and animals interpret and construct their respective worlds, though these constructions might be radically different. The alterity of the animals (to humans) is thus due to the difference between the humans' and animals' constructions of the world. The task of ethology is to understand this alterity by investigating the animals' construction of the world. The authors propose a bi-constructivist phenomenological approach to animal ethology, 'as the science of the human interpretation of animal interpretations' (Lestel *et al.*, 2014), thus moving the researcher's attention from studying animal behaviour to the interpretation of the animals' actions.

Ontological anthropology

Fundamental western ontological categories, such as the nature/culture dichotomy, were put into question in recent studies in ontological anthropology (e.g. Descola, 2009, Descola, 2013, Descola, 2014; Kohn, 2013, 2015; Viveiros de Castro, 2014), which found that 'a metaphysical distinction between nature and culture does not work in a region [such as that of Amazonian native groups], where nature and culture take on very different properties' (Kohn, 2015). Hence that dichotomy should not be seen as a universal constant, but as a social construction. Its deconstruction, in turn, weakens the perception of a fundamental divide between humans, as cultural beings, and animals as natural ones. Ontological anthropology recognizes that animals have selves and a 'point of view' and stresses human–animal mutual awareness (Kohn, 2013, 2015; Descola, 2014). Hence, it advocates the ethnographic study of 'how humans communicate with a host of nonhuman beings in a world that is itself communicative but not symbolic or linguistic' (Kohn, 2015).

The general inference from the parallel developments in diverse disciplinary fields is that, in recent times, the approach to animals underwent an ontological shift from viewing them as things or objects to regarding them as subjectivity-possessing persons (Lijmbach, 1998; Wolfe, 2009, p. 567). The difference between humans and non-human animals is not seen any more as one of kind, but rather one of degree. Humans and animals are considered to share some basic similarities, such as subjective experiences of well-being or suffering (Wemelsfelder, 1997), selfhood (Irvine, 2007) or personhood, consciousness, self-consciousness (Bekoff and Sherman, 2004), intentionality and agency (Pearson, 2017), although to different degrees (which have to be empirically established with respect to each particular species (Bekoff and Sherman, 2004, p. 176). This indicates an emergent paradigmatic shift in the treatment of animals and of their relations to humans, which necessitates a conceptual, theoretical and methodological reorientation of the study of human–animal (including tourist–animal) relations.

The shift entails the abandonment of the assumption of human exceptionalism (Pearson, 2017), adopted in the West from Aristoteles to Kant, namely of the existence of a basic and, in principle unbridgeable, human-animal divide (which, as we shall see below, is a pillar of the assumption of the 'Otherness' of animals in touristic imagery). This change further entailed a fundamental reorientation in the ethics of the human treatment of animals (in general and specifically in tourism), propounded in the work of contemporary animal ethicists (Fennell, 2012), while also leading to the emergence of a new, interdisciplinary sub-field, human–animal studies (HAS) (Shapiro and DeMello, 2010; Markwell, 2015a), accompanied by a growing engagement of researchers in the study of animals in 'real' social situations (rather than in laboratories). However, it is important to note that in none of the disciplines discussed above, or indeed in the emergent empirical studies of human–animal interaction, is the field of tourism explicitly brought up, despite the multiple engagements of tourists with a wide range of animals. This disregard, however, has been mutual; nor have tourism studies paid much attention to the developments in those disciplines, despite the growing interest in the role of animals in tourism and in tourist–animal interaction (e.g. Lovelock, 2007; Markwell, 2015b).

Deficiencies in the Approach to Animals in Tourism Studies

Studies of specific topics in the sub-field of tourism sometimes miss recent conceptual or theoretical developments in their broader disciplinary field. I have recently pointed this out with regard to current studies of ethnic tourism in South-east Asia (Cohen, 2016), which have missed significant recent theoretical approaches in the study of ethnicity. Studies of animals in tourism seem to suffer from a similar deficiency; with the exception of a few recent publications (to be presented below), most researchers of tourist–animal relations are apparently unaware of ongoing conceptual and theoretical developments in the fields of philosophy, biology, animal ethology and ontological anthropology, which have destabilized the basic assumptions of the modernist view of the human–animal relationship. Consequently, the studies are deficient in their treatment of the animal side of the relationship, which make them imbalanced, incomplete or, in some instances, insensitive to ethical issues in the use of animals in tourism.

The human–animal divide

Beirne (2007) has powerfully stated the common modernist assumption of a fundamental divide between humans and animals: 'The distinction between *Homo Sapiens a*nd non-human animals carries with it a cumbersome cultural baggage. Implicitly, it tends to be voiced as if humans were somehow not animals and as if all other animals were insapient ('dumb animals')...the distinction assumes that non-human animals were necessarily the Other' (Beirne, 2007). Most studies of animals in tourism continue tacitly to take-for-granted (or at least fail to question) this assumption of a fundamental divide, or abyss, between humans and animals, on which the image of the animals' inscrutable Otherness, promoted by the tourist industry, is based. In other words, students of animals in tourism mostly fail to distance themselves from the modernist image of the animal favoured by the industry, or fully grasp its ideological role in promoting animal tourism. Animals are perceived as attractive and fascinating, though potentially dangerous, objects to be gazed at (or hunted) by tourists or dominated, trained and tamed to perform for a tourist audience or to enable tourists to view, ride, touch or pet them. Tourism researchers looking at tourism as a business, including some from the sustainability camp, tend to disregard the conduct or plight of wild animals in their discussions of wildlife tourism (Rodger *et al.*, 2007), or of particular allegedly 'eco-friendly' activities, such as whale watching (Cunningham *et al.*, 2012).

While some psychologists have investigated the circumstances under which the closing of the human–animal divide affects moral concerns for animals (e.g. Bastian *et al.*, 2012), tourism studies have not yet directly addressed the relevance of the amelioration of the divide on tourism ethics. Many recent studies indeed take a strong ethical position, criticizing the exploitation of animals in tourist establishments as they are made into objects of the tourists' desire. But these studies (e.g. Duffy and Moore, 2011; Duffy, 2014; Kontogeorgopoulos, 2009) do not explicitly reject the human–animal divide or dwell upon the animals' autonomous conduct in their interaction with tourists. Their ethical advocacy of an amelioration of the plight of animals in tourist establishments is, thus, based on the researchers' humanist (rather than post-humanist) concern for the animals, but does not affect their exceptionalist image of human superiority over animals. This was also the case in my own critical studies of the treatment of animals in tourism, such as that of the Tiger Temple in Thailand (Cohen, 2013). A post-humanist animal ethics would go much further, conceding the animals an equal moral standing with humans and condemning their use in tourism as an infringement of their rights, rather than criticizing existing practices, owing to humanitarian compassion.

Anthropocentrism

Discussing the 'Animal Question', Franklin (2007, p. 32, empphasis in original) argues that 'because the entire issue is about how *we behave* and *how we should behave* the subjectivity and agency of the animal (always in question) is entirely lost or subsumed'. The study of human–animal relations is, thus, primarily concerned with the humans or, as Amiot and Bastian (2015) point out, it is 'anthropocentric'. This is equally the case in the specific field of tourism. The principal interest of studies of tourist–animal relations is in the tourists, their motivations, activities, experiences and satisfactions in interaction with animals in various settings, ranging from open nature (e.g. Lemelin *et al.*, 2008) to diverse establishments presenting animals to entertain tourists, such as zoos, theme parks, circuses and animal shows (Cohen, 2019b).

To be sure, many studies are concerned with the effects of the development of animal attractions on the animals; however, they discuss these effects solely from the human point of view, showing little if any awareness of the animals' 'point of view' regarding touristic activities. Thus, Knight (2010), in a study of the so-called 'wild monkey parks' in Japan, shows how the introduction of feeding was used to turn difficult-to-observe forest-dwelling macaque monkeys into 'ready-to-view-monkeys' for time-pressed Japanese visitors. Having access to the work of a Japanese primatologist, Knight was able to point to a series of changes the feeding engendered in the animals' behaviour. The author reports that 'given the conservatism of wild monkeys' feeding behaviour' it was difficult to 'get the monkeys to develop a taste for the food handouts' (Knight, 2010, p. 751). However, being interested primarily in the tourists rather than the animals, Knight does not further investigate the monkeys' adoption of the new diet and, in particular, does not ask whether the monkeys' switch to a new food should be taken simply as a behaviouristic conditioning process manipulated by humans, or whether it involved some autonomous agency or deliberate acquisition of a preference for a new taste on the monkeys' part, similar to the development of a taste for chillis in cats, noted by the phenomenological ethologists Lestel *et al.* (2014).

Some ecotourism researchers have paid detailed attention to the effects of the human interaction with animals in natural settings on the animals' behavioural reactions, health and well-being. It turned out that even in instances generally seen as eco-friendly, such as whale-watching, tourism can have significant detrimental effects on the animals. Thus, Higham and Neves (2015) reported that whales and other cataceans 'behave evasively when approached by whale-watching boats [and] engage in active avoidance maneuvers'. Repeated disturbances by whale-watching boats reduce the time the animals have for critical behaviours, such as resting and feeding, and cause 'altered behaviours that have broader biological and ecological consequences' (Higham and Neves, 2015). Higham *et al.* (2016) consequently argued that whale watching is not, as commonly believed, a harmless non-consumptive activity but, rather, a form of 'non-lethal consumptive exploitation'. While these are important insights into the consequences of human–whale interaction, they do not reveal the whales' 'point of view' regarding the boats and the humans observing them, or how the boats figure in the whales' construction of the world.

While these questions seem to be of little relevance for the study of whale tourism, they become highly relevant in cases of intimate embodied interaction between tourists and other sea animals during such activities as swimming with dolphins (Curtin, 2006; Wiener, 2015), whale sharks (Davis and Tisdell, 1999) or sea lions (Martinez, 2003). Human participants in these interactions are highly sensitive to the animals and imagine the encounter as a shared experience. According to Wiener (2015, pp. 155–156), 'swimmers report a stronger emotional relationship when they feel acknowledged by the participating dolphins...face-to-face interaction is desired to create closeness in human-dolphin relations'. However, Wiener claims that the dolphins' behaviour is often misinterpreted; swimmers anthropocentrically assume that 'dolphins are happy because of the upturned curve of their rostrum or a twinkle in their eye'

(Wiener, 2015, p. 156) but, in the absence of ethological studies, we do not know how the sea animals interpret the encounter or what role swimmers play in the construction of their world.

Various other instances could be offered to show that the studies emphasize the human point of view with respect to the animals with which the tourists interact, while the point of view of the animals is, if at all, only guessed at but, in fact, remains unresearched and unknown.

Mutuality

Most studies disregard the mutuality of the tourist–animal encounter. Derrida (2008), for example, has noted that we do not just look at animals but feel ourselves seen by them. That mutuality is hinted at (but unfortunately not further explored) in the title of Markwell and Cushing's (2009) article, 'The serpent's stare meets the tourist's gaze'). Even detailed studies, such as Lemelin's (2006) study of the 'ocular consumption' of polar bears, are silent about the animals' gaze on their observers, reflecting the common disregard of the mutuality of the tourists' and locals' gaze in tourist studies (Maoz, 2006). As in criminology, animals are often treated in tourism studies as 'passive, insentient objects acted upon by humans' (Beirne, 2007), but not as subjects with a will, agency and 'point of view' of their own.

Behaviourism

Studies of tourist–animal interaction are, at least implicitly, based on the conventional behaviouristic view of animals as reacting to humans rather than responding to them as in a process of interaction between independent subjects. They are not considered to possess the freedom of action which humans are endowed with, but to react 'instinctively' to human stimuli. In contrived settings, such as zoos or animal shows, professional keepers or performers handle, tame or train animals by controlling their behaviour and, in some instances, force them to behave in ways which are not part of their natural behavioural repertoire, causing them to suffer (Wilson, 2017). The professionalism of trainers consists of their ability to impose their will upon the animals' behaviour and suppress or preempt any resistance to it. This is particularly the case in training wild animals, such as big cats in circuses (Carmeli, 1999), and in human–animal contests, such as alligator-wrestling performances (Cattelino, 1968; Cohen, 2019a). But the suppression is rarely complete, and occasionally leads to successful acts of resistance on the part of the animals. The behaviouristic approach bars a full understanding of such instances of unsuccessful human control of the animal, ascribing them to professional incompetence or marking them as 'accidents', but failing to grasp their theoretical implications. 'Accidents' in which the animals manage to escape from their enclosures or cages in animal establishments, or attack, kill or hurt a trainer or a tourist, are interpreted as the outcome of insufficient precautions, skills or expertise on the part of the humans in the interaction (e.g. Duffy and Moore, 2011, p. 592), but not as signs of the animals' volition, agency, or capacity to invent new behaviours, emphasized by phenomenological ethology.

The prevailing, at least tacit, adoption by animal tourism researchers of a behaviouristic, essentially mechanistic and deterministic view of the animal behavioural repertoire, and of the conventional view of animal inferiority to humans, is considered deficient in light of the emergent new paradigm of the animal, according to which the behaviouristic approach offers an incomplete understanding of human–animal relations, since it fails to give sufficient weight to the animals' independent role and conduct in the interaction.

Rethinking the Animal in Tourism Studies

The study of animals in tourism has failed to take account of the emerging paradigmatic shift in the philosophical and scientific image of animals, while the burgeoning multi-disciplinary sub-field of human–animal relations has disregarded tourism as a significant instance of such relations. A major task facing the study of animals in tourism is, hence, to rethink its basic assumptions and theoretical approaches in light of the developments in those broader fields and, thus, to insert the particular insights gained in the study of tourist–animal relations into the current discussions of human–animal relations in the broader fields of animal ethology, anthropology and animal ethics. To achieve that several steps are necessary:

1. The first and crucial step toward this goal is to dissociate the image of the animal in tourism studies from the image promoted by the tourist industry, and to take account of the emergent new paradigm of the animal, shared by philosophy and several scientific disciplines, which denies the human–animal divide, emphasizes the basic similarities between humans and animals (rather than the divide separating them), and ascribes to animals a selfhood, personhood, consciousness, intentionality and a degree of freedom of action which enables them to *respond* to humans, rather than merely to *react* to human stimuli.
2. It follows from the first point that tourism studies should divest themselves of the anthropocentric perspective on animal–tourist relations which focuses on the desires, activities, emotions and satisfactions of the human participants in the interaction, and largely disregards those of the animal. I recognize that this anthropocentrism is in part due to the professional interest of tourism researchers, for whom animals are not the centre of interest. But a disregard for the animals' 'point of view' precludes the understanding of the process of interaction as shaped by the interplay between autonomous actors and makes the analysis of the tourists' conduct in the interaction one-sided and incomplete.
3. Tourism studies should take cognisance of Derrida's (2008) notion, that we not just look at animals but feel ourselves seen by them, which draws attention to the mutuality and reflexivity inherent in tourist–animal relations. It is the understanding of this mutuality, and of the pitfalls inherent in its misapprehension (some of which I have noted above), which animal ethology can help us to uncover and thereby enhance the benefits of tourist–animal interaction for both the tourists and the animals.
4. Consequently, more attention should be paid to the use of power by tourists, or the personnel of tourist establishments, by which the animals are constrained, dominated and their preferences and desires denied, as they are forced to entertain and serve the visitors by behaviours outside their natural repertoire. But attention should also be paid to the manner in which animals express resistance to human domination, leading to 'accidents', which, at least in some cases, are just unexpected, inventive or spontaneous deviations or counter-performances from their expected, trained behaviour.
5. Attention ought similarly to be paid to the 'back-stage' treatment of animals, particularly to the ways in which their volition is broken by training and suppression, so that they undergo a radical 'personality change', as in instances in which wild animals are made, by training, to lose their wildness or ferocity (Wilson, 2017; Cohen, 2019a), in order to make them amenable to perform for tourists or tame enough to come into embodied contact with them.
6. Finally, the study of tourist–animal interaction should not rely on commonsense interpretations of animal conduct by tourism researchers (which might be tainted by anthropomorphism), but involve, whenever possible, ethologists for a professional interpretation of the interaction. Such participation is advisable since some tourist behaviour, such as the attraction to young animals, apparently reflects an innate human biophilic predisposition (Borgi and Cirulli, 2016). Such an interdisciplinary approach to the study of tourist–animal interaction would be an important step forward in the research of tourist–animal relations.

Innovative Case Studies

In order to illustrate the advantages of a rethinking of the animal in tourism, I will present some recent case studies, which exemplify the insights that can be gained by the adoption of an innovative perception of animals, based on the emergent paradigmatic approach.

Danby (2018), taking 'a post-humanist approach towards human–animal relations within the leisure landscape...aims to explore innovative ways of being within post-modern society'. Danby emphasizes that 'post-human theory challenges the singular focus around human subjects [i.e. anthropocentrism], blurring boundaries between the human and nonhuman' (Danby, 2018). Drawing on post-humanist theory in her study of human–equine interaction, Danby seeks to develop 'innovative ways of understanding human–non-human relations where boundaries between the two [i.e. humans and horses] become blurred through lived leisure spaces' (Danby, 2018). She stresses that in her study, 'horses are viewed as key actors within the human–equine relationship, playing active roles' (Danby, 2018), and seeks to show how in equestrianism the 'close encounters, relations and experiences [of humans] with horses [are] contributing towards the mutual wellbeing of horses and humans' alike. She further claims that 'The sense of otherness is erased where human and nonhuman boundaries become blurred, allowing a sense of partnership, respect and equality within the relationship' (Danby, 2018). However, despite her declared anti-anthropocentrism, such interpretations nevertheless show traces of anthropocentric interpretation. Danby seems to be aware of this, by conceding that 'it is difficult to assess how horses...are perceiving these relationships [with humans]' (Danby, 2018).

In another work on the human–horse relationship, Schuurman and Franklin (2015, p. 20) state that recent research on the role of animals in society has 'turned to the actual practices involving humans and animals instead of looking at the animals...as essentially separate from the social and cultural worlds in which their lives are...embedded'; they see in this a 'shift of focus, [challenging] the rigid human–animal boundary by contextualizing it in empirical research'. In their study of the performance of expertise of horse trainers, they stress that in the study of the construction of expertise 'in the context of animal related activities it is essential that the actions of the animals are also accounted for' since 'animals have an effect on the construction of human expertise' (Schuurman and Franklin, 2015, p. 21). Their particular interest is in 'the variable roles of resistance on the part of the animal in creating incidences of "counter-performance"' (Schuurman and Franklin, 2015, pp. 20–21), and their impact on the 'overall performance of human expertise [in handling or training horses]'. The authors perceive 'counter-performance' as 'intentional resistance that challenges the hegemonic [human] definition of the situation' (Schuurman and Franklin, 2015, p. 24); hence, they argue that 'controlling counter-performance...is essential for the performance to succeed and reinforce the expertise of the trainer' (Schuurman and Franklin, 2015, p. 25). Unlike most earlier studies, Schuurman and Franklin (2015) thus ascribe an autonomous role to the animal in the construction of a performance, rather than seeing in manifestations of resistance mere 'accidents' in an otherwise smooth process. However, the absence of an independent ethological study of the trainer–horse interaction, we remain in the dark regarding the sources of the animals' intentions to 'counter-perform'.

Though enlightening about the interplay of autonomous human and animal behaviour demonstrated in these studies, they also seem to share a certain weakness: a fuller understanding of the horses' 'point of view' which is hard to unveil in the absence of an in-depth ethological study.

Conclusion

Tourism studies, particularly industry-oriented ones, have for obvious reasons prioritized the tourist over the animal in research on tourist–animal relations. However, some recent publications (e.g. Markwell, 2015b) have begun to pay increased attention to the animal, although they do not yet embrace the emergent new paradigm of human–animal relations based on a radically changed view of the animal proposed by contemporary philosophers and ethologists. The intention of this chapter was to draw attention to this new paradigmatic approach and, hence, introduce a new image of the animal into tourist–animal studies, thereby initiating a less anthropocentric and more balanced treatment of animals in tourism research, which will take account of both the human and the animal 'point of view'. This would also involve the abandonment or modification of the conventional behaviouristic approach to animals in tourist–animal research, and its substitution or supplementation with a phenomenological approach based on the ethological study of both tourists and animals in their interaction.

Contemporary research on tourist–animal relations in tourism is insular and has not drawn the attention of researchers outside the tourism fields. One indication of this is that tourism has been overlooked in Amiot and Bastian's (2015) massive review of the field of animal–human relations. One of the advantages of the proposed reorientation to the emergent new paradigm would be to put tourism–animal studies in the forefront of contemporary research and enable researchers to compare their study of tourist–animal relations with work on human–animal relations in other major fields, such as leisure (Young and Carr, 2018), therapeutics, industrialized agriculture and consumption, some of which, as we have seen above, have already started to embrace the new paradigm.

The proposed reorientation of the image of the animal in tourism studies would have significant implications for a critique of the ethics of the treatment of animals in tourist establishments in which they are often perceived as mere objects, things or just property. The ascription of personhood and other human-like traits to the animals would add weight to the condemnation of animal abuse in tourism and leisure establishments, help to enlighten or convince the public of the exploitation and suppression of the animals, and help educate university graduates in tourism studies of the standing and rights of animals in tourism.

References

Amiot, C.E. and Bastian, B. (2015) Toward a psychology of human–animal relations. *Psychological Bulletin* 141(1), 6–47. DOI: 10.1037/a0038147.

Badmington, N. (2003) Theorising posthumanism. *Cultural Critique* 53(1), 10–27. DOI: 10.1353/cul.2003.0017.

Bastian, B., Costello, K., Loughnan, S. and Hodson, G. (2012) When closing the human-animal divide expands moral concern. *Social Psychological and Personality Science* 3(4), 421–429. DOI: 10.1177/1948550611425106.

Beirne, P. (2007) Animal rights, animal abuse and green criminology. In: Beirne, P. and South, N. (eds) *Issues in Green Criminology*. Routledge, Abingdon, UK, pp. 55–83.

Bekoff, M. and Sherman, P.W. (2004) Reflections on animal selves. *Trends in Ecology & Evolution* 19(4), 176–180. DOI: 10.1016/j.tree.2003.12.010.

Borgi, M. and Cirulli, F. (2016) Pet-face: mechanisms underlying human-animal relationships. *Frontiers in Psychology* 7. DOI: 10.3389/fpsyg.2016.00298.

Brinkmann, S. (2017) Humanism after posthumanism: or qualitative psychology after the 'posts'. *Qualitative Research in Psychology* 14(2), 109–130. DOI: 10.1080/14780887.2017.1282568.

Carmeli, Y. (1999) Mimicry, display, and play of the real: a trainer and big cats. *Australian Drama Studies* 35, 109–119.

Cattelino, J. (1968) *High Stakes: Seminole Gaming and Sovereignty*. Duke University Press, Durham, North Carolina.

Cohen, E. (2013) 'Buddhist compassion' and 'animal abuse' in Thailand's Tiger Temple. *Society & Animals* 21(3), 266–283. DOI: 10.1163/15685306-12341282.

Cohen, E. (2016) Ethnic tourism in mainland Southeast Asia: the state of the art. *Tourism Recreation Research* 41(3), 232–245. DOI: 10.1080/02508281.2016.1188485.

Cohen, E. (2019a) Crocodile tourism: the emasculation of ferocity. *Tourism Culture & Communication* 19(2), 83–102. DOI: 10.3727/109830419X15536971539380.

Cohen, E. (2019b) Interacting with animals. In: Pearce, P. (ed.) *Tourist Behavior: The Essential Companion*.

Cunningham, P.A., Huijbens, E.H. and Wearing, S.L. (2012) From whaling to whale watching: examining sustainability and cultural rhetoric. *Journal of Sustainable Tourism* 20(1), 143–161. DOI: 10.1080/09669582.2011.632091.

Curtin, S. (2006) Swimming with dolphins: a phenomenological exploration of tourist recollections. *International Journal of Tourism Research* 8(4), 301–315. DOI: 10.1002/jtr.577.
Danby, P. (2018) Post-humanistic insight into human-equine interactions and wellbeing within leisure and tourism. In: Young, J. and Carr, N. (eds) *Domestic Animals, Human and Leisure: Rights, Welfare and Wellbeing*. Routledge, Abingdon, UK, pp. 158–176.
Davis, D. and Tisdell, C.A. (1999) Tourist levies and willingness to pay for a whale shark experience. *Tourism Economics* 5(2), 161–174. DOI: 10.1177/135481669900500203.
DeMello, M. (2012) *Animals and Society*. Columbia University Press, New York.
Derrida, J. (2008) *The Animal That Therefore I Am*. Fordham University Press, New York.
Descola, P. (2009) Human natures. *Social Anthropology* 17(2), 145–157. DOI: 10.1111/j.1469-8676.2009.00063.x.
Descola, P. (2013) *Beyond Nature and Culture*. University of Chicago Press, Chicago, Illinois.
Descola, P. (2014) All too human(still): A comment on Eduardo Kohn's *How Forests Think*. *HAU: Journal of Ethnographic Theory* 4(2), 267–273.
Duffy, R. (2014) Interactive elephants: nature, tourism and neoliberalism. *Annals of Tourism Research* 44, 88–101. DOI: 10.1016/j.annals.2013.09.003.
Duffy, R. and Moore, L. (2011) Global regulations and local practices: the politics and governance of animal welfare in elephant tourism. *Journal of Sustainable Tourism* 19(4-5), 589–604. DOI: 10.1080/09669582.2011.566927.
Edelman, D.B. and Seth, A.K. (2009) Animal consciousness: a synthetic approach. *Trends in Neurosciences* 32(9), 476–484. DOI: 10.1016/j.tins.2009.05.008.
Fennell, D. (2012) *Tourism and Animal Ethics*. Routledge, Abingdon, UK.
Ferrando, F. (2013) Posthumanism, transhumanism, antihumanism, metahumanism, and new materialisms. *Existenz* 8(2), 26–32.
Franklin, A. (2007) The 'Animal Question' and the consumption of wildlife. In: Lovelock, B. (ed.) *The Consumption of Wildlife: Hunting, Shooting and Sport Fishing*. Routledge, Abingdon, UK, pp. 31–44.
Higham, J. and Neves, K. (2015) Whales, tourism and manifold capitalist fixes: new relationships with the driving force of capitalism. In: Markwell, K. (ed.) *Animals in Tourism: Understanding Diverse Relationships*. Channel View Publications, Bristol, UK, pp. 109–127.
Higham, J., Beider, L., Allen, S., Corkeron, P. and Lusseau, D. (2016) Managing whale-watching as a non-lethal consumptive activity. *Journal of Sustainable Tourism* 24(1), 73–90.
Irvine, L. (2007) The question of animal selves: implications for sociological knowledge and practice. *Qualitative Sociology Review* 31(1), 5–22.
Iveson, R. (2010) Animals in looking-glass world: fables of überhumanism and posthumanism in Heidegger and Nietzsche. *Humanimalia* 1(2), 46–85.
Jackson, Z.I. (2013) Animal: new directions in the theorization of race and posthumanism. *Feminist Studies* 39(3), 669–685.
Knight, J. (2010) The ready-to-view wild monkey: the convenience principle in Japanese wildlife tourism. *Annals of Tourism Research* 37(3), 744–762.
Kohn, E. (2013) *How Forests Think: Toward an Anthropology Beyond the Human*. University of California Press, Berkeley, California.
Kohn, E. (2015) Anthropology of ontologies. *Annual Review of Anthropology* 44(1), 311–327. DOI: 10.1146/annurev-anthro-102214-014127.
Kontogeorgopoulos, N. (2009) Wildlife tourism in semi-captive settings: a case study of elephant camps in northern Thailand. *Current Issues in Tourism* 12(5-6), 429–449. DOI: 10.1080/13683500903042873.
Lemelin, R.H. (2006) The gawk, the glance, and the gaze: ocular consumption and polar bear tourism in Churchill, Manitoba, Canada. *Current Issues in Tourism* 9(6), 516–534. DOI: 10.2167/cit294.0.
Lemelin, R.H., Fennell, D. and Smale, B. (2008) Polar bear viewers as deep tourists: how specialised are they? *Journal of Sustainable Tourism* 16(1), 42–62. DOI: 10.2167/jost702.0.
Lestel, D., Bussolini, J. and Chrulew, M. (2014) The phenomenology of animal life. *Environmental Humanities* 5(1), 125–148. DOI: 10.1215/22011919-3615442.
Lijmbach, S. (1998) Animal Subjectivity: A Study into the Philosophy and Theory of Animal Experience. Doctoral thesis. Landbouwuniversiteit Wageningen.
Lovelock, B. (ed.) (2007) *Tourism and the Consumption of Wildlife: Hunting, Shooting and Sport Fishing*. Routledge, Abingdon, UK.
Maoz, D. (2006) The mutual gaze. *Annals of Tourism Research* 33(1), 221–239. DOI: 10.1016/j.annals.2005.10.010.

Marchesini, R. (2016) Posthuman antispeciesism. *Angelaki* 21(1), 217–233. DOI: 10.1080/0969725X.2016.1163853.

Markwell, K. (ed.) (2015b) *Animals in Tourism: Understanding Diverse Relationships*. Channel View Publications, Bristol, UK.

Markwell, K. (2015a) Birds, beast and tourists: Human-animal relationships in tourism. In: Markwell, K. (ed.) *Animals in Tourism: Understanding Diverse Relationships*. Channel View Publications, Bristol, UK, pp. 1–23.

Markwell, K. and Cushing, N. (2009) The serpent's stare meets the tourist's gaze: strategies of display at the Australian reptile Park. *Current Issues in Tourism* 12(5-6), 475–488. DOI: 10.1080/13683500903042899.

Martinez, A. (2003) Swimming With Sea Lions: Friend or Foe? Thesis. Flinders University of South Australia, Bedford Park SA.

Marvin, G. and McHugh, S. (eds) (2014) *Routledge Handbook of Human-Animal Studies*. Routledge, Abingdon, UK.

Miller, M.R., Philosophy Documentation Centre (2013) Descartes on animals revisited. *Journal of Philosophical Research* 38, 89–114. DOI: 10.5840/jpr2013386.

Pearson, C. (2017) History and animal agency. In: Kalof, L. (ed.) *The Oxford Handbook of Animal Studies*. Oxford University Press, Oxford, UK, pp. 240–257.

Rodger, K., Moore, S.A. and Newsome, D. (2007) Wildlife tours in Australia: characteristics, the place of science and sustainable futures. *Journal of Sustainable Tourism* 15(2), 160–179. DOI: 10.2167/jost619.0.

Ruiz, M. (2017) What are the similarities between animals and humans? Quora. Available at: https://quora.com/What-are-the-similarities-between-animals-and-humans (accessed 6 May 2019).

Schuurman, N. and Franklin, A. (2015) Performing expertise in human-animal relationships: Performative instability and the role of counterperformance. *Environment and Planning D: Society and Space* 33(1), 20–34. DOI: 10.1068/d13014p.

Shapiro, K. and DeMello, M. (2010) The state of human-animal studies. *Society & Animals* 18(3), 307–318.

Singer, P. (1975) *Animal Liberation*. Ecco/Harper Collins, New York.

Staats, A.W. (2003) A psychological behaviorism theory of personality. In: Milton, T. and Lerner, M.J. (eds) *Personality and Social Psychology: Handbook of Psychology*. 5. John Wiley and Sons, Hoboken, New Jersey, pp. 135–138.

Staddon, J. (2014) *The New Behaviorism*. Psychology Press, New York.

Viveiros de Castro, E. (2014) *Cannibal Metaphysics*. Univocal Publishing, Minneapolis, Minnesota.

Wemelsfelder, F. (1997) The scientific validity of subjective concepts in models of animal welfare. *Applied Animal Behaviour Science* 53(1-2), 75–88. DOI: 10.1016/S0168-1591(96)01152-5.

Wiener, C. (2015) Dolphin tourism and human perceptions. In: Markwell, K. (ed.) *Animals and Tourism: Understanding Diverse Relationships*. Channel View Publication, Bristol, pp. 146–162.

Wilkie, R. (2015) Multispecies scholarship and encounters: changing assumptions at the human-animal nexus. *Sociology* 49(2), 323–339. DOI: 10.1177/0038038513490356.

Wilson, D.A.H. (2017) Circus animals and the illusion of wildness. *Early Popular Visual Culture* 15(3), 350–366. DOI: 10.1080/17460654.2017.1383018.

Wolfe, C. (2009) Human, all too human: 'Animal studies' and the humanities. *PMLA* 124(2), 564–575. DOI: 10.1632/pmla.2009.124.2.564.

Young, J. and Carr, N. (2018) *Domestic Animals, Human, and Leisure: Rights, Welfare and Wellbeing*. Routledge, Abingdon, UK.

22 Conclusion: Tourism and Development in the Face of Uncertainty and Chaos

Richard Sharpley[1]*, Brian King[2] and Peter U.C. Dieke[3]
[1]University of Central Lancashire, Preston, UK; [2]The Hong Kong Polytechnic University, Kowloon, Hong Kong; [3]Makerere University Business School, Kampala, Uganda

Introduction

When this book was first conceived, the tourism sector around the world was operating relatively normally. Certainly, increasing concerns were being expressed about its environmental impacts in general and both so-called overtourism (Milano *et al.*, 2019) and the contribution of aviation to global warming in particular; however, although some destinations, such as Barcelona, were beginning to implement policy responses to the challenges of overtourism (Sharpley and Harrison, 2017), the environmental spotlight was yet to be focused on flying. Greta Thunberg was still an unknown Swedish schoolgirl and the concept of *flygskam*, or 'flight-shaming' had yet to emerge. In other words, it was, by and large, business as usual for tourism. Admittedly, the seeds of Thomas Cook's eventual collapse in September 2019 were perhaps being sown through this period (Calder, 2019) while, across the sector, businesses and destinations were experiencing varying degrees of success. Some low-cost airlines were, for example, expanding rapidly, claiming an increasing share of the equally rapidly growing air travel market (Statista, 2020); others, conversely, were less fortunate, not least Monarch Airlines. Its collapse in October 2017 was then the largest airline failure in UK aviation history. Nevertheless, international tourism was continuing its inexorable growth – in 2017, international arrivals increased by 7.0%, the highest annual growth since 2010 (UNWTO, 2018) – and consequently, tourism was maintaining its role in contributing to economic growth, development and regeneration in destinations regions and countries around the world.

In stark contrast, at the time of writing this chapter (late March 2020), tourism globally faces an existential crisis. In response to the coronavirus pandemic, countries around the world are closing their borders to international visitors and imposing restrictions on the mobility of their own citizens to a greater or lesser extent. The consequences have been immediate and significant. Demand for travel, tourism and hospitality services has plummeted; airlines are grounding aircraft, cruises have been cancelled, hotels are closing and the survival of the myriad businesses that directly and indirectly meet the needs of tourists is in doubt. As a result, millions of people around the world who depend on

*Corresponding author: rajsharpley@uclan.ac.uk

 Tourism in Development: Reflective Essays (eds P.U.C. Dieke *et al.*)
DOI: 10.1079/9781789242812.0022

tourism for their income and employment face an uncertain future, at least in the short term. According to the World Travel and Tourism Council, for example, up to 50 million jobs are at risk – 14% of all travel and tourism employment globally (WTTC, 2020). Moreover, the economic future of many destinations has been thrown into doubt, highlighting the dependency of many countries not only tourism as a development option but also on particular markets.

It is not possible to predict the actual longer-term consequences of the coronavirus on the tourism sector and how it responds at the time of writing. Indeed, to attempt to do so at the height of the crisis would be both speculative and inappropriate. Much will depend upon the duration of the crisis and, as noted in this volume by Alastair Morrison (Chapter 19), a glimmer of hope can be found in reports that China is beginning to relax its severe restrictions on social activity. Elsewhere, however, the opposite is the case in many countries as they go into lockdown to contain the spread of the virus. Much will also depend on government policies about providing financial support for businesses within the tourism sector, both large and small. These may reflect the varying ability of national economies to weather the coronavirus storm. It is hoped, however, that by the time this book is published, the worst of the crisis will have passed. And by then, two things are certain. First, significant academic attention will be devoted to both the consequences of the pandemic for the tourism sector and to how it is responding. Considerable research was devoted to the impacts on tourism of the SARS (severe acute respiratory syndrome) outbreak in 2002. Given the much greater scale and impact of the coronavirus, it is likely that the crisis will also generate greater volumes of research. And second, the tourism sector will be returning to some form of normality.

Importantly, however, it is most likely to be a 'new normal'. That is, as will be proposed in this chapter, the coronavirus crisis and its consequences can be considered from the perspective of chaos theory. As will be discussed, this proposes that chaotic systems – including tourism (McKercher, 1999) – although subject to periods of disorder or chaos, inevitably self-organize into a new form of order and structure. The chapter then proceeds to expand on remarks by Richard Butler in Chapter 4, to reflect on the potential nature of this 'new normal' in the context of tourism and development, highlighting a variety of potentially influential actors.

Chaos Theory and Tourism

According to Levy (1994), chaos theory originated in the 1970s in the natural sciences, specifically in the work of Lorenz (1993). Applications soon followed in the social sciences, notably in management disciplines in general and the management of crises in particular. By the late 1990s chaos theory was being proposed as relevant 'to such a broad array of disciplines that some view it as a scientific version of postmodernism; a scientific metaphor for late-20th century cultural values of relativism, plurality, and chance' (Murphy, 1996, p. 96). Inevitably, perhaps, tourism is one of many disciplines and fields that have drawn upon chaos theory. However, given the nature, scope and scale of tourism and, more specifically, its vulnerability to a variety of both internal and external (to the tourism system) influences, the extent of the contribution has been less than might have been expected. As McKercher, 1999, p. 425) observes, 'tourism is an inherently non-linear, complex and dynamic system that is well described within the chaos paradigm'. Lamenting the lack of academic attention that tourism scholars have given to the issue, he proposes a new 'chaos model of tourism' (McKercher, 1999, p. 430) which conceptualizes the inter-relationships between nine distinctive elements of the tourism system, with the key being that a disruption to any one element in the system can bring change to the system as a whole.

In a similar vein, Russell and Faulkner (1999, 2004) adopted the lens of chaos theory to challenge the predictability of the destination life cycle model (Butler, 1980). Zahra and Ryan (2007) offered a more nuanced perspective, arguing that tourism does not necessarily display the characteristics of unpredictability and chaos. More specifically, and of particular relevance to this chapter, chaos theory has also provided a conceptual framework for the analysis of tourism crisis management. For example, Ritchie (2004, p. 672) argues that 'chaos theory may provide some insights into crisis and disaster management for organizations in the tourism industry'. The H1N1 influenza ('swine flu') outbreak in Mexico in 2009 resembled the coronavirus crisis in many ways, albeit on a much smaller scale. This earlier crisis was the context for Speakman and Sharpley's (2012) chaos theory-informed analysis of the response to a tourism crisis.

What precisely is chaos theory? An in-depth consideration is beyond the scope of this chapter (see, for example, Murphy, 1996; Seeger, 2002; Sellnow *et al.*, 2002). Essentially, however, chaos theory is based on the premise that systems, whether natural or human, such as societies or business organizations, do not follow predictable, linear paths. In other words, chaos theory rejects the belief that any system can be understood in terms of its constituent elements and linear, predictable cause-and-effect relationships between those elements. It rejects the Newtonian principle that 'small changes in original conditions will create consistent changes in their effects' within a system (Murphy, 1996, pp. 96–7). Rather, proponents of chaos theory argue that 'multidirectional causality, the disproportionate impact of minor variance, and unforeseen, unmeasurable, unknowable variances affect system behavior in unpredictable ways' (Seeger, 2002, p. 331) to the extent that a small change in one element of a system may eventually bring about major changes to the system as a whole.

This is not to say, of course, that systems are in a constant condition of chaos and change. They may enjoy prolonged periods of equilibrium and also, following a period of disruption, demonstrate a capacity to regain stability, structure and order; 'chaotic systems can self-organize and self-renew, with periods of order broken by sudden transformations whose direction has elements of chance and cannot be reversed' (Murphy, 1996, p. 96). However, two points should be emphasized. First, the outcome of change cannot be predicted – namely the form of the renewed system. This prompts the previous suggestion that discussions regarding the nature of 'post-virus' tourism are inappropriate and at best speculative as the coronavirus crisis is unfolding. And second, instability and change are inherent characteristics of any system; systems exist on the 'edge of chaos' and may be disrupted or tipped into chaos by apparently small or insignificant events within the system, a phenomenon which has been described as the 'butterfly effect'. The eruption of the Eyjafjallajökull volcano in Iceland in 2010 and the subsequent disruption caused by the ash cloud is one such example (Guiver and Jain, 2011), while what is generally believed to have been the animal–human transfer of the coronavirus in a so-called wet market in Wuhan, China, as the origin of a global pandemic, is another.

Importantly, such events are often predictable. It has been argued, for example, that the coronavirus outbreak had long been foreseen (Quick, 2020). As has already been suggested, however, within the tenets of chaos theory it is the response of the system to the event that is unpredictable. This happens when it enters the stage of what is referred to in the language of chaos theory as 'bifurcation'. It is this moment when a system is thrown into disequilibrium and chaos, when its 'direction, character, and/or structure is fundamentally disrupted and departs from the previous path' (Seeger, 2002, p. 332). Yet this disorder is not unending. Perhaps most controversially (Levy, 1994), chaos theorists propose that systems possess an innate ability to self-organize, to recreate order and structure out of the chaos. And fundamental to this is the existence of so-called 'strange attractors', that is, particular forces or influences – 'islands of rationality...and stability' (Thiétart and Forgues, 1995, p. 19) – which serve to regulate the behaviour of a system. In social systems, these may be the set of beliefs and values with which a particular society or a human system, such as a business organization, identify. For Zahra and Ryan (2007), p. 855) they represent a 'common vision, sense of meaning, strategy or value system that drives people to achieve a common goal'. Equally, it may be institutions or organizations within a social system, for example, emergency services, which act as 'strange attractors' during a crisis. Either way, they operate as a form organizational force that influences the post-bifurcation state of a system.

As an inherently chaotic system, tourism has consistently revealed its ability to recover and self-organize following periods of disruption or chaos. As has been argued by Wang (2009, p. 75), the sector 'has always managed to resume or exceed its former production values within a period of just one or 2 years'. A similar point has been made in Chapter 4 of this volume by Richard Butler. He observed, as have others elsewhere, that tourism has weathered a number of chaotic events globally over the past half-century or more. With few exceptions, and most notably the year following the 2008 global financial crisis, international arrivals continued to grow year-on-year. It should of course be observed that, at the national level, the picture can be less positive; for example, since reaching a peak in 2010, tourism in Egypt has been adversely affected by political and terrorist events, only beginning to recover in 2017. Nevertheless, the apparent ease and speed with which the tourism sector has, in the past, bounced back from crises of one form or another is already being used by some as a basis for predicting the post-coronavirus trajectory of international tourism. For example, some are using the length of time taken for airlines and hotels to rebuild volumes and revenues following previous major crises as a yardstick to predict the recovery of the tourism sector from the impacts of the pandemic, although concerns are being expressed about widespread failures among tourism businesses.

Importantly, however, in chaos theory terms, different 'strange attractors' might have been at play in previous major tourism crises. Other factors may also lead to different responses to and outcomes of the coronavirus crisis for tourism and development. Specifically, and as was documented in Chapter 2, the development and promotion of tourism in particular and (sustainable) development policies in general have long reflected and been influenced by the predominant guiding principle of economic growth. Even the much-heralded Sustainable Development Goals (SDGs) controversially embrace economic growth; to an extent they may can considered as inherently contradictory – that is, unsustainable (Adelman, 2017). Consequently, the constant, or strange attractor, in the (chaotic) tourism system has long been a belief in economic growth, resulting in a drive for profit from the tourism sector, fed by ever-increasing tourist demand. Hence, in previous periods of chaos, tourism arguably self-organized post-crisis around this attractor (economic growth) to continue its overall growth trajectory. The authors now suggest, however, that the coronavirus crisis, and how nations might respond, is distinct from previous crises that have impacted on tourism. On this occasion, there is potential for different strange attractors to influence both the recovery of the tourism system – that is, in its return to a new equilibrium – and also its contribution to development.

The Future of Tourism and Development: Which Way Now?

Few would contest the coronavirus pandemic as the most serious crisis that the world has faced, at least since the end of the Second World War. Indeed, some have equated the fight against the pandemic as a global war, although the context and outcomes are very different (Elliott, 2020). It is unquestionable that the consequences of the pandemic and the transformation that it brings to global society will become evident only in the longer term, though the immediate impact on tourism in particular is unlike anything previously experienced by the industry. Certainly, previous crises have had a downward influence on tourism demand and /or limited the ability of some sectors to operate, often as a consequence of responses to the immediate problem that catalysed the tourism crisis. For example, following the outbreak of foot-and-mouth disease in the UK in 2001, the government effectively closed the countryside to prevent its spread, with immediate and serious consequences for rural tourism operators (Blake *et al.*, 2003; Miller and Ritchie, 2003; Sharpley and Craven, 2001). As is usually the case, however, this was a crisis that impacted primarily on just one part of the tourism sector within a national context. It was also relatively short-lived, as was the last worldwide tourism crisis. Although international arrivals and receipts declined by 4% and 6% in 2009 respectively, following the global financial crisis, the figures for both substantially exceeded pre-crisis levels by the following year. In contrast, as a result of the coronavirus pandemic, the global tourism industry has, in effect, been obliged to close down; not only has international travel almost ceased following restrictions on cross-border movements but also, in many countries, domestic tourism has also collapsed. Uniquely in the history of modern tourism, people around the world have temporarily been denied freedom of movement – in late March 2020, reports indicate that one third of the world's population is under lockdown (for example, Osborne, 2020). No clearer example can be provided of a system on the edge of and falling into chaos. In 2019, overtourism was the greatest concern for many destinations; just months later, the omnipresent reality is no tourism.

What then are the implications for tourism and development? From a practical perspective, and as reported by Richard Butler (Chapter 4), the tourism sector is likely to respond as it has always done, by attempting to rebuild through extensive marketing campaigns, special deals and other means. These may tempt back tourists who, as Butler suggests, may have a pent-up desire for tourism experiences following an extended period of immobility. Moreover, Zurab Pololikashvili, Secretary-General of the United Nations World Tourism Organization (UNWTO), has stated that 'By staying home today, we can travel tomorrow... [because] ... tourism is uniquely placed to lead future recovery. Our sector will provide the jobs people need to bounce back and will drive economic growth that will help whole communities and countries to recover' (hospitalitynet, 2020). In other words, the future for the UNWTO means business as usual, promoting the (re)growth of tourism to stimulate development. However, various factors may come into play. For example, air travel may be more expensive as airlines seek to rebuild their balance sheets and capitalize on reduced capacity, thereby dampening potential demand, while the cruise industry in particular may face significant barriers to shorter-term recovery as a result of the negative perceptions associated with the coronavirus pandemic onboard large vessels (Gulland, 2020). Moreover, this might be compounded by the pre-existing backlash against the cruise sector as a driver of overtourism, with notable popular destinations already reducing the number of ships docking. In a sense, cruising may become a negative 'strange attractor', perhaps a 'strange detractor'. More generally, tourists may become more wary of exposing themselves to the risk of infection and, hence, initially confine their travel domestically or avoid certain countries.

Significantly, however, aside from its immediate impacts on all sectors of the travel and tourism industry, the coronavirus pandemic has revealed the alarming dependency of many destinations and national economies on tourism. In short, this highlights the potential unsustainability (in the broader sense of the word) of tourism as an agent of development. Three decades ago, it was suggested that 'Tourism must be a recognized sustainable economic development option, considered equally with other economic activities when jurisdictions are making development decisions' (Cronin, 1990, p. 14); its potential should be recognized but, implicitly, also balanced with other economic sectors. This guiding principle cannot always be attained; it has long been acknowledged that tourism is a 'Hobson's choice' for some countries or an option of last resort (Lee, 1988). Nevertheless, one outcome of the crisis may be that destinations proceed to reassess their tourism policies, particularly as they apply to promotion and development. This might build a more balanced economy with less dependence and avoid a reversion to the overtourism phenomenon which was gathering pace prior to the spread of the virus.

The response to the coronavirus has also revealed the unsustainability of tourism in the environmental sense. As many reports have suggested, there has been an immediate and beneficial impact on air quality in many countries as a consequence of the lockdown and, in particular, a substantial reduction in car and air travel (for example, Amos, 2020). Levels of pollution have fallen significantly. The drop is not confined to air pollution; the physical environment in many overburdened destinations is also benefitting. There have been claims, for example, that the canals of Venice are cleaner than they have been for decades (Brunton, 2020) while, more generally, there will be a welcome absence of the detritus of large-scale tourism in destinations around the world. In some destinations, the reduction in 'people pollution' will be welcomed.

Clearly, a balance must be achieved between environmental benefits and socio-economic need; tourism must and will return and its role in development will continue. However, the coronavirus crisis might, in chaos theory terms, stimulate the emergence of an alternative (to economic growth) strange attractor – that of sustainable de-growth (essentially, a reduction in both production and consumption on a global scale along with a fundamental shift in society's understanding of the relationship between consumption, wealth and well-being). Though controversial, not least because for some it seems to contradict the very foundations of capitalism (Foster, 2011) and, for others, represents an unpalatable radicalism, the concept of de-growth is gaining traction within the tourism academy (Andriotis, 2018). A recognition of excessive tourism dependence and the environmental benefits flowing from reduced tourism, as described above, might encourage a consideration of de-growth and, optimistically, its adoption in tourism policy and planning. Indeed, given the likelihood that the next major tourism crisis will be climate change-related, such a move would be pre-emptive.

It is, however, debatable whether the coronavirus crisis will bring about such a shift in tourism policy and planning. In reality, the odds favour a return to 'business as usual' in terms of both supply and demand, certainly in the shorter term. However, after more than half a century of tourism for development, the future for tourism is uncertain, as has been reported in many of the preceding chapters of this book. For those confronting the survival of the planet and of humanity, global warming will quickly regain its position at the top of the agenda. Along with other challenges, this will demand a radical rethink about the production and consumption of tourism, about the nature and objectives of development and, indeed, the relationship between the two.

And Finally...

The primary purpose of this volume has been to honour Professor Carson (Kit) Jenkins' contributions to tourism scholarship and practice. In the light of current global challenges, this merits some retrospection. As has been mentioned previously, many contributors to this book first encountered Kit as colleagues or students at the Scottish Hotel School, University of Strathclyde. An example is Kit's successor as Head of School, (Professor) Roy Wood, whose profile is a valuable reference documenting Kit's contributions and qualities, both personal and professional (Wood, 2014).

Various chapter contributors to the current book supported Kit in hosting the *Tourism: The State of the Art* conference in 1994 to celebrate the 50th anniversary of the School and the 25th anniversary of the Scottish Tourism Board (King, 1994). Ten years later Kit was also integral to *Tourism 2004: State of the Art II* (King, 2004). A quarter-century has elapsed since the first conference, though many fundamental tourism and development phenomena remain contested. Though the sustainability concept is central to global discourse about the planet's future, its achievement appears increasingly out of reach in the face of the climate crisis. Closer to home, public sector accountability for the implementation of tourism development and capacity management (whether under- or over- tourism) remains equally elusive, despite UK devolution, a Scottish Parliament (in 1999) and a dedicated Tourism Minister.

Fully two decades ago, Kit observed that the gap between tourism academics and tourism practitioners 'is too wide and probably getting wider' (Jenkins, 1999). Fortunately, his efforts have made a difference. At national level, collaborations were evident between academics at *Tourism 2004: State of the Art II* – all Scottish universities involved in tourism education were represented on the 'scientific committee'. And at the international level the tourism scholarly community increasingly accepts that we should treasure evidence-based knowledge about tourism and its implications for development. Many doctoral delegates at *Tourism 2004: State of the Art II* originating in the Global South helped to advance our field through their insights and perspectives. This is consistent with Kit Jenkins' role in coalescing a 'global network of pioneers of tourism research and education' (Wood, 2014). Given the global reach of tourism and development, it is not coincidental that the editorial team for this volume is based in three continents – Nigeria (Africa), Hong Kong (Asia) and the UK (Europe). Such pooling of knowledge and insights is needed more than ever; UN Secretary General Antonio Guterres has described Covid-19 as the world's greatest test since World War II. Consistent with Kit Jenkins' long-held values, addressing problems that span the Global North and South will require an international perspective based on trust in scientific and scholarly insights, drawing upon resources from the public and private sectors. This will involve a consistent focus on how tourism impacts on people and on the natural world. Kit's distillation of the central issues facing tourism in the developing world, notably enshrined through his contributions to legislation in many African countries, provide a potential reference point as the world emerges from the current crisis with a new equilibrium.

References

Adelman, S. (2017) The sustainable development goals: Anthropocentrism and neoliberalism. In: French, D. and Kotzé, L. (eds) *Sustainable Development Goals: Law, Theory and Implementation*. Edward Elgar, Cheltenham, UK, pp. 15–40.

Amos, J. (2020) Coronavirus: Lockdowns continue to suppress European pollution. BBC News. 27 March (accessed 29 March 2020).

Andriotis, K. (2018) *Degrowth in Tourism: Conceptual, Theoretical and Philosophical Issues*. CAB International, Wallingford, UK.

Blake, A., Sinclair, M.T. and Sugiyarto, G. (2003) Quantifying the impact of foot and mouth disease on tourism and the UK economy. *Tourism Economics* 9(4), 449–465. DOI: 10.5367/000000003322663221.

Brunton, J. (2020) 'Nature is taking back Venice': wildlife returns to tourist-free city. *The Guardian*. 20 March. Available at: https://www.theguardian.com/environment/2020/mar/20/nature-is-taking-back-venice-wildlife-returns-to-tourist-free-city (accessed 29 March 2020).

Butler, R. (1980) The concept of a tourism area cycle of evolution. *Canadian Geographer* 24, 5–12.

Calder, S. (2019) Thomas Cook: a timeline of the slow and painful death of the travel firm. *The Independent*. 15 October. Available at: https://www.independent.co.uk/travel/news-and-advice/thomas-cook-collapse-timeline-directors-staff-business-committee-peter-fankhauser-a9156941.html (accessed 20 March 2020).

Cronin, L. (1990) A strategy for tourism and sustainable developments. *World Leisure & Recreation* 32(3), 12–18. DOI: 10.1080/10261133.1990.10559117.

Elliott, L. (2020) The coronavirus crisis may lead to a new way of economic thinking. *The Guardian*. 22 March. Available at: https://www.theguardian.com/business/2020/mar/22/the-coronavirus-is-leading-to-a-whole-new-way-of-economic-thinking (accessed 28 March 2020).

Foster, J.B. (2011) Capitalism and degrowth: an impossibility theorem. *Monthly Review* 62(8), 26–33. DOI: 10.14452/MR-062-08-2011-01_2.

Guiver, J. and Jain, J. (2011) Grounded: impacts of and insights from the volcanic ash cloud disruption. *Mobilities* 6(1), 41–55. DOI: 10.1080/17450101.2011.532651.

Gulland, A. (2020) Coronavirus: are cruise ships really 'floating Petri dishes'? *The Telegraph*. 19 February. Available at: https://www.telegraph.co.uk/global-health/science-and-disease/coronavirus-cruise-ships-really-floating-petri-dishes/ (accessed 29 March 2020).

hospitalitynet (2020) Press release: COVID-19 Statement from World Tourism Organization (UNWTO) – By staying home today, we can travel tomorrow. Hospitalitynet. 17 March. Available at: https://www.hospitalitynet.org/news/4097565.html (accessed 29 March 2020).

Jenkins, C.L. (1999) Tourism academics and tourism practitioners – bridging the great divide. In: Pearce, D.G. and Butler, R.W. (eds) *Contemporary Issues in Tourism Development*. Routledge, Abingdon, UK, pp. 52–64.

King, B.E.M. (1994) Tourism: the state of the art. *Journal of Vacation Marketing* 1(1), 107–108.

King, B.E.M. (2004) Tourism 2004: state of the art II. *Anatolia* 15(2), 206–208. DOI: 10.1080/13032917.2004.9687159.

Lee, J. (1988) *Tourism and Development in the Third World*. Routledge, Abingdon, UK.

Levy, D. (1994) Chaos theory and strategy: theory, application, and managerial implications. *Strategic Management Journal* 15(S2), 167–178. DOI: 10.1002/smj.4250151011.

Lorenz, E. (1993) *The Essence of Chaos*. University of Washington Press, Seattle, Washington.

McKercher, B. (1999) A chaos approach to tourism. *Tourism Management* 20(4), 425–434. DOI: 10.1016/S0261-5177(99)00008-4.

Milano, C., Cheer, J.M. and Novelli, M. (eds) (2019) *Overtourism: Excesses, Discontents and Measures in Travel and Tourism*. CAB International, Wallingford, UK.

Miller, G.A. and Ritchie, B.W. (2003) A farming crisis or a tourism disaster? An analysis of the foot and mouth disease in the UK. *Current Issues in Tourism* 6(2), 150–171. DOI: 10.1080/13683500308667949.

Murphy, P. (1996) Chaos theory as a model for managing issues and crises. *Public Relations Review* 22(2), 95–113. DOI: 10.1016/S0363-8111(96)90001-6.

Osborne, H. (2020) As a third of mankind enters coronavirus lockdown, the impact on cities can be seen from space. *Newsweek*. 26 March. Available at: https://www.newsweek.com/coronavirus-lockdown-space-third-mankind-1494412 (accessed 29 March 2020).

Quick, J. (2020) China's complacency has led to this predictable coronavirus outbreak. *The Telegraph*, 25 January. Available at: https://www.telegraph.co.uk/global-health/science-and-disease/chinas-complacency-has-led-predictable-coronavirus-outbreak/ (accessed 27 March 2020).

Ritchie, B.W. (2004) Chaos, crises and disasters: a strategic approach to crisis management in the tourism industry. *Tourism Management* 25(6), 669–683. DOI: 10.1016/j.tourman.2003.09.004.

Russell, R. and Faulkner, B. (1999) Movers and shakers: chaos makers in tourism development. *Tourism Management* 20(4), 411–423. DOI: 10.1016/S0261-5177(99)00014-X.

Russell, R. and Faulkner, B. (2004) Entrepreneurship, chaos and the tourism area lifecycle. *Annals of Tourism Research* 31(3), 556–579. DOI: 10.1016/j.annals.2004.01.008.

Seeger, M.W. (2002) Chaos and crisis: propositions for a general theory of crisis communication. *Public Relations Review* 28(4), 329–337. DOI: 10.1016/S0363-8111(02)00168-6.

Sellnow, T.L., Seeger, M.W. and Ulmer, R.R. (2002) Chaos theory, informational needs, and natural disasters. *Journal of Applied Communication Research* 30(4), 269–292. DOI: 10.1080/00909880216599.

Sharpley, R. and Craven, B. (2001) The 2001 foot and mouth crisis – rural economy and tourism policy implications: a comment. *Current Issues in Tourism* 4(6), 527–537. DOI: 10.1080/13683500108667901.

Sharpley, R. and Harrison, D. (2017) Conclusion: mass tourism in the future. In: Sharpley, R. and Harrison, D. (eds) *Mass Tourism in a Small World*. CAB International, Wallingford, UK, pp. 232–240.

Speakman, M. and Sharpley, R. (2012) A chaos theory perspective on destination crisis management: evidence from Mexico. *Journal of Destination Marketing & Management* 1(1-2), 67–77. DOI: 10.1016/j.jdmm.2012.05.003.

Statista (2020) Low cost carriers' worldwide market share from 2007 to 2018. Statista. Available at: https://www.statista.com/statistics/586677/global-low-cost-carrier-market-capacity-share/ (accessed 31 January 2020).

Thiétart, R.A. and Forgues, B. (1995) Chaos theory and organization. *Organization Science* 6(1), 19–31. DOI: 10.1287/orsc.6.1.19.

UNWTO (2018) *UNWTO Tourism Highlights, 2018 Edition*. UN World Tourism Organization, Madrid.

Wang, Y.-S. (2009) The impact of crisis events and macroeconomic activity on Taiwan's international inbound tourism demand. *Tourism Management* 30(1), 75–82. DOI: 10.1016/j.tourman.2008.04.010.

Wood, R.C. (2014) A portrait of Carson L. Jenkins. *Anatolia – An International Journal of Tourism and Hospitality Research* 25(1), 161–167.

WTTC (2020) Coronavirus puts up to 50 million travel and tourism jobs at risk says WTTC. World Travel & Tourism Council. Available at: https://www.wttc.org/about/media-centre/press-releases/press-releases/2020/coronavirus-puts-up-to-50-million-travel-and-tourism-jobs-at-risk-says-wttc/ (accessed 23 March 2020).

Zahra, A. and Ryan, C. (2007) From chaos to cohesion – complexity in tourism structures: an analysis of New Zealand's regional tourism organizations. *Tourism Management* 28(3), 854–862. DOI: 10.1016/j.tourman.2006.06.004.

Index

Note: Page numbers in **bold** type refer to **figures**
Page numbers in *italic* type refer to *tables*

This book is published by **CABI**, an international not-for-profit organisation that improves people's lives worldwide by providing information and applying scientific expertise to solve problems in agriculture and the environment.

CABI is also a global publisher producing key scientific publications, including world renowned databases, as well as compendia, books, ebooks and full text electronic resources. We publish content in a wide range of subject areas including: agriculture and crop science / animal and veterinary sciences / ecology and conservation / environmental science / horticulture and plant sciences / human health, food science and nutrition / international development / leisure and tourism.

The profits from CABI's publishing activities enable us to work with farming communities around the world, supporting them as they battle with poor soil, invasive species and pests and diseases, to improve their livelihoods and help provide food for an ever growing population.

CABI is an international intergovernmental organisation, and we gratefully acknowledge the core financial support from our member countries (and lead agencies) including:

UKaid from the British people

Ministry of Agriculture People's Republic of China

Australian Government Australian Centre for International Agricultural Research

Agriculture and Agri-Food Canada

Ministry of Foreign Affairs of the Netherlands

Schweizerische Eidgenossenschaft Confédération suisse Confederazione Svizzera Confederaziun svizra

Swiss Agency for Development and Cooperation SDC

Discover more

To read more about CABI's work, please visit: **www.cabi.org**

Browse our books at: **www.cabi.org/bookshop**,
or explore our online products at: **www.cabi.org/publishing-products**

Interested in writing for CABI? Find our author guidelines here:
www.cabi.org/publishing-products/information-for-authors/